BRADY

Paramedic National Standards Self-Test
Fifth Edition

MS. CHARLY D. MILLER, Paramedic

Emergency Medical Services
Educator/Author/Consultant

Based on the
The 1998 Emergency Medical Technician-Paramedic
National Standard Curriculum
as set forth by the
U.S. Department of Transportation,
National Highway Traffic Safety Administration
and
The 2005 American Heart Association Guidelines
for Cardiopulmonary Resuscitation and Emergency
Cardiovascular Care (*Circulation*, Volume 112,
Issue 24, Supplement: December 13, 2005)

PEARSON

Prentice
Hall

Upper Saddle River, New Jersey 07458

Library of Congress Cataloging-in-Publication Data

Miller, Charly D.
Paramedic national standards self test / Charly D. Miller.--5th ed.
  p. ; cm.
  ISBN-13 978-0-13-199987-9
  ISBN-10 0-13-199987-7
1. Medical emergencies--Examinations, questions, etc. 2. Emergency
medical personnel--Examinations, questions, etc.
[DNLM: 1. Emergencies--United States--Examination Questions. 2.
Emergency Medical Services--United States--Examination Questions. 3.
Allied Health Personnel--United States--Examination Questions]
I. Miller, C. D. (Charly D.), EMT-paramedic national standards review
self test. II. Title.
RC86.7.B596 2008
616.02'5'076–dc22

Publisher: Julie Levin Alexander
Publisher's Assistant: Regina Bruno
Executive Editor: Marlene McHugh Pratt
Senior Managing Editor
 for Development: Lois Berlowitz
Editorial Assistant: Jonathan Cheung
Director of Marketing: Karen Allman
Executive Marketing Manager: Katrin Beacom
Marketing Specialist: Michael Sirinides
Managing Editor for Production: Patrick Walsh

Production Liaison: Faye Gemmellaro
Production Editor: Jessica Balch, Pine Tree Composition
Manufacturing Manager: Ilene Sanford
Manufacturing Buyer: Pat Brown
Senior Design Coordinator: Christopher Weigand
Cover Designer: Kevin Kall
Interior Designer: Laserwords Pte. Ltd.
Composition: Pine Tree Composition
Printing and Binding: Bind-Rite Graphics
Cover Printer: Phoenix Color Corp.

Notice: The author and the publisher of this volume have taken care to make certain that the information given is correct and compatible with the standards generally accepted at the time of publication. Nevertheless, as new information becomes available, changes in treatment and in the use of equipment and procedures become necessary. The reader is advised to carefully consult the instructions and information material included in each piece of equipment or device before administration. Students are warned that the use of any techniques must be authorized by their medical adviser, where appropriate, in accord with local laws and regulations. The author and publisher disclaim any liability, loss, injury, or damage incurred as a consequence, directly or indirectly, of the use and application of any of the contents of this book.

Pearson Education Ltd.
Pearson Education Singapore, Pte. Ltd.
Pearson Education, Canada, Ltd.
Pearson Education—Japan

Pearson Education Australia Pty. Ltd.
Pearson Education North Asia Ltd.
Pearson Educación de Mexico, S.A. de C.V.
Pearson Education Malaysia, Pte., Ltd.
Pearson Education, Upper Saddle River, New Jersey

10 9 8 7 6 5
ISBN-13 978-0-13-199987-9
ISBN-10   0-13-199987-7

Contents

TEST SECTION 14 (45 QUESTIONS; 45 MINUTES ALLOTTED FOR COMPLETION) **377**

Multiple-lead ECG analysis (4-lead and 12-lead ECGs)

APPENDIX

Acknowledgments

I continue to be extremely grateful for, and deeply appreciative of, the extraordinary Brady editors I have worked with throughout the years: Natalie Anderson, Susan B. Katz, Judy Streger, and Marlene Pratt—as well as Brady's talented Monica Moosang and Production Editor Jessica Balch at Pine Tree Composition. Most importantly, Brady's remarkable staff of sales reps and marketing folks are the "best evah"! I thank each of them for all their hard work.

I continue to extend my appreciation to the 1993 staff of Rocky Mountain Adventist Healthcare, Prehospital Services, for their assistance in obtaining the majority of ECG strips still used in this text.

I also gratefully acknowledge the efforts expended and the valuable feedback provided by the following reviewers of my manuscript: Jonathan Hockman, NREMTP, EMSIC, Goldrush Consulting; B. J. Hutchinson, EMS Programs Manager, Hospital Consortium Education Network, Pleasanton, CA; Michael S. Murrow, Paramedic Instructor; Sandra Hartley, MS, CP, Paramedic Program Director, Pensacola Junior College, Pensacola, FL; John L. Beckman, FF/PM, EMS Instructor, Affiliated with the Addison Fire Protection District, Addison, IL; Timothy Duncan, RN, CCRN, CEN, CFRN, EMTP, St. Vincent Life Flight, Toledo, OH; Malcolm Miller, CCEMT-P, Deputy Director, Boyle County Emergency Medical Services, Danville, KY; Captain Donna Sparks, RN, EMT-P, EMS Coordinator, Erlanger Fire/EMS, Erlanger, KY; Attila J. Hertelendy, MHSM, CCEMT-P, NREMT-P, ACP, University of Mississippi Medical Center, School of Health Related Professions, Department of Emergency Medical Technology, Jackson, MS; Brittany Martinelli, MHSc, BSRT, NREMT-P, Assistant Professor EMS, Santa Fe Community College, Gainesville, FL; and Joyce Foresman-Capuzzi, RN, BSN, CEN, CPN, EMT-P, Business Development Representative, Temple Health System Transport Team, Philadelphia, PA.

Most importantly: **BLESS YOU, JJ!** James J. Wohlers is an incredibly talented and well-educated paramedic firefighter in Grand Island, Nebraska. I met JJ back in the '90's when we both were working as "Denver General" paramedics. JJ has been helping me in many aspects of the educational and Expert Witness work I've done since I left "the streets" in 1998. And, now he's helping me with my Brady self-test textbooks.

Lastly, to my parents, Ray and Carol Miller (and *all* my sibling units), I extend a resounding "Thank you for all you've done for me!!!"

Charly D. Miller

About the Author

Charly D. Miller is a nationally known Emergency Medical Services author, instructor, and consultant, who lives in Lincoln, Nebraska. A paramedic since 1985 (including nine years as a Denver General paramedic, in Denver, Colorado), Charly is a seasoned prehospital emergency care provider. With her additional experience as a psychiatric medical technician and an Army National Guard combat helicopter medic, Charly is one of the United States' most exciting and entertaining EMS educators.

Charly's personal-specialty educational and consultation topics include:

"All Tied Up & No Place To Go": Patient Restraint Issues & Techniques

"Restraint Asphyxia–Silent Killer": Understanding & Avoiding Restraint-Related Positional Asphyxia

"Because I Said So!": Patient Communication Techniques

"Need The Info!": Patient Interview Techniques

In addition to this text, Charly's other Brady-published texts include the *EMT-Basic National Standards Review Self-Test* and (as lead co-author) *Taigman's Advanced Cardiology (In Plain English)*.

To learn more about Charly or to contact her, visit her web site at *www.charlydmiller.com*.

About This Text

Dear Paramedic:

This text was written *by* a Paramedic, specifically *for* Paramedics. It was designed to assist you in your preparation for, and execution of, *any* Paramedic written examination. Specifically written to challenge you and help you to identify subjects requiring additional study, this text will also significantly improve your skill at reading and responding to "tricky" test questions.

Like its 4th edition, the questions and answers presented in this text's 5th edition are primarily based on the 1998 EMT-Paramedic National Standard Curriculum as set forth by the National Highway Traffic Safety Administration's Department of Transportation (DOT). However, many of the 4th edition text's questions and answers were affected (altered) by the recommended changes published in the December, 2005, American Heart Association Guidelines for Cardiopulmonary Resuscitation and Emergency Cardiovascular Care. So, some 4th edition questions and answers were removed. Also, all 4th edition questions and answers that remain in this edition have been updated to reflect the 2005 AHA guideline changes.

Additionally, this text's 5th edition includes 96 entirely NEW QUESTIONS, related to entirely NEW SUBJECTS—such as multiple lead ECG analysis (4-lead and 12-lead), fibrinolytic therapy eligibility considerations, capnometry, and the new AHA BLS guidelines. Thus, the Paramedic National Standards Self-Test 5th edition offers 14 Test Sections instead of 13 Test Sections, and 1696 questions instead of 1600 questions.

Each Test Section's Answer Key identifies each question's SUBJECT, in addition to providing reference page numbers related to that question's subject for a variety of texts:

Brady's *Paramedic Care: Principles & Practice,* 2nd edition, Volumes 1 through 5, © 2006
Brady's *Essentials of Paramedic Care,* 2nd edition, © 2007
Brady's *Prehospital Emergency Pharmacology,* 6th edition, © 2005
Brady's *Taigman's Advanced Cardiology (In Plain English),* © 1995
The 2005 American Heart Association Guidelines for Cardiopulmonary Resuscitation and Emergency Cardiovascular Care (*Circulation,* Volume 112, Issue 24, Supplement: December 13, 2005)

If you don't have access to the Brady textbooks, simply use the index of the texts you *do* have to seek out information on any subjects you have difficulty with.

PDF files of *Circulation*'s special 2005 AHA Guidelines supplement are freely available on the World Wide Web. To obtain them, go to:
http://circ.ahajournals.org/
Click on the "Circulation Supplements" link.
Then, click on the link to:
"December 13, 2005; 2005 American Heart Association Guidelines for Cardiopulmonary Resuscitation and Emergency Cardiovascular Care; Volume 112, Issue 24, Supplement: December 13, 2005"

The questions in this text are specifically designed to be more difficult and demanding than "average" tests' questions. If you can pass each of these fourteen self-test sections with a score of 90% or better, you will easily achieve an even higher score on any actual written exam.

Be sure to read the "How To Use This Text" section. These directions will help you use this text to your greatest advantage. All too frequently, paramedics fail written tests only because they haven't developed good test-taking skills, because they haven't learned how to read test questions carefully, or because they haven't reviewed critical information. Used as suggested, this text will enable you to avoid failure for any of those reasons.

Also, be sure to read my "Suggestions for Written Examination Preparation and Execution." This section provides important tips for mental and physical preparation before taking any actual written exam. Again, all too frequently, paramedics fail written tests even when they know the material, because they haven't developed good test-taking skills.

Lastly, I strongly suggest that you also read my "Suggestions for Practical Examination Preparation and Execution." Paramedics who operate with exceptional competency in the "real world," often fail practical (situational) examinations only because they are ill prepared to successfully perform in the simulated-emergency testing environment. This text will help you excel during the written exam. However, your performance during practical exams is also vitally important to your successful certification or recertification.

I sincerely hope that this text helps you to excel in all aspects of your paramedic examination.

I *do not* wish you "Good Luck!" Luck is simply not a consideration when it comes to successful written or practical test taking. Your dedicated preparation and review are the only factors that will help you to achieve the high scores you seek.

Sincerely yours,

Ms. Charly D. Miller

www.charlydmiller.com

How to Use This Text

This "self-test" text was developed to help you determine the strengths and weaknesses of your knowledge base, to determine which subjects you have difficulty with (which subjects you should especially study), and to practice your written test-taking skills. Here are some important suggestions for using this text to PRACTICE taking paramedic exams:

Read each question carefully. Knowledgeable and experienced EMTs often fail written exams simply because they don't read the questions thoroughly.

- *Pay very close attention to the kind of answer the question requires.* Is the question asking you to identify a correct or "true" answer, an incorrect or "false" answer, or the "best" answer?
- *Do not read more into the question or the answers than what is presented.* Do not ask yourself, "Well, what if ...?" Each question and the answers offered (whether it's a basic-information question or a scenario-based question) clearly presents all the "clues" you need to select the correct answer—if you know the subject.

Read each answer carefully. Then, after selecting an answer, go back and reread the question, inserting that answer.

- *Does the answer you selected still seem correct?* If not, or if you're not sure, you should review this question's subject. Circle this question's answer-sheet number lightly, in pencil, so you'll be reminded to review it later.
- *Do not guess when you don't know the answer.* When taking an actual written exam, a guess is always better than an unanswered question (unanswered questions are always an "error"). However, when practicing with this Self-Test text, do not guess! If you guess during a Self-Test, you may correctly answer the question purely by accident, and thus you'll not be alerted to review that subject. Without review, when answering a similar

question on an actual exam, there is no guarantee that you'll make the same, accidentally correct, guess. When practicing with Self-Tests, if you cannot answer a question with confidence, circle that question's answer-sheet number, skip it, and move on. Later, your circled question number will remind you to review the subject identified by the Answer Key.

The Self-Test Answer Key is your "structured review guide."   When you have completed each Self-Test section, compare your answers to the Answer Key provided at the back of this text. Note the reference page numbers and/or SUBJECT for each question you skipped or answered incorrectly. Go to the referenced texts (or the paramedic text you have) and review that subject's material. Then, retake the Self-Test. If you again encounter difficulty with one or more of those same subjects, repeat your review.

The more often you repeat this process, the better you will fare on your actual examination.

Time yourself as you take the Self-Tests.   Most examinations allow one minute per question. For instance: 60 questions = a one-hour time limit; 120 questions = a two-hour time limit; 180 questions = a three-hour time limit; and so on.

- Timed practice sessions help you learn to improve your test-taking speed.
- This text's 5th edition consists of 14 Test Sections, containing a total of 1696 questions. Even if you use each Test Section only once, you will have put in over 28 hours of test-taking practice!

Use group study sessions.   During the weeks before the test, get together with co-workers or friends who are taking the same exam and study as a group, especially in preparation for practical examinations. "Teaching" is a great way to "learn."

The National Registry of Emergency Medical Technicians (NREMT) Paramedic Skills and Oral Station Evaluation Sheets are reproduced in the Appendix of this text. These Evaluation Sheets represent the National Standard Performance requirements for the order in which specific practical skills (and oral responses) should be performed. Memorize them! Practice with them!

To obtain additional "practice" copies of these skills sheets, free of charge, see the instructions included in this text's Appendix, for how to go to the National Registry of Emergency Medical Technicians Web site and access and print out as many copies of these skills sheets as you need. Then, as your group practices the practical skills (and oral test) stations, use the NREMT skills to score each other prior to taking the exam.

In addition to using this text as I have suggested here, be sure to read my "Suggestions for Written Examination Preparation and Execution," and my "Suggestions for Practical Examination Preparation and Execution."

Suggestions for Written Examination Preparation and Execution

Most of us consider written examinations as delightful as a tooth extraction accomplished without the use of anesthetic! However, when taking a written examination, anesthesia is entirely inappropriate and unhelpful. As long as you've *prepared* well, you'll *do* well!

Test preparation should start long before exam day. First, if you haven't already, read the "How To Use This Text" section. Then, using this Self-Test text, determine your strengths and weaknesses, study the subjects you have difficulty with, and practice your test-taking skills.

TEST-DAY PREPARATION TIPS

Get a good night's sleep before taking the actual test. Losing sleep by furiously studying the night before an exam is sure to cause you more harm than good. Even if you've delayed your review and can't avoid last-minute studying, you *still* need to get plenty of sleep before the exam!

Go to bed earlier than usual on the night before, and then get up early enough to study on the morning of the test. Or, if you are one of those "night shift" people who absolutely cannot think efficiently in the early morning hours, take a good, long nap the previous afternoon or evening. Then study into the wee hours of the morning, and take another nap before the test.

Eat a GOOD breakfast on Test Day. Contrary to popular EMT behavior (of all certification levels), a good breakfast does *not* consist of coffee and donuts. The donut's sugar-energy burst will quickly peak, thereafter resulting in a rebound lethargy – probably during the time you're taking the test. The coffee's caffeine provides artificial stimulation, but doesn't improve your concentration or mental acuity, either.

Whether or not you adopt good nutritional habits in your normal day-to-day living, it will be a great advantage for you to do so just prior to an event as stressful and demanding as a written or practical paramedic examination. A good meal (breakfast or lunch), high in complex carbohydrates and protein, eaten more than an hour before the exam is the key to fueling the sustained energy levels and mental acuity you'll need during the high-stress test situation. Complex carbohydrates are found in things like fresh fruits and juices, whole-grain breads, or rice products. Lean meats, eggs, and milk products (yogurt, cheeses, and low-fat milk) are good sources of protein.

Allow yourself at least one hour between a good meal and the test time.   Immediately after eating a meal, a large proportion of your body's blood supply is diverted from your brain to your stomach, to aid in food digestion. By eating at least an hour before taking the test, your brain will not be robbed of fuel during the test. Instead, it will be fueled slowly and efficiently, helping you perform at your absolute best.

TIPS FOR TAKING THE ACTUAL WRITTEN EXAMINATION

Remember: Read each question carefully.   Is the question asking you to identify a correct or "true" answer, an incorrect or "false" answer, or the "best" answer? Don't read more into the question or answers that is stated. Avoid "what ifs" when considering your answer. Each question (or case scenario) should clearly present all the "clues" you need to select the correct response from the answers provided.

Read each answer option carefully.   Then, after making your selection, go back and reread the question, inserting that answer. Does the answer you selected still seem correct? If not, try another answer. If you're unsure, consider skipping the question (lightly circling the answer sheet's question number) and coming back to it later.

Budget your test-taking time.   Do not dwell upon any question that you can't confidently and quickly answer. Spending too much time on one difficult or confusing question gives you less time to answer questions you can confidently, correctly, and quickly complete. Skip any troublesome question and come back to it later. In this way, even if you run out of time before completing the entire test, you've confidently answered the maximum number of questions you possibly can.

Lightly circle (in pencil) any skipped question numbers on your answer sheet.   Test proctors will strongly warn you against making any "unnecessary stray marks or erasures" on your answer sheet. However, when you skip a question, you must be sure to also skip that question's answer-sheet selection! It is far better to *lightly* circle the number of a skipped question's answer set than to cause all subsequent questions to be incorrectly answered because you accidentally lost your place on the answer sheet.

Make sure that you gently, but completely, erase all extra marks or circles before submitting your completed answer sheet. "Unnecessary stray marks" or heavy erasures made on a computerized answer sheet may cause the computer to interpret your nearby answer selection as being wrong, even if you marked the correct answer! When making light pencil marks, avoid crossing over any of the form's print. When erasing pencil marks, erase them gently. Avoid dimming or blurring any of the answer sheet's print. Print damage may also interfere with the computer's answer-checking function, causing computer-interpretation errors that might count right answers as wrong answers!

As a last resort, GUESS—but only after you've "come back" to a skipped question.   An unanswered question is always counted as an incorrect answer. If you can't confidently (and quickly) answer a question, lightly circle its answer-sheet number and come back to it later. Sometimes, subsequent questions or answers will help you recognize the correct answer to a question you skipped! Even if that is not the case, it is best not to "dwell" upon a question you can't confidently and quickly answer. Move on.

Once you've been through the entire test and answered all of the questions you are confident of, if there is time left to do so, go back to your skipped questions. If you're still in doubt, make your best guess; a guess is always better than no answer! Again, be sure to gently erase the pencil circle around the answer-sheet number!

What about case-based questions?   Sometimes, medics with little "field" experience approach case-based "scenario" questions with extreme self-doubt and trepidation. Such worry is completely unnecessary! "Field" experience has very little to do with correctly answering written test questions, whether they are "straight" informational questions, or case-based "scenario" questions.

Case-based "scenario" questions simply test your ability to integrate your training and knowledge when considering situational information, as provided to you by the test questions. They test how well you are able to solve mysteries, given a limited amount of information. Approach case-based "scenario" questions with confidence, and play "EMS Detective." You have the knowledge that is being tested. You are not "running a call." You're simply figuring out a mystery, based *only* upon the information provided by the test questions. Read the scenario question and answers carefully, and answer it based only upon the information provided.

SUMMARY

Structured review, adequate rest, and efficient nutrition will give you an undeniable advantage and vastly improve your performance in any testing situation. Following the tips I've provided will help you confidently answer the largest number of questions possible, in the shortest amount of time. After that, return to the questions that require more thought (or a guess). Once you get to the "return" point, simply *relax* and have *faith* in yourself! You've already done well!

Suggestions for Practical Examination Preparation and Execution

Throughout your entire EMS career you will periodically be subjected to the dreaded "practical skills," "assessment," or "situational" examination stations. Practical examinations accompany all levels of EMS training courses, local and state certification and recertification tests, and all National Registry examinations. More and more often, practical examinations also accompany continuing-education workshops or courses (such as Advanced Cardiac Life Support or Pre-Hospital Trauma Life Support programs).

However, sometimes even the most "veteran" EMS providers tremble in their boots when faced with the ordeal of performing in the practical examination environment. There are a number of entirely understandable reasons for this trepidation:

- A written test evaluates only your medical knowledge and test-question/answer-reading skills. A practical examination places your medical knowledge, your equipment-use skills, and your physical performance under the strictest scrutiny. It's unnerving to be so closely observed and evaluated, especially when failure to "pass" and obtain certification threatens your livelihood!

- Even the most elaborately re-created practical examination simulations contain situational components that you must recognize and remember, without actually seeing, feeling, hearing, or smelling them.

- You may be facing the practical exam station alone, or you may be teamed with people you have never worked with before.

- The equipment you are given during a practical exam is rarely, if ever, presented in a manner you are accustomed to dealing with. Consequently, you may fumble around, hunting for instruments that, in a "real-life" situation – with your "real-life" equipment by your side – your hands would normally find on their own. Often, the equipment is also of a different brand, make, or model than what you're familiar with using.

- In addition to all of the above-stated factors, in their effort to remain unbiased and objective, even the friendliest practical examination evaluators often appear cold, even hostile. That, in itself, is enough to cause someone to become stressed!

Throughout my years of EMS service, I've participated in practical examinations as a test-taker, a "patient," and an evaluator. From these experiences, I've developed several suggestions for how to dramatically diminish your practical-exam-approach trepidation, and how to significantly improve your performance during practical examinations.

PREPARING FOR A PRACTICAL EXAMINATION

Periodic practice and review are essential. Regardless of the volume or variety of calls you've experienced, it is imperative that you remain skilled at operating all EMS equipment, imperative that you frequently refresh your knowledge of patient assessment and treatment protocols, and *especially* imperative that you refresh your knowledge of skills performance prior to performing in a practical examination situation.

We all know that "real life" situations often require us to deviate from, alter, or augment any "established" order of skills performance. But, in the practical examination situation, following a precise order of assessment or skills performance is a major factor in how you're scored. Even though you complete every step, if the order in which you complete the steps deviates from the "established" order of skills performance, your score will be diminished. Thus to prepare for a practical examination you must memorize the precise order of assessment and skills performance steps required for each station. Any testing authority's performance-order requirements should be available to you for study purposes. (In the Appendix of this text you will find copies of the skills station score sheets used by the National Registry of EMTs.) Get them. Memorize them. Practice following them!

Prior to the practical examination, get together with a small group of your peers and practice the skills, using the performance-order guidelines you are provided. Take turns being an examination evaluator, a "patient," and a skills performer. Experience in all three of these roles is enormously helpful in developing your awareness of common performance mistakes and in training yourself to avoid these common mistakes.

APPROACHING THE PRACTICAL EXAMINATION

First, please review the notes about sleep and nutrition from my "Suggestions for Written Examination Preparation and Execution." Those suggestions improve practical examination performance, as well.

PERFORMING IN THE PRACTICAL EXAMINATION

"Dress for Success!"   Your goal should be to clothe and equip yourself in a manner that makes you feel as "normal" and comfortable as possible, a manner that will help make your surroundings as familiar as possible.

Actually changing into your uniform before performing in the practical examination is ideal. This may appear "gung-ho" to others, but those others are not grading your performance. (They may laugh right up until the moment when they discover how far you out-performed them in the practical examination! Then, who'll have the "last laugh"?!)

At the very least, wear your own equipment holster and "cargo pants." This puts vital and frequently used items such as scissors, penlight, pens, etc., at your fingertips, just where you are used to finding them. Your own stethoscope ought to be draped, bundled, or strapped about whatever it is usually draped, bundled, or strapped. Indeed, if you have a personal emergency-care kit that you routinely use, bring it and request permission to use it!

Again, your goal is to clothe and equip yourself in a manner that makes you feel as normal and comfortable as possible, making your surroundings as familiar as possible. This will enormously improve your performance in any practical examination.

RULES SPECIFICALLY FOR EXCELLING IN "ASSESSMENT" OR "SITUATIONAL" PRACTICAL EXAMINATIONS

Practical examinations that involve manipulating or applying a single piece of equipment are relatively easy to prepare for and perform in. It is the "Assessment" or "situational" practical examinations that presents the greatest challenge to any EMS provider. To improve your performance in these situations, the most vital "secret for success" is to train yourself in some very specific ways to perform during the test:

- **Fix your eyes and your attention *only* on the evaluator as you begin the practical examination.** When you first enter the station and the evaluator is giving you an "introduction," do not preoccupy yourself with "eyeballing" the patient or scene, thinking you'll get "a head start" on the situation. *Good listening requires complete concentration.* Eyeballing the patient or scene will significantly distract you from hearing (and processing) the evaluator's introductory information, causing you to miss important details! Upon entering the station, fix your eyes and your attention on the evaluator, and concentrate solely upon the instructions and information she/he is providing you.
- **If you are offered time to examine the equipment before your evaluation begins, *do it!*** Even if you don't think you need to, carefully look at all the equipment, especially if anything differs from equipment you are accustomed to using. As you examine it, arrange the equipment in a manner that is familiar to you. Certainly, do *not* leave it in a messy pile at the corner of the room or beside the

patient! Use the entire time offered. Remember the old rule of "take *your* pulse first!" when responding to an emergency call? Anyone who discounts this rule is a fool. Anytime you're offered time to do something, take the *entire* amount of time and *do* it. At the very least, it will give you time to let your anxiety-related epinephrine burst wear off, so that you can function at your best.

- **When the evaluator describes the situation and scene, be especially alert for indications of the mechanism of injury, or the nature of illness, or descriptions of the scene itself—things that cannot be simulated in the testing environment.** If you are told that the steering wheel is bent, or the windshield fractured, or that items of furniture are broken, you are being told that for a reason. Take time to consider these important clues.

- **If this type of information is not offered, ask for it.** Scene Size-Up is important to every call, and especially important to every practical examination station. Ask yourself questions like; What is the condition of the apartment? Is it tidy? Messy? What does it smell like? What does the car look like? Where was the point of impact, and where was the patient sitting? Is there compartmental intrusion involving the patient's space? What does the patient look like on my approach? and so on. If you can't observe or sense the answers to these questions, ask the evaluator.

- **As you are directed to actually begin your performance, no matter what the situation is, the first question you should voice is "*Is the scene safe*?"** In real life you can usually sense potential dangers on approach. You can see smoke, smell a gas leak, hear a domestic altercation, or observe that the police have not yet arrived. Practical examinations are rarely able to simulate such normally apparent indications of danger. So, when beginning the Practical Examination, the very first words out of your mouth should be, "Is the scene safe?" It is important to *verbalize* this consideration, to let the evaluator know that you are considering it.

- **From this moment on, there is *no longer any reason for you to look at the evaluator!*** This is very, very important. Continually glancing back and forth between the evaluator and the patient is extremely distracting. It interrupts your concentration and the continuity of your performance. Once you've begun, focus your eyes and attention on the scene and the patient. After the introductory scene and patient information has been provided, your eyes and hands should never leave the patient and the simulated scene environment, except to obtain equipment. The evaluator *does not need to see your face* to answer your questions! She/he can listen to what you say, respond to your questions, and watch your skills performance without ever having eye contact with you.

- **Never stop talking.** Whether you are questioning the evaluator, addressing or questioning the patient, or describing your thoughts and actions, you should never stop talking. Frequently, even the best evaluators miss silent skills performances while they are making notes on a performance score sheet. If you aren't verbalizing every single thought and activity, while the evaluator is making notes she/he may miss things such as your sweep of the patient's clothing for gross bleeding. Thus, even though you performed it, you won't be credited with having done so. However, if you verbalize every single thing you *do* and verbalize everything you are *thinking*, you will be credited with all the activities and skills you are performing, whether or not the evaluator actually *sees* your performance.

- **Continuously talk to your patient.** This may sound difficult, but it is easily combined with the previous suggestion. Focus your attention on the patient, and tell the *patient* what you are thinking and doing. This will help to calm the patient "actor," as well as help to calm YOU, which will in turn increase the effectiveness of your performance. (BTW: This rule is true for practical examinations as well as "real" situations!) Also, address all your questions, explanations, and descriptions *to the patient*. The evaluator should provide you with any information that the patient is unable to verbalize in response to your direct questions, and you don't need to look at the evaluator to receive this information!

- **Ask for information.** If the patient doesn't exhibit or give you the answers you need, the evaluator should provide the information. If the evaluator doesn't volunteer the information, it's because She/he is waiting for you to ask for it. Again, you should not break your eye contact with the patient to ask for information. The evaluator is in the same room. She/he can respond to your questions without you looking at her/him.

The following is a written example of what a practical examination performance should *sound* like. Except for the first couple of cues, only the medic being tested (the performer) can be "heard."

Evaluator: You may begin.

Performer: Is the scene safe?

Evaluator: The scene is safe.

Performer: I am observing the scene as I approach. Is the mechanism of injury apparent?
I can see that my patient has some blood on his left thigh. Is the blood spurting as though it is an arterial bleed? It's not? Good.
First, I place my hand on his head and assess his level of consciousness.
Sir? Sir? Can you hear me?
Hi. My name is Mork. I'm a medic and I'm here to help you. I need you to stay very still and not move your head, please. Thank you! What's your name?
Great, Endor, thank you. Any pain in your neck as I run my fingers down it, Endor?

My invisible partner, Mindy, is going to hold your head to help you keep it absolutely still while I examine you. She will not stop holding your head until we have secured you to a long backboard.

First, I'd like to check your airway. I'm going to look inside your mouth and make sure everything's cool there. Anything loose or bothering you in there, Endor?

Good. How does your breathing feel?

As I listen to Endor's chest with this stethoscope, what do I hear?

How does your chest feel, Endor? How about when I compress it, like this?

It feels even in excursion, and I don't feel any crepitus. I don't see any wounds or deformities.

Endor, my invisible partner, Mindy, will observe your airway and respiratory effort as she continues to maintain immobilization of your c-spine—that's your neck—and she'll alert me to any changes while I examine you further.

I'm going to check your pulses now, Endor. Do I feel Endor's radial pulse?

What quality and rate do I feel?

Is Endor sweaty?

What's Endor's skin color?

Temperature?

Endor, I'm going to give you some oxygen to help you feel better. This mask may seem a little confining, but I'm going to run oxygen into it at fifteen liters per minute, and it will help you feel much, much better.

Yes: an evaluator could be scoring this medic's performance over the telephone!

A continuous narration of your thoughts and activities, whether delivered to your patient or to the evaluator (while still focusing on the patient), assists your score in a number of ways. It keeps you focused on the situation and your task performance, improves your mental concentration, and ensures that you proceed without forgetting things. It also helps to calm you and improve your physical performance. Most importantly, though, verbalizing everything you do and think will make it almost impossible for the evaluator to miss any of the things you do, even if she/he doesn't actually see your performance of something.

PARTNER OR GROUP APPROACHES TO A PRACTICAL EXAMINATION

All of the previous suggestions apply here as well. However, now you are teamed with other participants—people who may have been total strangers only moments before the practical examination. The old adage "Too many cooks spoil the broth" applies here in quadruplicate. Few things are as debilitating and disastrous as having two, three, or four EMTs crawling all over each other, each trying to treat the patient and "run the call" at the same time.

The secret to group success is organization and assignment of tasks. Each group member must have an assigned role, with one group leader assigned to be in charge of directing the entire group's performance. Obviously, these assignments must be made and agreed upon before the group enters the practical examination station. And, as each new station is approached, the roles should be clearly reassigned, allowing each participant an opportunity to rotate through each different role.

Groups of Two are Easy

For trauma practical examinations, one partner is group leader, and the other is the c-spine/airway monitor. The group leader introduces self and partner, directing the partner to maintain immobilization of the c-spine continuously and to monitor for changes in airway status after the initial airway/breathing examination. The group leader continues examination of the patient and performs all other necessary treatments. The group leader is also the team member in charge of verbalizing all the activities.

The c-spine/airway monitor maintains spinal immobilization and observes airway/respiratory status, no matter how tempted she/he is to help with other treatment. While doing so, (unless prohibited from "cuing" the group leader by the evaluator), she/he should provide verbal cues whenever the group leader appears to have forgotten something.

For medical practical examinations, one partner is group leader and the other is the airway monitor/equipment operator. The airway monitor/equipment operator's primary task is to manage oxygen equipment and delivery. If the patient does not require active airway maintenance and ventilation, this person is free to take vital signs, apply the ECG monitor (if available), start IVs, position the wheeled stretcher, and so on.

The group leader is still responsible for patient examination and assessment. However, when the airway monitor is free to also be an equipment operator, the group leader should direct the application of treatments, periodically participating in equipment operation.

Groups of three responders can be broken down into assignments of group leader, c-spine/airway monitor, and equipment operator.

Groups of four responders can be broken down into assignments of group leader, c-spine immobilizer, airway monitor, and equipment operator. If continued airway maintenance is not required, the airway monitor may also act as a secondary equipment operator (yet never forgetting to watch the airway).

If you have the misfortune of operating in a team consisting of five or more members, send all but four team members to direct traffic, boil water, anything to get them out of the way!

The most important key to group performance, no matter how you designate the tasks or roles, is that the assignments (and assignment responsibilities) are clearly understood by each member before you begin the practical examination.

IN SUMMARY

A brief recap of the essential points to remember for the practical examination:

- **Wear your uniform and bring your own equipment holster, stethoscope, and jump-kit if you have them and regularly work with them!** Make the practical examination's environment as normal, comfortable, and familiar as you possibly can.
- **As you first enter the station, fix your eyes and attention on the evaluator.** Carefully listen to the clues provided by the evaluator. Take all of the time you are offered to consider the clues. Ask for any and all scene or situation information if it is not offered. If you're offered time to check provided equipment before starting the exam, *do it.*
- **Ask, out loud, "Is the scene safe?"**
- **After beginning your performance, after the scene size-up, never let your eyes or hands leave the patient.** There certainly is no need to look at the evaluator again.
- **Never stop talking.** Talk to the patient, and ask the patient lots of questions. Verbalize everything you are doing and everything you are thinking. When asking questions of the evaluator, do not look away from the patient; the evaluator does not need your eye contact to answer you.
- **Clearly assign specific tasks to group members before beginning a group exam, and take turns with roles and task performance.**
- **When you are not the group leader, do not deviate from your assigned task.** However, if you are allowed to, provide reminders of business that should be performed (or questions that should be asked) when the group leader seems to have forgotten something.
- **Above all, enjoy yourself!** Really. Why not? When examination day arrives, set aside your fears. They won't help you. Take pride in the performance level you have worked so hard to achieve. Have confidence in your abilities. And, have a good time!

A Reminder: The National Registry of Emergency Medical Technicians (NREMT) Paramedic Skills and Oral Station Evaluation Sheets are reproduced in the Appendix of this text. These evaluation sheets represent the National Standard Performance requirements for the order in which specific practical skills should be performed. Memorize them! Practice with them!

To obtain additional copies of these skills sheets free of charge, see the instructions included in this text's Appendix, for how to go to the NREMT Web site, to access and print out as many copies as you need. Then, use them to score each other, as your group practices prior to taking the exam.

Test
Section
One

1

Test Section 1's subjects:

- EMS Systems
- Paramedic Roles and Responsibilities
- The Well-Being of the Paramedic
- Illness and Injury Prevention
- Medical/Legal Aspects of Advanced Prehospital Care
- Ethics in Advanced Prehospital Care
- Life-Span Development
- Therapeutic Communications
- Care-Provider Communication
- Documentation

Test Section 1 consists of 90 questions and is allotted 1.5 hours for completion.

1. Which of the following statements regarding ethics is false?
  (a) Ethics are principles, rules, and standards governing paramedic conduct.
  (b) Ethics are not laws.
  (c) Ethics deal with the relationship of a paramedic to her/his peers and to society at large.
  (d) Ethics are closely related to "morals"—personal standards regarding right and wrong human conduct.
  (e) Ethics require that personal considerations be placed before patient, public, or financial considerations.

2. Which of the following statements regarding professionals and professionalism is false?
  (a) Only an individual receiving monetary compensation for her/his work can legally be considered a "professional."
  (b) A professional has certain special skills and knowledge in a specific area.
  (c) A professional conforms to the standards of conduct and performance in a specific field of knowledge.
  (d) Professionalism promotes high-quality patient care and personal pride, and earns the respect of other team members.
  (e) A professional places concern for the patient before concern for personal ego.

3. As a paramedic, your roles and responsibilities include all of the following, except
  (a) initiating and continuing prehospital emergency care and providing appropriate invasive and noninvasive treatments, as dictated by protocols and medical control.
  (b) becoming involved in community education and training, helping citizens with learning basic life support, proper child-safety-seat use, and the like.
  (c) attempting rescue of a patient from an immediately life-threatening situation, even though you are not appropriately trained to do so, when appropriately trained rescue personnel are not available.
  (d) ensuring the maintenance, preparation, and stocking of emergency-care equipment and supplies.
  (e) recording details related to each incident and each patient's emergency care.

4. The process by which an agency or association grants recognition to an individual who has met certain predetermined qualifications for performing a specific activity is called
  (a) reciprocity.
  (b) certification.
  (c) licensure.
  (d) Answers (a) and (b).
  (e) Answers (a), (b), and (c).

5. Upon determining an individual's degree of occupational competency, a governmental agency grants the applicant permission to engage in that occupation. This process is called obtaining
   (a) reciprocity.
   (b) certification.
   (c) licensure.
   (d) Answers (a) and (b).
   (e) Answers (a), (b), and (c).

6. When an individual is automatically awarded similar certification or licensure because she or he currently holds the same certification or licensure validated by another agency or locality, this process is called
   (a) reciprocity.
   (b) co-certification.
   (c) co-licensure.
   (d) Answer (b) or (c).
   (e) Answer (a), (b), or (c).

7. Protocols are performance policies and procedures for all individuals operating within an EMS system. Protocols are
   (a) laws of operation that must never be altered or deviated from.
   (b) guidelines for usual approaches, subject to improvisation or adaptation, as required by unusual circumstances.
   (c) rules of operation, specifying the only actions that may be performed by EMS providers.
   (d) vague suggestions only, and do not constitute "required" actions in any situation.
   (e) Answer (a) or (c).

8. Protocols address all components of an EMS system, including
   (a) treatment of patients.
   (b) transfer and transport of patients.
   (c) triage of patients.
   (d) Answers (a) and (b) only.
   (e) Answers (a), (b), and (c).

9. An Emergency Medical Services (EMS) system is composed of a variety of components, including
   (a) citizen-initiated care.
   (b) care received in the emergency department.
   (c) public utilities (electric and gas companies).
   (d) Answers (a) and (b).
   (e) Answers (a), (b), and (c).

10. Performing patient-care procedures according to a medical director's previously established "standing orders" without contacting medical direction prior to performing them is an example of operating with
     (a) on-line (direct) medical direction.
     (b) off-line (indirect) medical direction.
     (c) intervening (may be direct or indirect) medical direction.
     (d) All of the above.
     (e) None of the above.

11. Using a radio or telephone to obtain verbal orders prior to performing patient-care procedures is an example of operating with
     (a) on-line (direct) medical direction.
     (b) off-line (indirect) medical direction.
     (c) intervening (may be direct or indirect) medical direction.
     (d) All of the above.
     (e) None of the above.

12. Auditing, peer reviews, and other post-emergency-response quality-assurance processes are considered part of
     (a) on-line (direct) medical direction.
     (b) off-line (indirect) medical direction.
     (c) intervening (may be direct or indirect) medical direction.
     (d) All of the above.
     (e) None of the above.

13. When at an emergency scene, who should be in charge of emergency medical care?
     (a) The EMS responder who arrived on-scene first, regardless of her/his EMS service rank or the rank of those subsequently arriving.
     (b) Any physically present licensed physician. (Whether or not on-line medical direction is available, any physically present licensed physician is ultimately responsible for on-scene care provision.)
     (c) The EMS health-care provider with the most knowledge and experience in the delivery of prehospital emergency care, regardless of her/his EMS service rank.
     (d) The highest-ranked EMS official (such as a fire chief).
     (e) Law-enforcement personnel.

14. The legal classification that applies to contractual wrongs committed by one individual against another individual is called
     (a) criminal law.
     (b) litigation.
     (c) civil (tort) law.
     (d) Any of the above.
     (e) None of the above.

15. The legal classification that applies to crimes and associated punishments related to wrongs committed by an individual against society is called

    (a) criminal law.

    (b) litigation.

    (c) civil (tort) law.

    (d) Any of the above.

    (e) None of the above.

16. A medical malpractice or negligence claim is an example of

    (a) criminal law.

    (b) litigation.

    (c) civil (tort) law.

    (d) Answer (a) or (b).

    (e) None of the above.

17. You are enroute to the emergency department with a critically ill patient. You call medical direction, providing all pertinent patient information, and request an order to administer a medication vital to your patient's survival. The medical director orders you to give the medication but directs you to administer a dose that you feel is contrary or inappropriate to national or local emergency-care protocol doses for this patient. You immediately explain why you consider the ordered dose to be inappropriate, and request a dose you know to be correct. The medical director denies your dose suggestion without counter-explanation, and repeats the order for a dose you still feel is contrary or inappropriate to your patient's survival. Thereafter, you should

    (a) notify the medical director that you refuse to follow the order as directed, and pursue alternative methods of appropriately treating the patient.

    (b) follow the medical director's order, regardless of your opinion of its appropriateness. Because the medical director ordered you to give it, you are not legally responsible for any untoward reactions that may occur.

    (c) thump the radio microphone or telephone receiver on the floor, complain about "transmission interference," discontinue contact (hang up), and ignore the order.

    (d) destroy all medication related to the order, and later document your "inability" to follow the order.

    (e) call the patient's private physician and obtain a second opinion about the medical director's order.

18.   As a paramedic, you may be held responsible for supervising the activities of individuals less medically qualified than you are, be they bystanders or co-responders. Should you allow less medically qualified individuals to perform in a medically inappropriate manner in your presence, without acting to stop and correct their performance, you may be liable for charges of negligence due to litigation generated by the
   (a) Underling Performance Act.
   (b) Paramedic In Charge Act.
   (c) Borrowed Servant Doctrine.
   (d) Answer (a) or (b).
   (e) None of the above. Such legal responsibility does not exist.

19.   If you fail to render appropriate care to an individual because of discrimination or bias on your part against the individual (such as a distaste for touching an AIDS or tuberculosis patient), you may be held liable for both medical negligence and
   (a) breech of the Responsibility To Care doctrine.
   (b) violation of the individual's civil rights.
   (c) commission of the Failure To Act doctrine.
   (d) Answer (a) or (c).
   (e) None of the above. Your personal rights to self-determination supersede the rights of any other individual and you are not required to act in a manner that significantly opposes your personal opinions or biases.

20.   Which of the following statements regarding "Good Samaritan" laws is true?
   (a) Good Samaritan laws differ from state to state. Some states' Good Samaritan laws do not protect trained prehospital care providers, whether paid or volunteer.
   (b) Good Samaritan laws provide protection to people who render care in an emergency situation but only when the care-providing individuals do not receive payment for their care.
   (c) Good Samaritan laws provide protection to all persons (paid and unpaid) who render care in an emergency situation, even when they perform in a manner later determined to be "grossly negligent," as long as it can be shown that the care-providers acted in "good faith" at the time of the emergency.
   (d) All of the above are true.
   (e) None of the above is true.

21.   In most states, the paramedic has an obligation to report (to appropriate local authorities) suspicions of all the following, except
   (a) abuse or neglect of children or the elderly.
   (b) sexual assault.
   (c) gunshot or stab wounds.
   (d) alcohol or drug abuse.
   (e) animal bites.

22. You are called to a private residence where you are met by a woman who tells you that her husband has "passed on." She reports that her husband had terminal cancer and had signed "legal papers" requesting that no resuscitation attempts be made when he died. The papers she is describing may be
   (a) Advanced Directives.
   (b) a Living Will.
   (c) Do Not Resuscitate orders.
   (d) Answer (a) or (b).
   (e) Answer (a), (b), or (c).

23. *(Preceding scenario continued)* The woman cannot find these "legal papers," but she assures you that they were written by her husband's lawyer, signed by her husband, and "even notarized!" The patient is lying in a hospital-style bed in the residence's front room (living room). You assess the patient, finding him apneic and pulseless, warm and dry, jaundiced and emaciated. In the absence of a legal document clearly identifying the patient's wishes, you should
   (a) recognize that the patient was obviously suffering from a terminal illness, respect his wishes by withholding resuscitation, and have the woman sign your trip report sheet as having "refused" care.
   (b) allow the woman a minimum of 15 more minutes to find the papers prior to initiating any kind of resuscitation measures.
   (c) delay resuscitation until the woman has called the patient's attorney and you have phone confirmation from a third party of the patient's wishes.
   (d) initiate and perform "slow code" protocols (only BLS), or "chemical code only" protocols (ACLS without intubation), until you contact medical direction and receive alternative orders.
   (e) aggressively initiate all aspects of ACLS resuscitation, call medical direction, and discuss your findings while preparing to transport the patient.

24. Conduct or care performance that fails to meet, or deviates from, the paramedic standard of care is considered
   (a) willful disobedience.
   (b) negligence.
   (c) reckless endangerment.
   (d) Any of the above.
   (e) None of the above.

25. The elements that must be proven in order to find a paramedic guilty of malpractice in a court of law include all of the following, except
   (a) that the patient asked the paramedic for help.
   (b) that the paramedic had a duty to act.
   (c) that a breach of the duty to act occurred.
   (d) that damages resulted from a breach of duty.
   (e) that the paramedic's action (or lack of action) was the proximate cause of the damage that occurred.

26. Paramedic John intubated an apneic, pulseless patient. He immediately secured the endotracheal tube, initiated ventilation, and thereafter proceeded to follow standard ACLS care protocols on scene and while transporting the patient to the emergency department. The patient was not resuscitated. Upon the emergency department's examination, it was discovered that Paramedic John had placed the endotracheal tube in the patient's esophagus. Because he failed to auscultate the patient's lungs and gastric area after intubation, failed to discover the esophageal intubation, and take appropriate action, Paramedic John committed a breach of duty called

   (a) malfeasance.

   (b) misfeasance.

   (c) nonfeasance.

   (d) Answers (a) and (c).

   (e) None of the above. Failing to auscultate does not constitute a breach of duty.

27. A patient who is unconscious may be treated under the principle of

   (a) informed consent.

   (b) expressed consent.

   (c) implied consent.

   (d) Answer (a) or (b).

   (e) Answer (b) or (c).

28. An adult patient who is alert and oriented to person, place, and time may be treated only after the patient demonstrates

   (a) informed and expressed consent.

   (b) expressed consent only.

   (c) implied consent only.

   (d) Answer (a) or (c).

   (e) Answer (b) or (c).

29. You are transferring a nursing-home resident to the emergency department for the nonemergency evaluation of a minor urinary tract infection. As you arrive at the hospital, your dispatcher notifies you of a 911 call waiting. All of the emergency department nurses are busy, so you give your verbal and written report to the admissions clerk who assures you she will inform the receiving nurse. With the clerk's assistance, you transfer the patient to an empty emergency department bed and respond to the emergency call. Which of the following statements regarding this situation is true?

   (a) Your transfer of care responsibility to the emergency department was legally completed when the admissions clerk "received" the patient.

   (b) Your transfer of care responsibility to the emergency department was legally completed when the patient was placed on the emergency department bed.

(c) You have abandoned your patient and may be liable for malpractice or negligence charges.

(d) Only if the clerk has EMT-Basic qualifications may she legally receive a non-emergency-care patient.

(e) Only if the clerk has EMT-Intermediate qualifications may she legally receive a nonemergency-care patient.

30. Which of the following statements regarding refusal of treatment is true?

(a) An alert adult who is oriented to person, place, and time may refuse treatment only after the potential consequences of refusal have been thoroughly explained (in "plain English"), and the adult has demonstrated a clear understanding of the consequences.

(b) Any patient who obviously requires treatment should receive it regardless of attempts to refuse. Involuntary treatment orders can be generated by law-enforcement personnel at your request.

(c) Even in the absence of a parent or legal guardian, an alert and oriented minor may refuse treatment but only after signing a "release from liability" statement.

(d) Once a patient signs a "release from liability" statement, no further action is required on the part of the paramedic.

(e) A person with an altered level of consciousness may refuse treatment only if over 18 years of age and only after signing a "release from liability" statement.

31. Touching a patient without consent may, technically, be considered

(a) assault.

(b) battery.

(c) false imprisonment.

(d) Any of the above.

(e) None of the above.

32. Creating apprehension of immediate bodily harm may, technically, be considered

(a) assault.

(b) battery.

(c) false imprisonment.

(d) Any of the above.

(e) None of the above.

33. Intentional and unjustifiable detention of a person may, technically, be considered

(a) assault.

(b) battery.

(c) false imprisonment.

(d) Any of the above.

(e) None of the above.

34. The paramedic who breaches patient confidentiality by discussing details of a patient contact with individuals unrelated to the patient's immediate care requirements may be charged with

    (a) invasion of privacy.

    (b) libel.

    (c) slander.

    (d) Answers (a) and (c).

    (e) Answer (a), (b), or (c).

35. A paramedic's best protection when involved in malpractice litigation is

    (a) a good EMS lawyer.

    (b) a complete and accurately written patient-care report.

    (c) a photographic memory.

    (d) repeatedly insisting, "I do not remember."

    (e) police testimony.

36. Stress is best defined as

    (a) an acute hostile reaction to an unusual or disturbing event.

    (b) a psychological or physical state of arousal caused by a stimulus.

    (c) a nonspecific emotional or physical response of the body to any demand made upon it.

    (d) Answer (a) or (b).

    (e) Answer (b) or (c).

37. Any stimulus or situation that causes stress is called

    (a) an aggressor.

    (b) an aggravator.

    (c) a stressor.

    (d) Answer (a) or (b).

    (e) Answer (a), (b), or (c).

38. Stress may result when an individual experiences

    (a) frustrations or disappointments.

    (b) pleasant situations or successes.

    (c) poor health or nutrition.

    (d) Answers (a) and (c).

    (e) Answers (a), (b), and (c).

39. The physiological effects of stress include all of the following, except

    (a) decreased or inhibited production of sympathetic hormones.

    (b) enlargement of the outer portion (cortex) of the adrenal gland.

    (c) atrophy of the thymus gland and other lymphatic structures.

    (d) development of ulcers in the stomach and duodenum.

    (e) alteration of the body's homeostasis.

40. In response to stress, the hypothalamus produces corticotropin-releasing factor (CRF), which stimulates the sympathetic nervous system and endocrine system. Hormones released because of CRF stimulation include all of the following, except
   (a) norepinephrine and epinephrine.
   (b) growth hormone and prolactin.
   (c) cortisol.
   (d) delta endorphins (μ_3 & μ_4).
   (e) beta (β) endorphins.

41. Stress can suppress the immune system and has been linked to all of the following immune-related diseases or conditions, except
   (a) CAD, hypertension, CVA, and dysrhythmias.
   (b) rheumatoid arthritis.
   (c) asthma and hay fever.
   (d) asphyxia.
   (e) impotence, eczema, and acne.

42. The "adaptation" stage of stress response, in which the individual begins to adapt (cope) and physiologic parameters of the body return to normal, is also called the
   (a) alarm stage of stress.
   (b) resistance stage of stress.
   (c) denial stage of stress.
   (d) hostility stage of stress.
   (e) exhaustion stage of stress.

43. The stress-response stage that follows long, continued exposure to the cause(s) of stress is called the
   (a) alarm stage of stress.
   (b) resistance stage of stress.
   (c) denial stage of stress.
   (d) hostility stage of stress.
   (e) exhaustion stage of stress.

44. Physiological and emotional stress responses are greatest during the
   (a) alarm stage of stress.
   (b) resistance stage of stress.
   (c) denial stage of stress.
   (d) hostility stage of stress.
   (e) exhaustion stage of stress.

45. Common physical signs or symptoms exhibited by individuals suffering excessive stress may include all of the following, except

    (a) elation.

    (b) sleep disturbances.

    (c) nausea and vomiting.

    (d) diarrhea.

    (e) vision problems.

46. Common cognitive signs or symptoms exhibited by individuals suffering excessive stress may include all of the following, except

    (a) confusion or disorientation.

    (b) memory deficiencies.

    (c) increased attention span.

    (d) concentration or decision-making difficulties.

    (e) nightmares or distressing dreams.

47. Common emotional signs or symptoms exhibited by individuals suffering excessive stress may include all of the following, except

    (a) elation or a sense of security.

    (b) anxiety, fearfulness, or panic.

    (c) hopelessness or suicidal ideation.

    (d) feeling overwhelmed, abandoned, or lost.

    (e) anger or hostility.

48. Common behavioral signs or symptoms exhibited by individuals suffering excessive stress may include all of the following, except

    (a) acting in an unusual manner (such as being excessively humorous).

    (b) hyperactivity or hypoactivity.

    (c) alcohol or drug abuse.

    (d) loss of appetite.

    (e) "masking" signs and symptoms by acting absolutely "normal" at all times.

49. Circadian rhythms are best defined as

    (a) biological cycles that repeat themselves approximately every 24 hours.

    (b) biological cycles of hormonal and body-temperature fluctuations.

    (c) biological cycles effecting appetite and fatigue.

    (d) All of the above.

    (e) None of the above.

50. All of the following statements about sleep deprivation are true, except

    (a) sleep deprivation is common among EMS individuals who work at night.

    (b) sleep deprivation increases the risk of motor-vehicular accidents involving vehicles operated by emergency personnel.

   (c) sleep deprivation is most effectively prevented by vigorously exercising just prior to taking a nap.

   (d) sleep deprivation often accompanies the disruption of circadian rhythms.

   (e) sleep deprivation may be caused by stress, or may, in and of itself, produce stress signs and symptoms.

51.   EMS shifts often require individuals to sleep during the day. All of the following techniques will minimize the stress of daytime sleeping, except

   (a) taking the phone off the hook or putting the answering machine volume to zero.

   (b) vigorously exercising just prior to sleeping.

   (c) adjusting your sleep environment to mimic nighttime conditions: blocking all light from the windows, turning down the heat, and so on.

   (d) sleeping at the same time on your days off that you sleep on days that you work.

   (e) avoiding consumption of a large meal prior to sleeping.

52.   Which of the following events can, by themselves, be classified as a **critical incident**—one that should trigger the intervention of a Critical Incident Stress Debriefing (CISD) team, for all involved emergency-care providers?

   (1) The death or injury of a child.

   (2) The death or injury of someone known to the EMS provider.

   (3) Incidents that are prolonged in time.

   (4) Incidents that attract a great deal of media attention.

   (5) Mass Casualty Incidents (MCI's).

   (6) The off-duty death or injury of a co-worker.

   (7) The on-duty death or injury of a co-worker.

   (8) The suicide of a co-worker.

   (9) A civilian death caused by a co-worker.

  (10) An emergency call that threatened the safety of a co-worker.

     (a) Only number 5; all others are truly unfortunate incidents, but, by themselves, they do not cause "critical" stress reactions.

     (b) Numbers 5, 7, and 8.

     (c) Numbers 5, 7, 8, and 9.

     (d) Numbers 5, 7, and 10.

     (e) Numbers 1 through 10.

53. Which of the following statements about Critical Incident Stress Debriefing (CISD) is true?

    (a) Because of its current popularity, CISD is overused.

    (b) CISD should be moderated by a trained team that includes peers of the individuals who are being debriefed, as well as mental-health professionals.

    (c) CISD is best accomplished by peers and co-workers, individuals known by those being debriefed; strangers should not be involved in CISD.

    (d) CISD should be limited to one meeting only, so as to minimize the amount of attention paid to the stressful incident, allowing involved individuals to distance themselves from the incident as soon as possible.

    (e) None of the above statements about CISD is true.

54. The Moro reflex occurs when a healthy newborn is surprised, causing the arms to be thrown wide, with fingers spread apart, and is often followed by a grabbing motion. Another term for this reflex is the

    (a) sucking reflex.

    (b) rooting reflex.

    (c) palmar grasp reflex.

    (d) startle reflex.

    (e) protective reflex.

55. When his cheek is touched by a cloth or hand, the hungry infant's head automatically turns to the side, either left or right. This is called the

    (a) sucking reflex.

    (b) rooting reflex.

    (c) palmar grasp reflex.

    (d) startle reflex.

    (e) protective reflex.

56. Which of the following statements about the psychosocial development of an infant is false?

    (a) Because of their underdeveloped senses (especially vision), most infants automatically have a positive response to any individual who securely holds them and speaks softly.

    (b) In response to being separated from a parent or a familiar care-taker, most infants will protest.

    (c) Infants can experience despair or hopelessness.

    (d) In response to prolonged separation from parents or familiar care-takers, infants can become detached or withdrawn.

    (e) Infants experiencing a parental-separation reaction may protest for up to a week.

57. The importance of interaction with peers (individuals of similar age) begins during the _____ Life-Span Development stage.
   (a) Infancy (birth to 12 months old)
   (b) Toddler (12 to 36 months old)
   (c) Preschool Age (3 to 5 years old)
   (d) School Age (6 to 12 years old)
   (e) Adolescence (13 to 18 years old)

58. In most individuals, vital sign "norms" reach that of adult ranges during the _____ Life-Span Development stage.
   (a) Infancy (birth to 12 months old)
   (b) Toddler (12 to 36 months old)
   (c) Preschool Age (3 to 5 years old)
   (d) School Age (6 to 12 years old)
   (e) Adolescence (13 to 18 years old)

59. All of the following techniques positively affect your ability to develop a trusting and therapeutic rapport with the patient, except for
   (a) using the patient's name as often as possible.
   (b) speaking slightly slower than a normal rate of speech.
   (c) speaking loudly, so that miscommunication is minimized.
   (d) controlling your facial expression and keeping it calm.
   (e) moving slowly in a confident and controlled manner.

60. Unfortunately, EMS personnel sometimes fail to communicate effectively with patients. Reasons for communication failure include all of the following, except
   (a) personal prejudice on the part of the paramedic related to the specific patient or the patient's specific illness or injury.
   (b) failure to ensure patient privacy (asking questions in the presence of individuals that the patient does not want to expose information to).
   (c) practicing patience and flexibility during communication (allowing the patient freedom to determine what and how much information to divulge).
   (d) external distractions such as television noise, traffic, crowds, and even EMS radios.
   (e) internal distractions on the part of the paramedic (such as thinking ahead to the "next question to ask" rather than listening to the patient's response).

61. Which of the following statements about nonverbal-communication physical-positioning techniques ("body language") is false?

    (a) Positioning yourself so that your eye level is just above the patient's eye level communicates your authority and expertise, which automatically reassures emergency patients.

    (b) Positioning yourself so that your eye level is the same as the patient's eye level communicates your equality with the patient.

    (c) Positioning yourself so that your eye level is just below the patient's eye level communicates your willingness to allow the patient some control, and is especially helpful with pediatric and geriatric patients.

    (d) Positioning yourself with arms crossed in front of your chest sends negative signals to your patient.

    (e) Nodding your head up and down while listening to the patient sends positive signals to your patient.

62. During a patient interview, the use of **leading questions**

    (a) will help the patient to provide you with the most accurate information.

    (b) will encourage the patient to describe their complaint more completely.

    (c) may result in obtaining inaccurate information.

    (d) Answers (a) and (b).

    (e) Answers (a), (b), and (c).

63. During a patient interview, the use of **open-ended questions**

    (a) will help the patient to provide you with the most accurate information.

    (b) will encourage the patient to describe their complaint more completely.

    (c) may result in obtaining inaccurate information.

    (d) Answers (a) and (b).

    (e) Answers (a), (b), and (c).

64. Which of the following statements about **listening** to a patient is false?

    (a) Listening requires your complete attention.

    (b) The art of listening requires constant practice, and the provider's development of an ability to cease his inner dialogue while listening to the patient.

    (c) Professional, experienced care providers are able to effectively listen to the patient while simultaneously formulating their next question.

    (d) In order to listen well, the provider must stop doing other things and focus upon the patient's response.

    (e) Never finish the patient's responses for him.

65. When dealing with loss, death, and dying, which of the following statements regarding the denial stage is true?

    (a) Denial is employed by almost all patients and is a temporary stage.

    (b) Denial may be limited to the individual's initial reaction, or may be recurrent.

   (c) Denial is a defense mechanism, acting as a buffer and allowing the individual to put off dealing with the loss.

   (d) Answers (a) and (c) are true.

   (e) Answers (a), (b), and (c) are true.

66. When dealing with loss, death, and dying, which of the following statements regarding the anger stage is true?

   (a) The individual or family may displace their feelings of anger by being hostile toward the paramedic.

   (b) A show of tolerance in response to this hostility will encourage escalation of the anger and may promote violence.

   (c) Avoid discussion of the individual's or family's anger and discreetly summon the police for protection.

   (d) All of the above are true.

   (e) None of the above is true.

67. The bargaining stage of loss (dealing with death and dying) is

   (a) an attempt to postpone or alter the inevitable.

   (b) a normal defense mechanism–a normal response.

   (c) an attempt to reach an agreement or bargain that will prevent death.

   (d) Answers (a) and (c).

   (e) Answers (a), (b), and (c).

68. When dealing with the death of a patient, the paramedic

   (a) may go through the same grief stages as the family.

   (b) may need to express feelings of guilt or helplessness.

   (c) should recognize that expressions of personal emotions felt for a stranger's death are inappropriate for a professional.

   (d) Answers (a) and (b).

   (e) Answers (a), (b), and (c).

69. Which of the following statements about telling family members or friends that the patient is dead is false?

   (a) Be sure to avoid using the terms "dead" or "died;" instead, use non threatening phrases such as "passed on," "not with us any longer," and the like.

   (b) Assure the family and friends that everything possible was done for the patient.

   (c) Avoid religious statements or phrases such as "God's will" or "gone to heaven," and the like.

   (d) Avoid assumptive statements or phrases about "relief from pain" or "had a long life," and the like.

   (e) Offer to summon assistance for the survivors, such as a Victims' Advocate member, if your service has the ability to do so.

70. Which of the following statements regarding UHF is true?
   (a) UHF stands for "ultrahigh frequency."
   (b) UHF travels further and better over varied terrain than low-band frequencies, and is more frequently used by county agencies than by city agencies.
   (c) UHF has better penetration of the concrete and steel found in urban areas than low-band frequencies do.
   (d) Answers (a) and (b) are true.
   (e) Answers (a) and (c) are true.

71. A _____ radio transmission communication system uses one frequency and does not allow for simultaneous two-way communications.
   (a) simplex
   (b) monoplex
   (c) biplex
   (d) duplex
   (e) triplex

72. Simultaneous two-way radio communication (much like a telephone conversation) requires two frequencies and is called a _____ transmission system.
   (a) biplex
   (b) duplex
   (c) triplex
   (d) quadraplex
   (e) multiplex

73. The transmission of data (ECG strips) while simultaneously engaged in two-way radio communication on the same channel can occur using a _____ transmission system.
   (a) biplex
   (b) duplex
   (c) triplex
   (d) quadraplex
   (e) multiplex

74. The responsibilities of an experienced emergency medical dispatcher (EMD) include all of the following, except
   (a) professionally representing the EMS system, often being the public's first contact with it.
   (b) directing the appropriate emergency vehicle(s) to the correct address.
   (c) recognizing a "prank" call and immediately disconnecting that caller to free-up the line for real emergencies.
   (d) recognizing the basic nature of many emergencies purely from the caller's description, and instructing the caller to perform appropriate emergency-care measures until assistance arrives.
   (e) maintaining written and phone-recorded records of emergency calls and responses.

75. Which of the following statements regarding radio communication is false?
  (a) Using plain English is often better than using radio 10-codes.
  (b) Radio 10-codes are useless when communicating with individuals or services who don't use them.
  (c) Using specific radio-code responses with dispatchers familiar with them can shorten radio air time while still providing clear information.
  (d) 10-codes especially speed the report of patient medical information (speeding the delivery of medical care) by minimizing the time required to obtain medication orders or directions from any medical advisor answering the hospital's radio or phone.
  (e) The paramedic must be equally familiar with radio codes and medical terminology to facilitate the most rapid and most accurate radio or phone communication with both dispatch and medical control.

76. Radio or land-line communication of patient information to medical control (prior to emergency department arrival) should include all of the following, except
  (a) emergency-response-unit call name, number, and name of the paramedic.
  (b) the patient's name and address.
  (c) the patient's age and sex.
  (d) the patient's approximate weight.
  (e) the patient's chief complaint and associated symptoms.

77. Radio or land-line communication with medical control (prior to emergency department arrival) may also include all of the following, except
  (a) whether the patient is lethargic, obtunded, or comatose.
  (b) vital signs, neurological status, and ECG findings, if applicable.
  (c) a brief, pertinent description of the current emergency (OPQRST).
  (d) a brief, pertinent past medical history, medications, and allergies (SAMPLE).
  (e) treatment rendered prior to calling and requests for orders, if applicable.

78. All of the following statements regarding paramedic communications with medical control (prior to emergency department arrival) are true, except
  (a) Always repeat ("echo") orders given by the physician.
  (b) Always question orders that are not clear to you.
  (c) Always question orders that do not seem appropriate for the patient's condition.
  (d) Always contact the physician, unless the patient has signed the appropriate forms for refusal of treatment or transport.
  (e) Always consult with the physician when you are uncertain of what course to take.

79. Which of the following statements regarding written prehospital patient-care reports (PCRs) is false?

    (a) PCRs should include a written description of the patient's initial condition and information provided to the prehospital care providers upon their arrival.

    (b) PCRs are a legal record of the care rendered to the patient in the prehospital phase of emergency care.

    (c) PCRs provide prehospital care providers with absolute immunity from abandonment litigation, because they are a legally recognized "refusal" form for the patient to sign when refusing care or transportation.

    (d) PCRs are often used for medical audits, quality control, and data collection.

    (e) PCRs must be complete, legible (easily read by anyone), and signed by the PCR author.

80. At 10:00 P.M. on June 5th, you run a call and deliver the patient to the emergency department. At 3:00 A.M. on June 6th, you run another call and deliver that patient to the same emergency department. While there, you notice that your earlier patient has been admitted. Suddenly, you realize you forgot to note something on the earlier patient's care report (PCR). Which of the following statements describes the correct action for you to take?

    (a) No action is required. Once a patient has been admitted, nothing you have to add will make a difference to the patient's care.

    (b) Write in the information on your copy of the patient's care report, as though it had been included when the call was originally run. Do not change the hospital's copy; you'll no longer have access to it, anyway.

    (c) Write in the information on your copy of the patient's care report, as though it had been included when the call was originally run. Be sure to find the hospital's copy of the patient care report, and make the same additions to that copy.

    (d) Write an entirely new PCR, using the same dates and times as the first one but including information you earlier forgot to note. Replace both your copy and the hospital's copy of the PCR, and then destroy all copies of the first one.

    (e) Using another PCR form, write an "amendment" to your first PCR, identifying the patient, the emergency-call-identification information, and noting the information that you earlier forgot to include. Identify the actual date and time that you wrote the PCR amendment. Submit a copy of the PCR amendment to the hospital, and submit a copy of the PCR amendment to your service.

81. While writing a PCR, you make a mistake in your description of something. Which of the following statements describes the correct action for you to take?

    (a) Thoroughly blacken-out the erroneous description, being sure to press firmly to blacken-out that portion of all carbon copies beneath the top copy. Then, write in the correct information.

    (b) Throw away the incorrect PCR and start a new one, being sure you don't make any further mistakes. You cannot submit a PCR if any corrections were made during its writing.

(c) Draw a single line through the erroneous description, so that the underlying text can still be read, place your initials above or beside the lined-through portion, and then write in the correct information.

(d) Throw away your PCR and have your partner write the new one, dictating the correct information to be included.

(e) Thoroughly blacken out the erroneous description, being sure to press firmly to blacken out that portion of all carbon copies beneath the top copy. Then place your initials above or beside the blackened-out portion, and write in the correct information.

82. The medical abbreviation or symbol that means "after" is ___.

(a) $\bar{a}$

(b) $\bar{p}$

(c) $\bar{c}$

(d) $\bar{s}$

(e) $\bar{w}$

83. The medical abbreviation or symbol that means "before" is ___.

(a) $\bar{a}$

(b) $\bar{p}$

(c) $\bar{c}$

(d) $\bar{s}$

(e) $\bar{w}$

84. The medical abbreviation or symbol that means "with" is ___.

(a) $\bar{a}$

(b) $\bar{p}$

(c) $\bar{c}$

(d) $\bar{s}$

(e) $\bar{w}$

85. The medical abbreviation or symbol that means "without" is ___.

(a) $\bar{a}$

(b) $\bar{p}$

(c) $\bar{c}$

(d) $\bar{s}$

(e) $\bar{w}$

86. The medical symbol that means "woman" or "female" is ___.

(a) $\emptyset$

(b) β

(c) ♀

(d) Ψ

(e) ♂

87. The medical symbol that means "man" or "male" is ___.

   (a) Ø

   (b) β

   (c) ♀

   (d) Ψ

   (e) ♂

88. The medical symbol that means "change" is ___.

   (a) ≈

   (b) Δ

   (c) ≅

   (d) ±

   (e) ↔

89. The medical symbol that means "approximately" is ___.

   (a) ≈

   (b) Δ

   (c) ≅

   (d) ±

   (e) ↔

90. The medical abbreviation BS means

   (a) breath sounds.

   (b) blood sugar.

   (c) bovine scat.

   (d) Answer (a) or (b).

   (e) Answer (a) or (c).

The Answer Key to Test Section 1 is on page 402.

Test
Section
Two

2

Test Section 2's subjects:

- General Principles of Pathophysiology
- History Taking
- Physical Exam Techniques
- Patient Assessment
- Clinical Decision Making
- Assessment-Based Management
- Trauma Systems and Mechanism of Injury
- Hemorrhage and Shock

Test Section 2 consists of 150 questions and is allotted 2.5 hours for completion.

1.   The largest organ in the human body is the
     (a) liver.
     (b) lungs.
     (c) small intestine.
     (d) large intestine.
     (e) skin.

2.   The body is continuously synthesizing or breaking down substances; thus the
     internal environment is always changing, yet always tending toward a state of
     balance. Homeostasis is defined (and often referred to) as
     (a) a static state of balance.
     (b) a dynamic steady state.
     (c) a constant state of synthesis.
     (d) a state of synthesis that alternates between being active and inactive.
     (e) a continuous-activity state of balance.

3.   The thick, viscous fluid that fills and gives shape to each cell of the body, is called
     (a) exoplasm.
     (b) protoplasm.
     (c) cytoplasm.
     (d) either exoplasm or protoplasm.
     (e) either protoplasm or cytoplasm.

4.   Structures within each cell that perform specific functions are called
     (a) monocytes.
     (b) cytokines.
     (c) organelles.
     (d) phagocytes.
     (e) granulocytes.

5.   The conversion of nutrients into energy is a function of the cell's
     (a) nucleus.
     (b) endoplasmic reticulum.
     (c) Golgi apparatus.
     (d) mitochondria.
     (e) lysosomes.

6.   Protection against disease, production of nutrients (such as sugars and amino
     acids), and breakdown of bacteria and organic debris are functions of the cell's
     (a) nucleus.
     (b) endoplasmic reticulum.

(c) Golgi apparatus.

(d) mitochondria.

(e) lysosomes.

7. Deoxyribonucleic acid (DNA), the genetic material that determines our inherited traits, is stored within the cell's

(a) nucleus.

(b) endoplasmic reticulum.

(c) Golgi apparatus.

(d) mitochondria.

(e) lysosomes.

8. Energy created within the cell is often stored in the form of ATP, which yields energy when it is split by enzyme action. ATP stands for

(a) abdominal triproteinemia.

(b) adrenal triproteinemia.

(c) adrenal triphosphate.

(d) adenosine triproteinemia.

(e) adenosine triphosphate.

9. Which of the following are considered to be among the seven major cell functions?

(1) movement

(2) conductivity

(3) metabolic absorption

(4) secretion

(5) excretion

(6) respiration

(7) reproduction

(a) All of the above functions (1, 2, 3, 4, 5, 6, and 7).

(b) Only functions 1, 2, 3, 4, 5, and 6.

(c) Only functions 2, 3, 4, and 5.

(d) Only functions 2, 3, and 4.

(e) Only functions 1 and 2.

10. The skin, the mucous membranes, and the intestinal tract's lining are each examples of _____ tissue.

(a) nerve

(b) connective

(c) smooth muscle

(d) skeletal muscle

(e) epithelial

11. _____ is the most abundant type of tissue in the body.
    (a) Nerve tissue
    (b) Connective tissue
    (c) Smooth muscle tissue
    (d) Skeletal muscle tissue
    (e) Epithelial tissue

12. Which of the following types of muscle tissue has the ability to contract when stimulated?
    (a) Cardiac muscle tissue.
    (b) Smooth muscle tissue.
    (c) Skeletal muscle tissue.
    (d) Answers (a) and (c) only.
    (e) Answers (a), (b), and (c).

13. The most abundant form of muscle tissue in the human body is
    (a) cardiac muscle tissue.
    (b) smooth muscle tissue.
    (c) skeletal muscle tissue.
    (d) Answers (a) and (b) are equally abundant.
    (e) Answers (a) and (c) are equally abundant.

14. _____ has the ability to spontaneously contract (without external stimulation).
    (a) Cardiac muscle tissue
    (b) Smooth muscle tissue
    (c) Skeletal muscle tissue
    (d) Answers (a) and (b).
    (e) Answers (b) and (c).

15. Ductless glands that secrete hormones directly into the circulatory system are called _____ glands.
    (a) epithelial
    (b) exocrine
    (c) paracrine
    (d) endocrine
    (e) autocrine

16. Glands that secrete hormones which reach an epithelial surface via ducts are called _____ glands.
    (a) epithelial
    (b) exocrine
    (c) paracrine
    (d) endocrine
    (e) autocrine

17. The most common form of cellular injury is caused by
   (a) chemical insults.
   (b) infections.
   (c) hypoxia.
   (d) inflammatory insults.
   (e) nutritional imbalances.

18. A pathogen is best defined as a microorganism that
   (a) produces only bacterial diseases.
   (b) is capable of causing infection or disease.
   (c) combats viral or bacterial diseases.
   (d) produces only fungal infections.
   (e) is capable of defeating infection or disease.

19. Ischemia can result from all of the following, except
   (a) a blockage in the delivery of oxygenated blood to the cells.
   (b) carbon monoxide poisoning.
   (c) aerobic metabolism.
   (d) red blood cell deficiencies.
   (e) hemoglobin deficiencies.

20. On average, water accounts for ___ percent of an adult's total body weight.
   (a) 15
   (b) 30
   (c) 60
   (d) 75
   (e) 90

21. On average, water accounts for ___ percent of an infant's total body weight.
   (a) 15
   (b) 30
   (c) 60
   (d) 75
   (e) 90

22. The largest percentage of an individual's body water is contained within
   (a) the intracellular compartment.
   (b) the extracellular compartment.
   (c) the interstitial compartment.
   (d) the intravascular compartment.
   (e) skin.

23. Dehydration (an abnormal decrease in total body water) can result from which of the following?

(1) plasma losses from burns or open wounds   (4) hyperventilation
(2) vomiting or diarrhea   (5) bowel obstruction
(3) diaphoresis   (6) poor nutritional states

     (a) 1, 2, 3, 4, 5, or 6.
     (b) 1, 2, 3, 4, or 5.
     (c) 1, 2, 3, or 4.
     (d) 1, 2, or 3.
     (e) 1 or 2.

24. Electrolytes are substances that

     (a) dissociate (separate) into electrically charged particles when placed into water.
     (b) associate (combine) into electrically charged particles when placed into water.
     (c) have an equal number of protons and electrons.
     (d) Answers (a) and (c).
     (e) Answers (b) and (c).

25. Which of the following statements about a cation are true?

     (a) A cation is an ion with a positive charge.
     (b) A superscript plus sign ($^+$) indicates a cation.
     (c) Because it is positively charged, a cation will be attracted to a positive pole.
     (d) Answers (a) and (b).
     (e) Answers (a), (b), and (c).

26. Which of the following statements about an anion are true?

     (a) An anion is an ion with a negative charge.
     (b) A superscript minus sign ($^-$) indicates an anion.
     (c) Because it is negatively charged, an anion will be attracted to a negative pole.
     (d) Answers (a) and (b).
     (e) Answers (a), (b), and (c).

27. Electrolytes are usually measured in

     (a) percentages (%).
     (b) micrograms per liter (µg/L or mcg/L).
     (c) milliliters (ml).
     (d) milliequivalents per liter (mEq/L).
     (e) drops (gtt).

28. Open-ended questions are

     (a) the best questions to use when your patient is short of breath.
     (b) questions that require the patient to provide more than just a one-word answer.

(c) questions that require only a "yes" or "no" answer.
(d) Answers (a) and (b).
(e) Answers (a) and (c).

29. Which of the following are good examples of open-ended questions?
   (1) "Can you describe the pain in your chest?"
   (2) "What is bothering you?"
   (3) "Do you take diuretics?"
   (4) "Is your pain sharp or stabbing?"
   (5) "How does your head feel?"

      (a) 1, 2, and 3.
      (b) 1, 3, 4, and 5.
      (c) 2 and 5.
      (d) 1, 3, and 5.
      (e) 1, 2, 3, 4, and 5.

30. In the History of Present Illness information mnemonic OPQRST-ASPN, the first letter P stands for
   (a) provocation. (What makes it worse?)
   (b) palliation. (What makes it better?)
   (c) previous (previous health information).
   (d) Answers (a) and (b).
   (e) Answers (a) and (c).

31. At the end of the History of Present Illness information mnemonic OPQRST-ASPN, the letters P and N stand for
   (a) previously known (has had the problem before).
   (b) pertinent negatives (denials of complaints).
   (c) past notes (previous health information).
   (d) Answers (a) and (b).
   (e) Answers (a) and (c).

32. Using your sense of touch to examine the patient's body is a technique called
   (a) auscultation.
   (b) palpitation.
   (c) palpation.
   (d) percussion.
   (e) inspection.

33. The presence of a radial pulse is considered to suggest a systolic blood pressure of
   (a) 90 mmHg.
   (b) 60 mmHg.
   (c) 80 mmHg.
   (d) 40 mmHg.
   (e) 70 mmHg.

34.   The presence of a carotid pulse is considered to suggest a systolic blood pressure of
  (a) 90 mmHg.
  (b) 60 mmHg.
  (c) 80 mmHg.
  (d) 40 mmHg.
  (e) 70 mmHg.

35.   The systolic blood pressure measures the force of pressure against the
  (a) veins when the ventricles contract.
  (b) arteries when the ventricles contract.
  (c) veins when the atria relax.
  (d) veins when the ventricles relax.
  (e) arteries when the ventricles relax and fill with blood.

36.   The diastolic blood pressure measures the force of pressure against the
  (a) veins when the ventricles contract.
  (b) arteries when the ventricles contract.
  (c) veins when the atria relax.
  (d) veins when the ventricles relax.
  (e) arteries when the ventricles relax and fill with blood.

37.   A patient's pulse pressure is
  (a) the difference between the systolic and diastolic blood pressure measurements (diastolic blood pressure subtracted from systolic blood pressure).
  (b) the amount of change in systolic blood pressure when the patient changes position from lying to sitting.
  (c) the sum of three consecutive systolic blood pressure measurements, divided by 3 (the average of three systolic blood pressures).
  (d) the sum of three consecutive diastolic blood pressure measurements, divided by 3 (the average of three diastolic blood pressures).
  (e) determined by dividing the systolic blood pressure by the diastolic blood pressure.

38.   Which of the following emergencies is unlikely to cause hypertension?
  (a) Anaphylaxis.
  (b) Cerebrovascular accident.
  (c) Renal disease.
  (d) Closed head injury.
  (e) Arteriosclerosis.

39.   "Tilt test" vital signs are first measured when the patient is supine, and then measured again 30 to 60 seconds after the patient has changed to a sitting or standing position. Hypovolemia may be suggested if, after changing to a sitting or standing position, the patient's pulse rate has

    (a) increased by 4 to 6 beats per minute.

    (b) decreased by 4 to 6 beats per minute.

    (c) increased by 10 to 20 beats per minute.

    (d) decreased by 10 to 20 beats per minute.

    (e) remained the same.

40. The medical phrase for when vital sign changes occur after the patient changes from a supine to a sitting or standing position is

    (a) orthopnic vital sign changes.

    (b) polypostural vital sign changes.

    (c) range-of-motion vital sign changes.

    (d) orthostatic vital sign changes.

    (e) paradoxical vital sign changes.

41. Normal body temperature is approximately 98.6° Fahrenheit; the Celsius equivalent of this body temperature is

    (a) 30°C.

    (b) 31°C.

    (c) 37°C.

    (d) 39°C.

    (e) 44.8°C.

42. The metal end-piece of a quality stethoscope has two sides. One side is a rigid, flat, diaphragm. The other side is a cone-shaped, bell-like device. The "bell" side should be used when auscultating

    (a) lung sounds.

    (b) heart sounds.

    (c) bowel sounds.

    (d) Answers (a) and (c).

    (e) Answers (b) and (c).

43. Which of the following statements about accurate blood pressure measurement is false?

    (a) Using a blood pressure cuff wider than one-half to two-thirds of the patient's upper arm will result in an inaccurate blood pressure measurement.

    (b) If your patient is obese and you do not have an obese-size blood pressure cuff, use the adult blood pressure cuff on the obese patient's forearm, and auscultate the blood pressure at the radial pulse site.

    (c) If the adult cuff covers three-fourths or more of your adult patient's upper arm, consider using a pediatric blood pressure cuff.

    (d) If the adult cuff covers three-fourths or more of your adult patient's upper arm, consider auscultating a blood pressure with the adult cuff applied to the patient's thigh.

    (e) In renal failure patients, the arm containing a dialysis shunt will be the arm with the best circulatory perfusion. Use that arm in order to measure the most accurate blood pressure.

44. During your first attempt to auscultate the patient's blood pressure measurement, you accidentally deflate the cuff too fast to accurately determine a systolic reading, but manage to stop deflation before reaching the diastolic measurement. You should

    (a) immediately reinflate the cuff (prior to complete deflation) to beyond the point where you can hear the pulse beat, and more slowly deflate it to gain the accurate systolic measurement.

    (b) slowly continue the deflation and record the diastolic reading. Then, estimate the systolic reading you most likely measured.

    (c) completely deflate the cuff (without noting a diastolic measurement), and wait for at least 30 seconds before again inflating the cuff to auscultate an accurate systolic and diastolic blood pressure.

    (d) resume deflation and note the diastolic measurement, then immediately reinflate the cuff to measure the accurately corresponding systolic blood pressure measurement.

    (e) discontinue attempts to auscultate a blood pressure, and palpate one, instead.

45. At sea level, a pulse oximetry reading of between _____ indicates that your patient's blood has a normal oxygen saturation.

    (a) 76 and 80%

    (b) 81 and 85%

    (c) 86 and 90%

    (d) 91 and 95%

    (e) 96 and 100%

46. Which of the following statements about pulse oximetry assessment is false?

    (a) A pulse oximeter reading of below 90% indicates the need for high-flow oxygen administration, and consideration of aggressive airway management with positive pressure ventilation.

    (b) Because modern pulse oximeter devices are highly sensitive to the difference between oxygen molecule bound hemoglobin versus carbon monoxide molecule bound hemoglobin, pulse oximetry is especially helpful in evaluating the severity of carbon monoxide poisoning cases.

    (c) If the patient's actual pulse does not correspond with the digital pulse rate readout (or pulsation wave) displayed by your pulse oximeter model, your pulse oximeter is probably providing an inaccurate reading of the patient's oxygen saturation.

    (d) If your patient is hypovolemic, the pulse oximeter reading will probably be false.

    (e) If your patient is hypothermic, the pulse oximeter reading will probably be false.

47. Which of the following statements about blood glucose measurement devices (glucometers) is false?

    (a) All patients exhibiting an altered level of consciousness should have their blood glucose level evaluated.

    (b) Modern glucometers are entirely accurate, and require only infrequent (once- or twice-a-year) calibration to ensure reliable blood glucose level evaluations, especially when regularly used.

(c) All patients exhibiting signs or symptoms of cerebrovascular accident should have their blood glucose level evaluated.

(d) All glucometer models work differently. If using an unfamiliar glucometer, read and follow the directions carefully. The slightest deviation from any particular model's directions may produce a false reading.

(e) Hand-held glucometers are only moderately accurate, even when properly used and calibrated daily.

48. The medical term _____ refers to small red spots on the skin, from which irregular red lines ("legs") radiate in all directions.
  (a) ecchymosis
  (b) petechiae
  (c) purpura
  (d) venous star
  (e) spider angioma

49. The medical term _____ refers to blue spots on the skin, from which irregular blue lines ("legs") radiate in all directions.
  (a) ecchymosis
  (b) petechiae
  (c) purpura
  (d) venous star
  (e) spider angioma

50. The medical term _____ refers to small reddish-purple spots on the skin.
  (a) ecchymosis
  (b) petechiae
  (c) purpura
  (d) venous star
  (e) spider angioma

51. Facial bones include all of the following, except
  (a) the zygoma.
  (b) the maxilla.
  (c) the mandible.
  (d) the sphenoid bone.
  (e) the palantine bone.

52. The cranium consists of the frontal, temporal,
  (a) occipital, maxillary, and nasal concha bones.
  (b) sphenoid, occipital, and parietal bones.
  (c) ethmoid, sphenoid, and zygomatic bones.
  (d) parietal, zygomatic, and sphenoid bones.
  (e) occipital, parietal, and zygomatic bones.

53. The orbits of the eye (the "eye sockets") are formed by the
    (a) frontal bones.
    (b) zygomatic bones.
    (c) maxillary and nasal bones.
    (d) Answers (a) and (b).
    (e) Answers (a), (b), and (c).

54. The anterior chamber of the eye is filled with a clear liquid called
    (a) visceral humor.
    (b) aqueous humor.
    (c) vitreous humor.
    (d) aquial gel.
    (e) globular gel.

55. The posterior chamber of the eye is filled with a clear gel-like substance called
    (a) visceral humor.
    (b) aqueous humor.
    (c) vitreous humor.
    (d) aquial gel.
    (e) globular gel.

56. The anterior and posterior chambers of the eye are separated by the
    (a) retina.
    (b) cornea.
    (c) lacrimal ducts.
    (d) pupil.
    (e) lens.

57. A structure of the eye that does not have its own vasculature and is dependent upon lacrimal fluid for nutrients and oxygen is called the
    (a) iris.
    (b) cornea.
    (c) sclera.
    (d) conjunctiva.
    (e) lens.

58. The cervical spine consists of ___ vertebrae.
    (a) 5
    (b) 7
    (c) 9
    (d) 12
    (e) 13

59.   The thoracic spine consists of ___ vertebrae.

    (a) 5

    (b) 7

    (c) 9

    (d) 12

    (e) 13

60.   The lumbar spine consists of ___ vertebrae.

    (a) 5

    (b) 7

    (c) 9

    (d) 12

    (e) 13

61.   The spinal-structure abnormality involving an exaggerated inward curve of the lumbar spine is commonly called "swayback." The medical term for this abnormality is

    (a) kyphosis.

    (b) hyperphosis.

    (c) lordosis.

    (d) hypodosis.

    (e) scoliosis.

62.   The spinal-structure abnormality involving an exaggerated outward curve of the thoracic spine is commonly called "hunchback." The medical term for this abnormality is

    (a) kyphosis.

    (b) hyperphosis.

    (c) lordosis.

    (d) hypodosis.

    (e) scoliosis.

63.   The medical term for the spinal-structure abnormality involving an exaggerated lateral curvature of one or more spinal areas is

    (a) kyphosis.

    (b) hyperphosis.

    (c) lordosis.

    (d) hypodosis.

    (e) scoliosis.

64. There are ___ pairs of cranial nerves.
    (a) 5
    (b) 7
    (c) 9
    (d) 12
    (e) 13

65. The cranial nerves originate from the _____ of the brain.
    (a) base
    (b) cerebral cortex
    (c) bilateral temporal lobes
    (d) bilateral parietal lobes
    (e) frontal lobe

66. Which of the following activities should be performed during the Scene Size-Up portion of an emergency response?
    (1) Don body substance isolation personal protective equipment.
    (2) Assess the attitudes of bystanders. (Do they appear hostile, panicked?)
    (3) Assess the scene for safety risks.
    (4) If the scene is on a roadway, ensure safe traffic routing.
    (5) Determine the number of patients involved.
    (6) Evaluate the scene for signs of the mechanism of injury or the nature of illness.

    (a) 3, 4, and 5.
    (b) 1, 3, 4, and 5.
    (c) 2, 3, 4, and 5.
    (d) 1, 3, 4, 5, and 6.
    (e) 1, 2, 3, 4, 5, and 6.

67. From the following numbered activities, select only those steps that should be performed during the Initial Assessment phase of patient evaluation and examination. In addition, place your selection in the appropriate order of initial-assessment performance and management priority.
    (1) Determine the patient's "priority" assignment (need for rapid or delayed treatment or transport).
    (2) Stabilize the cervical spine if indicated by the mechanism of injury.
    (3) Assess circulation (check for a pulse).
    (4) Assess airway and breathing.
    (5) Assess mental status.
    (6) Measure the blood pressure.
    (7) Form a general impression of the patient's condition.

    (a) 4, 3, 5, 6, 7, 1.
    (b) 2, 4, 3, 5, 7, 1.

(c) 7, 2, 5, 4, 3, 1.

(d) 5, 4, 3, 2, 7, 1.

(e) 5, 2, 4, 3, 6, 7.

68. The A in the mental status (level of consciousness) AVPU assessment mnemonic stands for

(a) average patient response, whether well oriented or confused.

(b) adequate patient response, only if well oriented.

(c) apneic patient, whether well oriented or confused.

(d) alert (awake) patient, only if well oriented.

(e) alert (awake) patient, whether well oriented or confused.

69. The V in the mental status (level of consciousness) AVPU assessment mnemonic stands for

(a) the patient's violent response to stimulation.

(b) a patient who responds to verbal stimulation.

(c) a patient who responds to various stimuli.

(d) the patient's vague response to stimulation.

(e) a patient who has varied (inconsistent) response to stimulation.

70. The P in the mental status (level of consciousness) AVPU assessment mnemonic stands for a patient

(a) who responds to painful stimulation.

(b) with a partial response to stimulation.

(c) with a physical response to verbal stimulation.

(d) with pulses absent.

(e) with pulses present.

71. The U in the mental status (level of consciousness) AVPU assessment mnemonic stands for a patient

(a) you are unable to test with stimulation.

(b) who is unwilling to respond to stimulation.

(c) with uneven response to stimulation.

(d) who is unresponsive to any stimulation.

(e) with an unreliable (suspicious) response to stimulation.

72. An audible high-pitched screech on inspiration indicates an upper airway obstruction, and may be produced by any of the following, except

(a) a foreign body.

(b) severe swelling.

(c) a relaxed tongue.

(d) epiglottitis.

(e) an allergic reaction.

73. The medical term for the audible inspiratory high-pitched screech caused by an upper airway obstruction is
   (a) snoring.
   (b) wheezing.
   (c) stridor.
   (d) rales.
   (e) rhonchi.

74. To manually open the airway of an atraumatic medical patient, the _____ may be performed.
   (a) head-tilt/chin-lift maneuver
   (b) head-tilt/neck-lift maneuver
   (c) jaw-thrust maneuver
   (d) Answer (a) or (c).
   (e) Answer (a), (b), or (c).

75. To manually open the airway of a trauma patient, the _____ may be performed.
   (a) head-tilt/chin-lift maneuver
   (b) head-tilt/neck-lift maneuver
   (c) jaw-thrust maneuver
   (d) Answer (a) or (c).
   (e) Answer (a), (b), or (c).

76. A weak and thready pulse might indicate any of the following, except
   (a) hypertension.
   (b) heart failure.
   (c) hypovolemia.
   (d) systemic vasodilation.
   (e) dehydration.

77. The presence of poor capillary refill is a reliable sign of poor circulation in
   (a) infants only.
   (b) infants and young children only.
   (c) infants, young children, and adults.
   (d) young children and adults only.
   (e) infants and adults only.

78. Which of the following criteria indicates a patient who should be considered Top Priority for rapid transportation to an emergency department (a patient whose transport should not be delayed by on scene Detailed Assessment performance)?
   (1) Complaints of severe pain.
   (2) Signs and symptoms of complicated childbirth.
   (3) Complaints of chest pain with a systolic blood pressure below 100.
   (4) Difficulty breathing.

(5) Unresponsive patients.

(6) Conscious patients who cannot follow commands.

(7) Patients who generate a poor General Impression.

    (a) 2, 3, 4, and 5.

    (b) 2, 3, 4, 5, and 6.

    (c) 2, 3, 4, 5, and 7.

    (d) 2, 3, 4, 5, 6, and 7.

    (e) 1, 2, 3, 4, 5, 6, and 7.

79. In the absence of obvious external signs of traumatic injury, predictors of serious internal traumatic injury include

(1) A helmeted individual involved in a motorcycle crash.

(2) A restrained individual found in a vehicle compartment in which a dead, unrestrained individual was also found.

(3) Any occupant of a rollover accident.

(4) A restrained high-speed-collision passenger.

(5) An unrestrained high-speed-collision passenger.

(6) An individual ejected from a vehicle.

    (a) 2, 3, 5, and 6.

    (b) 1, 3, 5, and 6.

    (c) 1, 3, 4, 5, and 6.

    (d) 1, 2, 3, 5, and 6.

    (e) 1, 2, 3, 4, 5, and 6.

80. In the rapid trauma assessment mnemonic DCAP-BTLS, the letter A stands for

    (a) apnea.

    (b) absent pulse(s).

    (c) altered level of consciousness.

    (d) ataxia or aphasia.

    (e) None of the above.

81. In the rapid trauma assessment mnemonic DCAP-BTLS, the letter P stands for

    (a) pain.

    (b) penetration.

    (c) pulse absence.

    (d) priapism.

    (e) None of the above.

82. In the rapid trauma assessment mnemonic DCAP-BTLS, the letter T stands for

    (a) trauma.

    (b) tremulousness.

    (c) tenderness.

    (d) tinnitus.

    (e) None of the above.

83. In the rapid trauma assessment mnemonic DCAP-BTLS, the letter S stands for
   (a) swelling.
   (b) severe injury.
   (c) somatic complaints.
   (d) significant blood loss.
   (e) None of the above.

84. Your patient is lying supine on a flat surface. You can see the slight outline of his jugular veins. Which of the following statements about this observation is true?
   (a) It is not unusual to see the outline of jugular veins when someone is lying supine on a flat surface.
   (b) Observation of jugular vein outlines indicates abnormal jugular venous distention only if the patient is in a Trendelenburg position.
   (c) No matter how the patient is positioned, any observance of jugular vein outlines (slight or pronounced) indicates the abnormal presence of jugular venous distention.
   (d) All of the above are true.
   (e) None of the above is true.

85. Distended jugular veins may indicate that
   (a) something is inhibiting the patient's blood return to his chest.
   (b) the patient may have cardiac tamponade.
   (c) the patient may have a tension pneumothorax.
   (d) Answer (a) or (b) only.
   (e) Answer (a), (b), or (c).

86. You notice that, on inspiration, your patient's trachea seems to "tug" (shift) to one side, and then return to the midline. This observation suggests that your patient may have a
   (a) tension pneumothorax on the side that the trachea is tugging toward.
   (b) tension pneumothorax on the side that the trachea is tugging away from.
   (c) simple pneumothorax on the side that the trachea is tugging toward.
   (d) simple pneumothorax on the side that the trachea is tugging away from.
   (e) hemothorax on the side that the trachea is tugging toward.

87. You notice that your patient's trachea seems to be displaced from the midline (shifted to one side) during both inspiration and expiration. This observation suggests that your patient may have a
   (a) tension pneumothorax on the side that the trachea is shifted toward.
   (b) tension pneumothorax on the side that the trachea is shifted away from.
   (c) simple pneumothorax on the side that the trachea is shifted toward.
   (d) simple pneumothorax on the side that the trachea is shifted away from.
   (e) hemothorax on the side that the trachea is shifted toward.

88. "Cullen's sign" is a characteristic bruising pattern indicative of internal abdominal hemorrhage, and is recognized as

    (a) bruising of the umbilical area.

    (b) bruising of the flanks.

    (c) a sign that one or more hours have passed since the abdominal trauma occurred.

    (d) Answers (a) and (c).

    (e) Answers (b) and (c).

89. "Grey-Turner's sign" is a characteristic bruising pattern indicative of internal abdominal hemorrhage, and is recognized as

    (a) bruising of the umbilical area.

    (b) bruising of the flanks.

    (c) a sign that one or more hours have passed since the abdominal trauma occurred.

    (d) Answers (a) and (c).

    (e) Answers (b) and (c).

90. When using the SAMPLE patient history mnemonic, the letter A stands for

    (a) associated symptoms.

    (b) aggravation and alleviation.

    (c) allergies.

    (d) Any or all of the above.

    (e) None of the above.

91. When using the SAMPLE patient history mnemonic, the letter M stands for

    (a) mechanism of injury.

    (b) medications.

    (c) movement or motor findings.

    (d) Any or all of the above.

    (e) None of the above.

92. When using the SAMPLE patient history mnemonic, the letter P stands for

    (a) past medical history.

    (b) pain description.

    (c) patient-provided care, prior to your arrival.

    (d) Any or all of the above.

    (e) None of the above.

93. When using the SAMPLE patient history mnemonic, the letter L stands for

    (a) last ate.

    (b) last drank.

    (c) last oral intake of anything.

    (d) Any or all of the above.

    (e) None of the above.

94. Identify the most correct performance order for the following components of a full patient assessment:

(1) Detailed Physical Exam   (4) Ongoing Assessment
(2) Scene Size-Up   (5) Focused History and Physical Exam
(3) Initial Assessment

    (a) 2, 3, 1, 4, 5.
    (b) 2, 3, 5, 1, 4.
    (c) 2, 3, 4, 1, 5.
    (d) 3, 2, 5, 1, 4.
    (e) 3, 2, 4, 1, 5.

95. When should an Ongoing Assessment be conducted?

(a) Once, just prior to arrival at the emergency department.
(b) Every 5 minutes of transportation, for all patients (stable or unstable).
(c) Every 5 minutes for stable patients, and every 15 minutes for unstable patients.
(d) There are no guidelines for how often an Ongoing Assessment should be conducted.
(e) None of the above answers is correct.

96. The most important goal of developing good clinical decision-making and assessment-based management skills is to

(a) become adept at differentiating between critical life-threatening, potentially life-threatening, and non-life-threatening patient presentations.
(b) obtain the fullest and most complete account of the patient's past medical history to report to the emergency department physician.
(c) obtain the fullest and most complete account of the patient's current complaint to report to the emergency department physician.
(d) Answers (b) and (c).
(e) None of the above answers is correct.

97. A small medical facility serving a remote area, which acts only to stabilize and prepare patients with moderate or serious injuries before transferring them to a higher level trauma facility, would most likely have a _____ Trauma Center/Facility designation.

(a) Level I
(b) Level II
(c) Level III
(d) Level IV
(e) Level X

98. Which of the following procedures must be performed prior to initiating transport of a seriously injured trauma victim?

(a) Recognition of a critical mechanism of injury.
(b) Initiation of IV fluid replacement for patients with a systolic blood pressure less than 90.

(c) Obtaining a medical history and immobilizing extremity fractures.

(d) Answers (a) and (b).

(e) Answers (a), (b), and (c).

99.   Newton's first law of motion (inertia) states that

(a) a body in motion will remain in motion unless acted upon by an outside force.

(b) a body at rest will remain at rest unless acted upon by an outside force.

(c) force equals mass multiplied by velocity of acceleration or deceleration.

(d) Answers (a) and (b).

(e) Answers (a) and (c).

100.   An automobile driven by a restrained individual was traveling approximately 30 mph when it impacted, head-on, with a cement bridge abutment, coming to a very sudden stop. You arrive within five minutes of the accident. You see that the front end of the auto was crumpled and bent off to one side. The dashboard and steering wheel show no patterns of damage from driver impact. The driver denies striking his head, denies experiencing any loss of consciousness, and denies CTL spine pain. The driver's thick clothing prevented restraint abrasions or bruises from occurring. Which of the following statements regarding the energy of this accident is false?

(a) A destructive energy was newly created the instant the auto struck the abutment.

(b) Vehicle deformity and a bent frame indicate that the auto absorbed energy upon impact.

(c) The restrained driver of this auto also absorbed energy upon impact.

(d) Three impacts occurred: the auto impacted the abutment, the driver impacted the seatbelts, and the driver's internal organs impacted various internal structures or restraining ligaments.

(e) Even though the individual was restrained at the time of impact, and has no external signs of trauma, the transfer or conversion of energy may have caused internal organ injuries that have not yet created observable signs or symptoms.

101.   The branch of physics that deals with how force and mass affect objects in motion, and the energy exchanges that occur during collision or impact, is called

(a) forceology.

(b) motionology.

(c) massology.

(d) kinetics.

(e) dyanetics.

102.   Severe deceleration forces may result in life-threatening hemorrhage of the _____ secondary to laceration by the ligamentum teres.

(a) aorta

(b) liver

(c) kidneys

(d) spleen

(e) bladder

103. Severe deceleration forces may result in life-threatening hemorrhage of the _____ secondary to laceration by the ligamentum arteriosum.

   (a) aorta
   (b) liver
   (c) kidneys
   (d) spleen
   (e) bladder

104. Your patient, an unrestrained driver, was involved in a frontal impact accident. He traveled "down and under" on impact, his knees striking and damaging the lower margin of the driver's side dash. Based upon the dashboard damage, you should strongly anticipate all of the following, except

   (a) chest injuries.
   (b) femoral shaft injuries.
   (c) distal tibial injuries.
   (d) posterior hip fractures.
   (e) posterior hip dislocations.

105. Another frontal impact accident resulted in the unrestrained driver traveling an "up and over" pathway. There is a star fracture of the windshield (caused by patient contact) high above the steering wheel. Based upon this sign, you should anticipate

   (a) compression injuries of the cervical spine.
   (b) hyperflexion injuries of the cervical spine.
   (c) hyperextension injuries of the cervical spine.
   (d) Answers (b) and (c).
   (e) Answer (a), (b), or (c).

106. This same frontal impact, "up and over" travel pathway (with a star fracture of the windshield caused by patient contact high above the steering wheel), should also cause you to anticipate

   (a) skull or facial injuries.
   (b) abdominal hollow organ rupture and liver laceration.
   (c) bilateral femoral fractures.
   (d) Answers (a) and (b).
   (e) Answer (a), (b), or (c).

107. Which of the following statements regarding ejection injuries is false?

   (a) Ejection always involves multiple impacts and produces multiple injuries.
   (b) Ejection most commonly occurs in frontal impact accidents.
   (c) Although risk of multiple minor injuries is high, ejection often "saves" the unrestrained victim from more life-threatening injuries (such as drowning or being trapped in a burning vehicle).
   (d) Risk of ejection is significantly decreased by the wearing of seat belts and shoulder restraints.
   (e) Rollover accidents commonly result in partial or complete ejection of unrestrained vehicle occupants.

108. In an accident involving an off-center impact, causing rotation of the auto around the point of impact, you should anticipate
    (a) lateral and frontal injury patterns.
    (b) the occupant furthest from the point of impact to suffer significant deceleration or acceleration injuries.
    (c) the occupant nearest to the point of impact to suffer significant deceleration or acceleration injuries.
    (d) Answers (b) and (c).
    (e) Answers (a), (b), and (c).

109. Which of the following statements regarding restraint systems is false?
    (a) Even properly positioned, lap and shoulder belts may produce significant injuries during an accident.
    (b) Steering wheel and dash mounted airbags are effective only in frontal collisions.
    (c) Survival rates are greatly improved by the use of restraint systems.
    (d) When air bags are available, use of the old lap and shoulder belt system is no longer necessary.
    (e) Dash mounted air bags have killed infants and small children.

110. Suspected alcohol intoxication or substance abuse is an important factor when considering the mechanism of injury because
    (a) the relaxation produced by drugs or alcohol significantly lessens the severity of injury related to the mechanism.
    (b) blood testing is required by law in many states.
    (c) an intoxicated patient's altered ability to perceive pain may mask symptoms of serious injury.
    (d) Answers (a) and (b).
    (e) Answers (a) and (c).

111. Motorcycle accidents with frontal impact frequently cause
    (a) abdominal injury from handlebar contact.
    (b) pelvic injury from handlebar contact.
    (c) bilateral femoral fractures from handlebar contact.
    (d) Answers (b) and (c).
    (e) Answers (a), (b), and (c).

112. When involved in a motorcycle accident, if the victim was wearing a helmet the likelihood of
    (a) head injury is greatly diminished.
    (b) spinal injury is greatly diminished.
    (c) spinal injury is greatly increased.
    (d) Answers (a) and (b).
    (e) Answers (a) and (c).

113. Statistically, what is the most common type of impact in motor vehicle accidents?
   (a) Frontal impact.
   (b) Lateral impact.
   (c) Rear-end impact.
   (d) Rollover impacts.
   (e) There is no most common type of motor vehicle accident impact.

114. Which of the following statements about the various traumatic mechanisms associated with blast injuries is false?
   (a) As the compression/decompression pressure wave passes through an individual's body, air-filled structures (such as the lungs, auditory canals, sinuses, and bowel) sustain greater injury than do solid or fluid-filled structures.
   (b) Underwater detonation greatly diminishes the potential for injury; thus the deeper an individual is submerged during an underwater detonation, the less injury they are likely to sustain.
   (c) A blast wind follows behind the pressure wave created by the explosion.
   (d) Projectiles driven by the blast wind may cause injuries as severe as bullet wounds.
   (e) A strong enough pressure wave or blast wind may propel a victim, causing blunt or penetrating impacts with stationary objects.

115. The most common and most serious type of injury associated with explosions is
   (a) inner-ear injury (hearing loss).
   (b) closed head injury.
   (c) lung injury.
   (d) bowel rupture.
   (e) rescuer injury from detonation of additional terrorist booby traps.

116. Your patient jumped from a second-story window to escape from a person threatening assault. Her chief complaint is bilateral ankle and heel pain from first landing on both feet (erect) and then falling forward into the dirt. In addition to ankle and heel injuries, given this mechanism of injury, you must also suspect
   (a) spinal fracture.
   (b) abdominal or thoracic internal injuries.
   (c) upper extremity injuries.
   (d) Answers (a) and (c).
   (e) Answers (a), (b), and (c).

117. When evaluating a fall victim's mechanism of injury, which of the following points of information are vitally important to obtain immediately?
   (1) Cause of the fall.
   (2) The patient's systolic and diastolic blood pressure.
   (3) Type of landing surface (cement, dirt, grass, etc.).
   (4) Potential for objects struck during the fall.

(5) Body part(s) first impacting upon landing.
(6) Body part(s) impacted during the fall.
(7) Height of the fall.

    (a) 1, 3, 4, 5, 6, and 7.
    (b) 2, 3, 4, 5, 6, and 7.
    (c) 2, 3, 5, and 7.
    (d) 3, 4, 5, 6, and 7.
    (e) 1, 2, 3, 4, 5, 6, and 7.

118. Which of the following statements regarding bullet wounds is true?
    (a) Bullets designed to expand ("mushroom") on initial impact create entrance wounds that are larger than their exit wounds.
    (b) A higher-velocity (e.g., rifle) bullet produces less cavitation injury than a medium-velocity (e.g., handgun) bullet, because a faster-traveling bullet is less likely to tumble or fragment as it moves through body tissue.
    (c) All bullets create greater entrance wounds than exit wounds, because the velocity is slowed by the body tissue prior to exit.
    (d) Solid organs (such as the liver, spleen, or kidneys) have greater resiliency and absorb energy better than muscles and tendons; thus the extent of solid organ tissue damage along a projectile's pathway will be less than the damage to muscles and tendons.
    (e) A bullet passing though lung tissue creates less direct tissue damage than a bullet passing through any other body tissue.

119. Which of the following best defines shock?
    (a) A state of inadequate blood pressure (less than 100 systolic).
    (b) A state of inadequate pulse rate (less than 80 per minute).
    (c) A state of inadequate respirations (less than 20 per minute).
    (d) A state of inadequate hydration.
    (e) A state of inadequate cellular oxygenation.

120. Stroke volume is affected by all of the following, except
    (a) the volume of blood returned to the atria.
    (b) the volume of blood delivered to the ventricles from the atria.
    (c) the amount of atrial contraction that occurs during ventricular diastole.
    (d) the amount of ventricular muscle fiber stretching during ventricular diastole.
    (e) the resistance of arterial tone during ventricular systole.

121. At rest, the normal adult's average stroke volume is
    (a) 15 ml.
    (b) 25 ml.
    (c) 50 ml.
    (d) 70 ml.
    (e) 150 ml.

122. Cardiac output is defined as the amount of blood pumped by the heart in
  (a) one minute, and is computed as stroke volume X heart rate.
  (b) five minutes, and is computed as stroke volume X heart rate X 5.
  (c) a single cardiac contraction, and is computed as systolic blood pressure ÷ by diastolic blood pressure.
  (d) Answer (a) or (c).
  (e) Answer (b) or (c).

123. Failure of the Frank-Starling mechanism results in
  (a) greater cardiac output.
  (b) increased stroke volume.
  (c) decreased stroke volume.
  (d) Answers (a) and (b).
  (e) Answers (a) and (c).

124. The open space within the interior of a blood vessel is called the
  (a) vessel potential space.
  (b) vessel tubule.
  (c) orifice.
  (d) lumen.
  (e) interstitial space.

125. The outermost layer of a blood vessel's wall is called the
  (a) tunica exterior.
  (b) tunica adventitia.
  (c) tunica media.
  (d) tunica intima.
  (e) tunica interior.

126. The innermost layer of a blood vessel's wall (the layer that lines the open space within its interior) is called the
  (a) tunica exterior.
  (b) tunica adventitia.
  (c) tunica media.
  (d) tunica intima.
  (e) tunica interior.

127. The middle layer of a blood vessel's wall is called the
  (a) tunica exterior.
  (b) tunica adventitia.
  (c) tunica media.
  (d) tunica intima.
  (e) tunica interior.

128. The muscular layer of a blood vessel's wall is called the

    (a) tunica exterior.

    (b) tunica adventitia.

    (c) tunica media.

    (d) tunica intima.

    (e) tunica interior.

129. Which of the following blood vessel types has the greatest ability to vary the size of its interior open space?

    (a) Major veins.

    (b) Venules.

    (c) Major arteries.

    (d) Arterioles.

    (e) Capillaries.

130. All arteries carry

    (a) blood away from the heart.

    (b) oxygenated blood.

    (c) blood toward the heart.

    (d) oxygen-depleted blood.

    (e) Answers (a) and (b).

131. All veins carry

    (a) blood away from the heart.

    (b) oxygenated blood.

    (c) blood toward the heart.

    (d) oxygen-depleted blood.

    (e) Answers (c) and (d).

132. The peripheral vascular resistance against which the heart must pump is determined by the condition of an individual's

    (a) arteries.

    (b) veins.

    (c) capillaries.

    (d) Answers (a) and (c).

    (e) Answers (b) and (c).

133. The greatest percentage of the body's total blood volume (64%) is normally contained within which of the following portions of the vascular system?

    (a) The arterial system.

    (b) The venous system.

    (c) The lymphatic system.

    (d) The capillary system.

    (e) The renal system.

134. Normally, the most common blood cell type (composing about 45% of the total blood volume) is the

    (a) leukocyte.
    (b) erythrocyte.
    (c) plasmacyte.
    (d) lymphocyte.
    (e) platelet.

135. Oxygen is transported from the lungs to the cells by attaching itself to the hemoglobin molecules that are found in

    (a) leukocytes.
    (b) erythrocytes.
    (c) plasmacytes.
    (d) lymphocytes.
    (e) platelets.

136. Responsibility for blood clotting and blood vessel repair is attributed to

    (a) leukocytes.
    (b) erythrocytes.
    (c) plasmacytes.
    (d) lymphocytes.
    (e) platelets.

137. Inadequate cellular oxygenation causes cells to rely solely upon anaerobic metabolism, which

    (1) produces pyruvic acid, which soon converts into lactic acid.
    (2) produces lactic acid, which soon converts into pyruvic acid.
    (3) if uncorrected, can produce a state of acidosis, which will reduce the ability of hemoglobin to carry oxygen.
    (4) increases aerobic cellular utilization of glucose.
    (5) results in a high serum glucose (hyperglycemia).
    (6) results in a low serum glucose (hypoglycemia).

        (a) 1, 4, and 6.
        (b) 2, 4, and 6.
        (c) 2, 4, and 5.
        (d) 1, 3, and 5.
        (e) 2, 3, and 5.

138. Which of the following statements regarding baroreceptors is false?

    (a) Baroreceptors are sensory nerve endings located in the carotid-artery sinuses, atrial walls, the vena cava, and the aortic arch.
    (b) Baroreceptors are stimulated by changes in blood pressure.

(c) Baroreceptors excrete a hormone that stimulates the adrenal glands, affecting either increased or decreased catecholamine release.

(d) If baroreceptors detect an increase in blood pressure, the brain is cued to decrease the heart rate and decrease the preload or afterload.

(e) If baroreceptors detect a decrease in blood pressure, the brain is cued to activate the sympathetic nervous system.

139. A patient should be considered to have entered the Decompensated stage of shock when his

(a) pulse rate increases above 100 beats per minute (tachycardia).

(b) skin becomes cool, pale, and clammy.

(c) systolic blood pressure drops.

(d) attitude becomes anxious, restless, or combative.

(e) respiratory rate increases above 20 breaths per minute (tachypnea).

140. Hypovolemic shock can be caused by which of the following mechanisms?

(1) External or internal hemorrhage. (6) Burns.
(2) Bowel obstruction. (7) Ascites secondary to liver failure.
(3) Excessive vomiting or diarrhea. (8) Excessive sweating.
(4) Pancreatitis or peritonitis. (9) Head and spine injury.
(5) Diabetic ketoacidosis.

(a) 1, 3, 6, and 8.

(b) 1, 3, 4, 5, 6, 8, and 9.

(c) 1, 2, 3, 4, 5, 6, and 8.

(d) 1, 2, 3, 4, 5, 6, 7, and 8.

(e) 1, 2, 3, 4, 5, 6, 7, 8, and 9.

141. Feces described or observed as being black and tarlike in color, may indicate the presence of digested gastrointestinal bleeding. The medical term for black or tar-colored stool is

(a) melena.

(b) melanoma.

(c) hematemesis.

(d) hematochezia.

(e) hemoptysis.

142. The medical term for feces that exhibits the obvious presence of red blood ("frank blood") is

(a) melena.

(b) melanoma.

(c) hematemesis.

(d) hematochezia.

(e) hemoptysis.

143. Which of the following statements about septic shock is false?

    (a) Massive infection can produce toxins that increase blood vessel permeability, resulting in significant fluid loss from the vasculature.

    (b) Geriatric or pediatric septic shock is always accompanied by high fever.

    (c) In the early stage of septic shock, cardiac output is increased; however, toxin-produced vasodilation may prevent an increase in blood pressure.

    (d) The organ system most susceptible to septic shock is the respiratory system; thus dyspnea, or altered lung sounds, are often present.

    (e) Septic shock should be suspected in any emergency patient with a recent history of infection or illness.

144. Which of the following statements regarding neurogenic shock is false?

    (a) Peripheral vasoconstriction will be disabled in the areas below an injury site productive of neurogenic shock; thus a relative hypovolemia may occur.

    (b) Sympathetic nervous system cardiac stimulation may be disabled by neurogenic shock; thus the neurogenic shock patient may not be tachycardic.

    (c) Deprivation of oxygen or glucose to the medulla of the brain may produce neurogenic shock.

    (d) Neurogenic shock patients exhibit all the classic signs and symptoms of shock, except that their skin always remains warm, dry, and pink from head to toe.

    (e) Neurogenic shock is most frequently caused by severe spinal cord injury.

145. Shock caused by total-body anaphylaxis or sepsis is classified as _____ shock.

    (a) irreversible

    (b) cardiogenic

    (c) distributive

    (d) obstructive

    (e) respiratory

146. Shock caused by carbon monoxide poisoning is classified as _____ shock.

    (a) irreversible

    (b) cardiogenic

    (c) distributive

    (d) obstructive

    (e) respiratory

147. Shock caused by a tension pneumothorax, pulmonary emboli, or cardiac tamponade is classified as _____ shock.

    (a) irreversible

    (b) cardiogenic

    (c) distributive

    (d) obstructive

    (e) respiratory

148. Possible complications of treating a hypotensive patient by applying and inflating the pneumatic antishock garment (PASG) include all of the following, except
   (a) increased intracranial pressure.
   (b) decreased diaphragmatic excursion (interference with respirations).
   (c) increased pulmonary congestion.
   (d) worsening of cardiogenic shock.
   (e) increased mortality when used in cases of penetrating chest trauma.

149. Which of the following statements about helicopter transport of the emergency patient is false?
   (a) The decreased atmospheric pressure of helicopter transportation may increase the severity of asthma or COPD emergencies.
   (b) The decreased atmospheric pressure of helicopter transportation will improve the condition of patients with a suspected pneumothorax or tension pneumothorax.
   (c) Rapid changes in altitude will affect the air pressure of the PASG.
   (d) Rapid changes in altitude will affect the air pressure of an endotracheal tube cuff.
   (e) Rapid changes in altitude will affect the air pressure of an air splint.

150. Especially for trauma patients, anticipation of shock should first be considered
   (a) during your Scene Size-Up.
   (b) after formation of your General Impression.
   (c) after your Initial Assessment.
   (d) during your Rapid Trauma Assessment.
   (e) during your Detailed Physical Exam.

The Answer Key to Section 2 is on page 406.

Test
Section
Three

3

Test Section 3's subjects:

- General Pharmacology
- Venous Access and Medication Administration Techniques

Test Section 3 consists of 120 questions and is allotted 2 hours for completion.

1.   The chief extracellular ion is
    (a) sodium.
    (b) potassium.
    (c) calcium.
    (d) chloride.
    (e) magnesium.

2.   The chief intracellular ion is
    (a) sodium.
    (b) potassium.
    (c) calcium.
    (d) chloride.
    (e) magnesium.

3.   The symbol for sodium is ___.
    (a) S^+
    (b) So^+
    (c) N^+
    (d) Na^+
    (e) Na^-

4.   The symbol for chloride is ___.
    (a) C^-
    (b) C^+
    (c) Cl^-
    (d) Cl^+
    (e) Cl^{++}

5.   The symbol for magnesium is ___.
    (a) Ma^+
    (b) Ma^-
    (c) $Mang^+$
    (d) Mg^-
    (e) Mg^{++}

6.   The symbol for bicarbonate is ___.
    (a) B^+
    (b) Bi^-
    (c) HCO^+
    (d) BC^-
    (e) HCO_3^-

7.   The symbol for calcium is ___.
        (a) C^{++}
        (b) Ca^{++}
        (c) C^-
        (d) Ca^{--}
        (e) K^+

8.   The symbol for potassium is ___.
        (a) PO^+
        (b) P^+
        (c) K^+
        (d) P^-
        (e) PO^-

9.   The symbol for phosphate is ___.
        (a) HPO_4^-
        (b) $H_2O_3^+$
        (c) $H_2O_3^-$
        (d) HCO_2^-
        (e) HPO_2^+

10.   The principal buffer of the body is ___.
        (a) K^+
        (b) HCO_3^-
        (c) HPO_3^-
        (d) $CaCl_2^-$
        (e) HPO_2^+

11.   Fluid with osmotic pressure equal to that of normal body fluid is called
        (a) homeotonic.
        (b) hypertonic.
        (c) isotonic.
        (d) hypotonic.
        (e) homeostatic.

12.   Fluid with osmotic pressure less than that of normal body fluid is called
        (a) homeotonic.
        (b) hypertonic.
        (c) isotonic.
        (d) hypotonic.
        (e) homeostatic.

13. Fluid with osmotic pressure greater than that of normal body fluid is called
  (a) homeotonic.
  (b) hypertonic.
  (c) isotonic.
  (d) hypotonic.
  (e) homeostatic.

14. Water will move across a semipermeable membrane
  (a) in the direction of least resistance.
  (b) from an isotonic solution to a homeotonic solution.
  (c) from a homeostatic solution to an isotonic solution.
  (d) from an area of higher solute concentration to that of lower solute concentration.
  (e) from an area of lower solute concentration to that of higher solute concentration.

15. Sodium will move across a semipermeable membrane
  (a) in the direction of least resistance.
  (b) from an isotonic solution to a homeotonic solution.
  (c) from a homeostatic solution to an isotonic solution.
  (d) from an area of higher solute concentration to that of lower solute concentration.
  (e) from an area of lower solute concentration to that of higher solute concentration.

16. Movement of a solute across a semipermeable membrane is called
  (a) osmosis.
  (b) simple transfer.
  (c) diaphoresis.
  (d) diffusion.
  (e) suffusion.

17. Movement of a solvent across a semipermeable membrane is called
  (a) osmosis.
  (b) simple transfer.
  (c) diaphoresis.
  (d) diffusion.
  (e) suffusion.

18. Of the following ways in which substances move across a semipermeable membrane, the fastest is
  (a) osmosis.
  (b) simple transfer.

(c) diaphoresis.

(d) diffusion.

(e) suffusion.

19. Electrolytes move across a semipermeable membrane using

(a) osmosis.

(b) simple transfer.

(c) diaphoresis.

(d) diffusion.

(e) suffusion.

20. Water will move across a semipermeable membrane using

(a) osmosis.

(b) simple transfer.

(c) diaphoresis.

(d) diffusion.

(e) suffusion.

21. Which of the following statements regarding active transport is false?

(a) Energy is required to accomplish active transport.

(b) Active transport is slower than diffusion or osmosis.

(c) Larger molecules can move across semipermeable membranes with active transport.

(d) Molecules can move toward areas of higher concentration with active transport.

(e) Proteins can move across semipermeable membranes with active transport.

22. Facilitated diffusion

(a) employs "helper proteins" to cross the cell membrane.

(b) requires energy to occur.

(c) is a selective process, occurring only with certain molecules.

(d) All of the above.

(e) None of the above.

23. Protein-containing intravenous fluids are called

(a) isotonic osmolloids.

(b) colloids.

(c) hypertonic osmolloids.

(d) crystalloids.

(e) hypotonic osmolloids.

24. Intravenous fluids that do not contain proteins are called
    (a) isotonic osmolloids.
    (b) colloids.
    (c) hypertonic osmolloids.
    (d) crystalloids.
    (e) hypotonic osmolloids.

25. Prehospital intravenous fluid replacement is initiated with
    (a) isotonic osmolloid solutions.
    (b) colloid solutions.
    (c) hypertonic osmolloid solutions.
    (d) crystalloid solutions.
    (e) hypotonic osmolloid solutions.

26. A greater increase in intravascular fluid can be accomplished sooner with infusion of
    (a) isotonic osmolloid solutions.
    (b) colloid solutions.
    (c) hypertonic osmolloid solutions.
    (d) crystalloid solutions.
    (e) hypotonic osmolloid solutions.

27. Normal saline (NS) and 5% dextrose in water (D_5W) are examples of
    (a) isotonic osmolloid solutions.
    (b) colloid solutions.
    (c) hypertonic osmolloid solutions.
    (d) crystalloid solutions.
    (e) hypotonic osmolloid solutions.

28. Plasmanate and Hetastarch are examples of
    (a) isotonic osmolloid solutions.
    (b) colloid solutions.
    (c) hypertonic osmolloid solutions.
    (d) crystalloid solutions.
    (e) hypotonic osmolloid solutions.

29. Lactated Ringer's (LR or RL) is an example of
    (a) an isotonic osmolloid solution.
    (b) a colloid solution.
    (c) a hypertonic osmolloid solution.
    (d) a crystalloid solution.
    (e) a hypotonic osmolloid solution.

30.   Solutions containing higher solute concentrations than that within the cell are called
       (a) isotonic.
       (b) isotonic or hypertonic.
       (c) hypotonic.
       (d) isotonic or hypotonic.
       (e) hypertonic.

31.   Solutions with a similar solute concentration as that within the cell are called
       (a) isotonic.
       (b) isotonic or hypertonic.
       (c) hypotonic.
       (d) isotonic or hypotonic.
       (e) hypertonic.

32.   If a normally hydrated person receives an infusion of _____ solution, a shift of fluid will occur from the extracellular to the intracellular compartments.
       (a) an isotonic
       (b) an isotonic or hypertonic
       (c) a hypotonic
       (d) an isotonic or hypotonic
       (e) a hypertonic

33.   If a normally hydrated person receives an infusion of _____ solution, fluid will shift from the intracellular to the extracellular compartments.
       (a) an isotonic
       (b) an isotonic or hypertonic
       (c) a hypotonic
       (d) an isotonic or hypotonic
       (e) a hypertonic

34.   Solutions containing a lesser concentration of solutes than that within the cell are called
       (a) isotonic.
       (b) isotonic or hypertonic.
       (c) hypotonic.
       (d) hypertonic or hypotonic.
       (e) hypertonic.

35.   Acid-base balance refers to the concentration of
       (a) the chief extracellular ion of body fluids.
       (b) the chief intracellular ion of body fluids.
       (c) hydrogen ions in body fluids.
       (d) All of the above.
       (e) None of the above.

36. Which of the following statements regarding pH is false?

    (a) The term *pH* is used to express the hydrogen-ion concentration of a fluid.

    (b) The pH becomes lower as the hydrogen-ion concentration becomes greater.

    (c) At a pH of 7, a solution is neutral (pure water has a pH of 7).

    (d) The pH becomes higher as the hydrogen-ion concentration becomes greater.

    (e) The hydrogen-ion concentration in the body changes from second to second.

37. A pH measurement below 7 reflects

    (a) an increased concentration of H^+ and is called acidosis.

    (b) a decreased concentration of H^+ and is called acidosis.

    (c) an increased concentration of alkaline ions and is called alkalosis.

    (d) Answers (a) and (c).

    (e) Answers (b) and (c).

38. A pH measurement above 7.55 reflects

    (a) an increased concentration of H^+ and is called acidosis.

    (b) a decreased concentration of H^+ and is called alkalosis.

    (c) a decreased concentration of alkaline ions and is called alkalosis.

    (d) Answers (a) and (c).

    (e) Answers (b) and (c).

39. The normal human pH range is _____.

    (a) 7.0 to 7.6

    (b) 7.25 to 7.35

    (c) 7.35 to 7.45

    (d) 7.45 to 7.55

    (e) 7.80 to 7.95

40. The body's normal pH is maintained by

    (1) the respiratory system.   (4) the renal system.

    (2) the endocrine system.   (5) the digestive system.

    (3) the bicarbonate buffer system.   (6) the pituitary gland.

        (a) 3.

        (b) 1, 3, and 4.

        (c) 2, 3, and 4.

        (d) 2, 3, 5, and 6.

        (e) 1, 2, 3, 4, 5, and 6.

41. The system able to remove hydrogen ions in the fastest manner is

    (a) the bicarbonate buffer system.

    (b) the respiratory system.

    (c) the renal system.

    (d) the digestive system.

    (e) the endocrine system.

42. The slowest-acting system for removing hydrogen ions is
   (a) the bicarbonate buffer system.
   (b) the respiratory system.
   (c) the renal system.
   (d) the digestive system.
   (e) the endocrine system.

43. The bicarbonate buffer system has two components,
   (a) carbonic acid ($H_3CO_2^-$) and bicarbonate (HCO_3^+).
   (b) carbonic acid (H_2CO_3) and bicarbonate (HCO_3^-).
   (c) carbonic acid (HCO_3) and bicarbonate ($H_3CO_2^-$).
   (d) carbonic acid (HCO_2) and bicarbonate ($H_3CO_3^+$).
   (e) carbonic acid (HCO_2) and bicarbonate ($H_2CO_3^-$).

44. Which of the following statements about carbonic acid is false?
   (a) Because it is unstable, carbonic acid will eventually dissociate into water (H_2O) and carbon dioxide (CO_3).
   (b) Compared to pure hydrogen, carbonic acid is a weak acid.
   (c) Carbonic acid can dissociate into hydrogen and bicarbonate.
   (d) Erythrocytes contain an enzyme that can speed the breakdown of carbonic acid.
   (e) A decrease in hydrogen ions results in an increase in carbonic acid.

45. An increased respiratory rate results in increased
   (a) elimination of CO_2 and a decreased pH.
   (b) retention of CO_2 and an increased pH.
   (c) retention of CO_2 and a decreased pH.
   (d) elimination of CO_2 and an increased pH.
   (e) None of the above.

46. A decreased respiratory rate results in increased
   (a) elimination of CO_2 and a decreased pH.
   (b) retention of CO_2 and an increased pH.
   (c) retention of CO_2 and a decreased pH.
   (d) elimination of CO_2 and an increased pH.
   (e) None of the above.

47. Respiratory acidosis is caused by abnormal
   (a) elimination of CO_2 and a decreased pH.
   (b) retention of CO_2 and an increased pH.
   (c) retention of CO_2 and a decreased pH.
   (d) elimination of CO_2 and an increased pH.
   (e) None of the above.

48. Respiratory alkalosis is caused by abnormal
    (a) elimination of CO_2 and a decreased pH.
    (b) retention of CO_2 and an increased pH.
    (c) retention of CO_2 and a decreased pH.
    (d) elimination of CO_2 and an increased pH.
    (e) None of the above.

49. Anaerobic metabolism can cause
    (a) metabolic alkalosis.
    (b) respiratory acidosis.
    (c) respiratory alkalosis.
    (d) homeostasis.
    (e) metabolic acidosis.

50. Prolonged vomiting or excessive use of diuretics can cause
    (a) metabolic alkalosis.
    (b) respiratory acidosis.
    (c) respiratory alkalosis.
    (d) homeostasis.
    (e) metabolic acidosis.

51. Magnesium sulfate and calcium chloride are examples of medications originally derived from
    (a) animal sources.
    (b) vegetable (plant) sources.
    (c) mineral sources.
    (d) synthetic production.
    (e) None of the above.

52. Insulin and epinephrine are examples of medications originally derived from
    (a) animal sources.
    (b) vegetable (plant) sources.
    (c) mineral sources.
    (d) synthetic production.
    (e) None of the above.

53. Digitalis, morphine sulfate, and atropine are examples of medications originally derived from
    (a) animal sources.
    (b) vegetable (plant) sources.
    (c) mineral sources.
    (d) synthetic production.
    (e) None of the above.

54. Lidocaine and bretylium tosylate are examples of medications originally derived from
   (a) animal sources.
   (b) vegetable (plant) sources.
   (c) mineral sources.
   (d) synthetic production.
   (e) None of the above.

55. In 1970 an act was passed to regulate and control the manufacturing, distribution, and dispensing of drugs that have abuse potential. This act is called the
   (a) Federal Food, Drug, and Cosmetic Act.
   (b) Harrison Narcotic Act.
   (c) Controlled Substance Act.
   (d) Narcotic Control Act.
   (e) Pure Food and Drug Act.

56. The official name for epinephrine is
   (a) epi.
   (b) epinephrine hydrochloride.
   (c) adrenalin.
   (d) epinephrine hydrochloride, U.S.P.
   (e) beta-(3,4-dihydroxyphenl)-a-methylaminoethanol.

57. The chemical name for epinephrine is
   (a) epi.
   (b) epinephrine hydrochloride.
   (c) adrenalin.
   (d) epinephrine hydrochloride, U.S.P.
   (e) beta-(3,4-dihydroxyphenl)-a-methylaminoethanol.

58. The generic name for epinephrine is
   (a) epi.
   (b) epinephrine hydrochloride.
   (c) adrenalin.
   (d) epinephrine hydrochloride, U.S.P.
   (e) beta-(3,4-dihydroxyphenl)-a-methylaminoethanol.

59. The trade or proprietary name for epinephrine is
   (a) epi.
   (b) epinephrine hydrochloride.
   (c) adrenalin.
   (d) epinephrine hydrochloride, U.S.P.
   (e) beta-(3,4-dihydroxyphenl)-a-methylaminoethanol.

60. When a drug is completely dissolved in water, the resulting preparation is called a
   (a) syrup.
   (b) tincture.
   (c) solution.
   (d) suspension.
   (e) spirit.

61. When alcohol is used to chemically extract a drug, the resulting preparation is called a
   (a) syrup.
   (b) tincture.
   (c) solution.
   (d) suspension.
   (e) spirit.

62. Some liquid preparations contain drugs that do not dissolve. After resting, the solid portion separates and settles to the container bottom. These preparations require shaking or rolling before use, and are called
   (a) syrups.
   (b) tinctures.
   (c) solutions.
   (d) suspensions.
   (e) spirits.

63. Most drugs are prepared in solid form. Solid drug forms include all of the following, except
   (a) elixirs.
   (b) powders.
   (c) suppositories.
   (d) capsules.
   (e) tablets.

64. The term or phrase _____ refers to the action of two drugs combining to create a much stronger effect than either drug can achieve alone.
   (a) synergism
   (b) potentiation
   (c) cumulative effect
   (d) antagonism
   (e) therapeutic action

65. The term or phrase _____ refers to an increased drug effect that is achieved by repeated doses.
   (a) synergism
   (b) potentiation

(c) cumulative effect

(d) antagonism

(e) therapeutic action

66. The term or phrase _____ refers to one drug interfering with the body's response to another drug.

    (a) synergism

    (b) potentiation

    (c) cumulative effect

    (d) antagonism

    (e) therapeutic action

67. The term or phrase _____ refers to one drug's effect being increased or enhanced by another drug's effect.

    (a) synergism

    (b) potentiation

    (c) cumulative effect

    (d) antagonism

    (e) therapeutic action

68. The term or phrase that best describes any undesired or unintended response to a drug is

    (a) side effects.

    (b) tachyphylaxis.

    (c) contraindications.

    (d) idiosyncrasy.

    (e) hypersensitivity.

69. The term or phrase that best describes an allergic reaction to a medication is

    (a) side effects.

    (b) tachyphylaxis.

    (c) contraindications.

    (d) idiosyncrasy.

    (e) hypersensitivity.

70. Sometimes, a particular patient will experience an undesired and unavoidable drug effect that is not experienced by other patients who have taken the same medication. The term or phrase that best refers to this occurrence is

    (a) side effects.

    (b) tachyphylaxis.

    (c) contraindications.

    (d) idiosyncrasy.

    (e) hypersensitivity.

71.   The term or phrase _____ refers to specific physiological or medical reasons that a patient should not receive a particular medication.
       (a) side effects
       (b) tachyphylaxis
       (c) contraindications
       (d) idiosyncrasy
       (e) cumulative effects

72.   When a condition or complaint is resistant or unresponsive to normally successful treatment or therapeutic actions, it is said to be _____ to that treatment or action.
       (a) refractory
       (b) antagonistic
       (c) refractional
       (d) synergistic
       (e) reciprocal

73.   Examples of the _____ route of medication administration include oral, rectal, or nasogastric-tube administration.
       (a) injectable
       (b) noninjectable
       (c) enteral
       (d) parenteral
       (e) transenteral

74.   Examples of the _____ route of medication administration include IV, IM, or SQ administration.
       (a) injectable
       (b) noninjectable
       (c) enteral
       (d) parenteral
       (e) transenteral

75.   Examples of the _____ route of medication administration include intraosseous, inhalation, or transdermal administration.
       (a) injectable
       (b) noninjectable
       (c) enteral
       (d) parenteral
       (e) transenteral

76.   Vaginal administration of a drug is an example of the _____ route of medication administration.
       (a) injectable
       (b) noninjectable

(c) enteral

(d) parenteral

(e) transenteral

77. Which of the following factors may affect or alter an individual's response to a drug?

(1) Age   (5) Gender

(2) Body mass and weight   (6) Environmental influences

(3) Time of administration   (7) Genetic traits

(4) General health condition   (8) Psychological state

     (a) 1, 2, 3, 4, 5, 6, 7, and 8.

     (b) 1, 2, 3, 4, 5, and 6.

     (c) 1, 2, 3, 4, and 7.

     (d) 1, 2, 4, 6, and 7.

     (e) 1, 2, and 4.

78. Which of the following statements regarding drug administration routes is false?

(a) In emergency medicine, the most frequently preferred route of drug administration is IV.

(b) The endotracheal (ET) route delivers approximately the same blood concentration of drug as does the IV or IO route.

(c) The speed of drug absorption is approximately the same when given either IV or ET.

(d) Sublingual drug absorption is faster than subcutaneous drug absorption.

(e) If an IV cannot be established, the IO route of drug administration is preferred over the ET route.

79. Drug-elimination routes employed by the body include all of the following, except

(a) excretion via hemorrhage.

(b) excretion via urination.

(c) excretion via expiration.

(d) excretion via the feces.

(e) excretion via salivation.

80. The majority of medications achieve their effect by

(a) altering the physical properties of cells (especially cell walls).

(b) combining with chemicals within the body to neutralize their harmful effects.

(c) altering normal metabolic pathways or functions.

(d) combining with chemicals within the body and biotransforming into an entirely different substance that produces the desired effect.

(e) attaching to specific proteins present on cell membranes called drug-receptor sites.

81.   An "agonist" drug achieves its effect by
      (a) altering the physical properties of specific target-cell walls to stimulate a response.
      (b) binding with a drug-receptor site and stimulating a response.
      (c) altering normal metabolic functions so that a response does not occur.
      (d) combining with a body chemical and biotransforming into a substance that prevents an undesired response.
      (e) binding with a drug-receptor site and blocking the receptor's ability to respond to other substances.

82.   Vegetative functions are the responsibility of the
      (a) central nervous system.
      (b) peripheral nervous system.
      (c) parasympathetic nervous system.
      (d) sympathetic nervous system.
      (e) independent nervous system.

83.   Functions during stress are the responsibility of the
      (a) central nervous system.
      (b) peripheral nervous system.
      (c) parasympathetic nervous system.
      (d) sympathetic nervous system.
      (e) independent nervous system.

84.   The chemical mediator acetylcholine is the primary neurotransmitter employed by the
      (a) central nervous system.
      (b) peripheral nervous system.
      (c) parasympathetic nervous system.
      (d) sympathetic nervous system.
      (e) independent nervous system.

85.   The hormones epinephrine and norepinephrine are employed by the
      (a) central nervous system.
      (b) peripheral nervous system.
      (c) parasympathetic nervous system.
      (d) sympathetic nervous system.
      (e) independent nervous system.

86.   Constricted pupils, increased salivation, and decreased heart rate are effects that occur with
      (a) stimulation by acetylcholine release.
      (b) stimulation of alpha 1 receptors.

    (c) stimulation of beta 1 receptors.
    (d) stimulation of beta 2 receptors.
    (e) stimulation of dopaminergic receptors.

87.   Bronchodilation and vasodilation are effects that occur with
    (a) stimulation by acetylcholine release.
    (b) stimulation of alpha 1 receptors.
    (c) stimulation of beta 1 receptors.
    (d) stimulation of beta 2 receptors.
    (e) stimulation of dopaminergic receptors.

88.   Renal vasodilation occurs in response to
    (a) stimulation by acetylcholine release.
    (b) stimulation of alpha 1 receptors.
    (c) stimulation of beta 1 receptors.
    (d) stimulation of beta 2 receptors.
    (e) stimulation of dopaminergic receptors.

89.   Increased heart rate is an effect that occurs with
    (a) stimulation by acetylcholine release.
    (b) stimulation of alpha 1 receptors.
    (c) stimulation of beta 1 receptors.
    (d) stimulation of beta 2 receptors.
    (e) stimulation of dopaminergic receptors.

90.   Peripheral vasoconstriction and mild bronchoconstriction are effects that occur with
    (a) stimulation by acetylcholine release.
    (b) stimulation of alpha 1 receptors.
    (c) stimulation of beta 1 receptors.
    (d) stimulation of beta 2 receptors.
    (e) stimulation of dopaminergic receptors.

91.   Increased cardiac automaticity, conduction, and contractile force are effects that occur in response to
    (a) stimulation by acetylcholine release.
    (b) stimulation of alpha 1 receptors.
    (c) stimulation of beta 1 receptors.
    (d) stimulation of beta 2 receptors.
    (e) stimulation of dopaminergic receptors.

92. Administration of an adrenergic drug results in
   (a) sympathetic nervous system stimulation.
   (b) inhibition or blocking of sympathetic nervous system stimulation.
   (c) parasympathetic nervous system stimulation.
   (d) Answers (a) and (c).
   (e) Answers (b) and (c).

93. Administration of an antiadrenergic drug results in
   (a) sympathetic nervous system stimulation.
   (b) inhibition or blocking of sympathetic nervous system stimulation.
   (c) parasympathetic nervous system stimulation.
   (d) Answers (a) and (c).
   (e) Answers (b) and (c).

94. Administration of a sympatholytic drug results in
   (a) sympathetic nervous system stimulation.
   (b) inhibition or blocking of sympathetic nervous system stimulation.
   (c) parasympathetic nervous system stimulation.
   (d) Answers (a) and (c).
   (e) Answers (b) and (c).

95. Administration of a sympathomimetic drug results in
   (a) sympathetic nervous system stimulation.
   (b) inhibition or blocking of sympathetic nervous system stimulation.
   (c) parasympathetic nervous system stimulation.
   (d) Answers (a) and (c).
   (e) Answers (b) and (c).

96. Medications that stimulate parasympathetic nervous system actions are known as
   (a) parasympathomimetics.
   (b) cholinergics.
   (c) anticholinergics.
   (d) Answers (a) and (b).
   (e) Answers (a) and (c).

97. Medications that block parasympathetic nervous system actions are known as
   (a) cholinergic blockers.
   (b) parasympatholytics.
   (c) anticholinergics.
   (d) Answers (a) and (b).
   (e) Answers (a), (b), and (c).

98. A drug that influences the force of cardiac-muscular contractility is
   (a) an inotrope, or inotropic.
   (b) a phototrope, or phototropic.
   (c) a phenotrope, or phenotropic.
   (d) a chemotrope, or chemotropic.
   (e) a chronotrope, or chronotropic.

99. A drug that influences the heart rate is
   (a) an inotrope, or inotropic.
   (b) a phototrope, or phototropic.
   (c) a phenotrope, or phenotropic.
   (d) a chemotrope, or chemotropic.
   (e) a chronotrope, or chronotropic.

100. Your patient weighs 154 pounds. How many kilograms (kg) does your patient weigh?
   (a) 70 kg.
   (b) 85 kg.
   (c) 77 kg.
   (d) 93.5 kg.
   (e) 74.8 kg.

101. Your patient weighs 55 kilograms. How many pounds (lbs.) does your patient weigh?
   (a) 137.5 lbs.
   (b) 190 lbs.
   (c) 209 lbs.
   (d) 110 lbs.
   (e) 121 lbs.

The following seven questions are designed only to test your mathematics skills. They do not reflect common prehospital medication ratios of milligrams-to-milliliters. The answers may not reflect realistic infusion rates. All *supplies* referred to, however, *represent standard EMS equipment* (such as macrodrip and microdrip administration sets).

102. You are presented with an ampule that contains 200 mg of a medication in 40 ml of solution. What is the concentration of medication per milliliter?
   (a) 0.5 mg per ml.
   (b) 5.0 mg per ml.
   (c) 1.0 mg per ml.
   (d) 50 mg per ml.
   (e) 10 mg per ml.

103. You are presented with a tubex (preloaded syringe) that contains a medication concentration of 2 mg per milliliter of solution. There are 150 mg of medication in the tubex. The tubex contains
    (a) 30 ml.
    (b) 100 ml.
    (c) 75 ml.
    (d) 37.5 ml.
    (e) 300 ml.

104. If 1200 mg of a medication is injected into a 250 ml bag of D_5W, what concentration of micrograms of medication per milliliter is achieved?
    (a) 480 µg/ml.
    (b) 4800 µg/ml.
    (c) 0.48 µg/ml.
    (d) 4.80 µg/ml.
    (e) 300,000 µg/ml.

105. You have just spiked a 500 ml bag of normal saline with a macrodrip infusion set that delivers 15 gtt/ml. You are ordered to run the IV at 24 gtt/min. How much time will it require to infuse the 500 ml?
    (a) 5 hours and 12 minutes.
    (b) 3 hours and 6 minutes.
    (c) 6 hours and 12 minutes.
    (d) 45 minutes.
    (e) 34 minutes.

106. You are ordered to administer one liter of fluid at a rate of 250 ml/hr. Your macrodrip infusion set delivers 10 gtt/ml. How many drops per minute must you infuse?
    (a) 4 gtt/min.
    (b) 21 gtt/min.
    (c) 25 gtt/min.
    (d) 42 gtt/min.
    (e) 250 gtt/min.

107. You are ordered to administer a medication at a rate of 4 µg/kg/min to a 220-pound patient. You prepare the infusion using 750 mg of the medication, a 250-ml bag of D_5W, and a microdrip infusion set. How many drops per minute do you infuse?
    (a) 25 gtt/min.
    (b) 15 gtt/min.
    (c) 8 gtt/min.
    (d) 9 gtt/min.
    (e) 32 gtt/min.

108. You are ordered to administer 9 mg/kg of medication to a 176-pound patient. The medication is packaged as 1 gram in 5 ml of solution. How many milliliters must be administered to comply with the ordered dose?
    (a) 3.6 ml.
    (b) 3.9 ml.
    (c) 72 ml.
    (d) 7.2 ml.
    (e) 3600 ml.

109. Which of the following statements regarding colloidal IV fluids is most true?
    (a) Colloid IV solutions contain large proteins that cannot easily pass across the capillary cell membrane.
    (b) Colloids have osmotic properties that cause water to be drawn from the surrounding, extravascular spaces into the circulatory system.
    (c) Colloidal IV fluids remain in the circulatory system longer than crystalloidal solutions.
    (d) Answers (b) and (c) are true.
    (e) Answers (a), (b), and (c) are true.

110. Which of the following statements regarding crystalloidal IV fluids is false?
    (a) Because all crystalloid solutions contain glucose (which prevents effective fluid replacement), colloid solutions are preferred for prehospital fluid replacement.
    (b) Crystalloid IV solutions are available as isotonic solutions, hypertonic solutions, or hypotonic solutions.
    (c) 3 milliliters of crystalloid solution is required to effectively replace each milliliter of blood loss.
    (d) Consisting primarily of electrolytes and water, crystalloid solutions lack larger proteins.
    (e) Crystalloid solutions are the primary prehospital IV solutions.

111. Which of the following IV solutions is an isotonic electrolyte solution, containing sodium chloride, potassium chloride, calcium chloride, and sodium lactate in water?
    (a) 5% dextrose in water (D_5W).
    (b) Normal saline solution.
    (c) Lactated Ringer's solution.
    (d) Plasma protein fraction (Plasmanate®).
    (e) Hetastarch (Hespan®).

112. Macrodrip IV administration tubing sets have a drops-per-milliliter ratio
    (a) of 80 drops to 1 ml.
    (b) of 60 drops to 1 ml.
    (c) of 45 drops to 1 ml.
    (d) that may vary from 10 to 20 drops per ml, depending upon the manufacturer.
    (e) None of the above.

113. Microdrip IV administration tubing sets have a drops-per-milliliter ratio
    (a) of 80 drops to 1 ml.
    (b) of 60 drops to 1 ml.
    (c) of 45 drops to 1 ml.
    (d) that may vary from 10 to 20 drops per ml, depending upon the manufacturer.
    (e) None of the above.

114. Which of the following IV catheter sizes represents the largest diameter catheter?
    (a) 10-gauge.
    (b) 14-gauge.
    (c) 16-gauge.
    (d) 22-gauge.
    (e) 24-gauge.

115. Which of the following statements regarding catheter size and its effect upon fluid administration rates is false?
    (a) The larger the IV catheter lumen, the greater the rate of flow.
    (b) The longer the IV catheter length, the faster the rate of flow.
    (c) The smaller the IV catheter gauge number, the greater the IV catheter lumen.
    (d) The higher you position the bag of IV fluid (relative to the patient's level), the faster the rate of IV flow.
    (e) Compression of the IV fluid bag will increase the rate of IV flow.

116. Infiltration of fluid or medication into the tissues surrounding the IV site is called
    (a) extravasation.
    (b) intravenous shift.
    (c) third spacing.
    (d) Any of the above.
    (e) None of the above.

117. The cause of a pyrogenic reaction to IV cannulation is
    (a) irritation and inflammation of the cannulated vein by the IV solution, needle, catheter, or infused medication.
    (b) advancing the catheter incompletely and then withdrawing the catheter back over (or through) the needle.
    (c) the infusion of air from an incompletely flushed administration set.
    (d) the presence in the IV solution or administration set of foreign proteins capable of causing fever.
    (e) an antigen or antibody reaction to the IV solution.

118. The cause of thrombophlebitis from IV cannulation is
    (a) irritation and inflammation of the cannulated vein by the IV solution, needle, catheter, or infused medication.
    (b) advancing the catheter incompletely and then withdrawing the catheter back over (or through) the needle.

    (c) the infusion of air from an incompletely flushed administration set.

    (d) the presence in the IV solution or administration set of foreign proteins capable of causing fever.

    (e) an antigen or antibody reaction to the IV solution.

119.   Signs and symptoms of a pyrogenic reaction to IV cannulation include

    (a) complaints of pain along the path of the vein.

    (b) erythema and edema at the puncture site.

    (c) the abrupt onset of fever, chills, nausea, headache, or backache.

    (d) Answers (a) and (b).

    (e) Answers (b) and (c).

120.   Signs and symptoms of thrombophlebitis caused by IV cannulation include

    (a) complaints of pain along the path of the cannulated vein.

    (b) erythema and edema at the puncture site.

    (c) the abrupt onset of fever, chills, nausea, headache, or backache.

    (d) Answers (a) and (b).

    (e) Answers (a), (b), and (c).

The Answer Key to Test Section 3 is on page 412.

Test Section Four

Test Section 4's subjects:

- EMS-Administered Drugs
- activated charcoal
- adenosine
- albuterol
- aminophylline
- amiodarone
- aspirin
- atropine
- calcium chloride
- $D_{50}W$
- dexamethasone
- diazepam/Valium
- diphenhydramine
- dobutamine
- dopamine
- epinephrine 1:10,000
- epinephrine 1:1000
- flumazenil
- furosemide
- glucagon
- ipecac
- isoproterenol
- lidocaine
- magnesium sulfate
- methylprednisolone
- morphine sulfate
- naloxone
- nitroglycerin
- oxytocin
- procainamide
- propranolol
- racemic epinephrine
- sodium bicarbonate
- thiamine
- vasopressin
- verapamil

Test Section 4 consists of 135 questions and is allotted 2 hours and 15 minutes for completion.

Author's Note: *The correct answers for many questions in this section are based upon the 2005 American Heart Association Guidelines for Cardiopulmonary Resuscitation and Emergency Cardiovascular Care (Circulation, Volume 112, Issue 24 Supplement; December 13, 2005). If you notice a contradiction between the Answer Key and the information supplied by any textbook that hasn't been 2005 AHA-updated, the Answer Key identifies the currently correct answer.*

Within each of the December 2005, American Heart Association's protocols for provision of advanced cardiac life support, the various interventions possible for consideration are identified by a Class of Recommendation. The first three questions in this Test Section refer to these 2005 AHA ACLS intervention recommendation classes.

1. Interventions identified as _____ indicate unacceptable and possibly harmful interventions.
   (a) Class I
   (b) Class IIa
   (c) Class IIb
   (d) Class III
   (e) Indeterminate

2. Interventions considered always acceptable, safe and effective, are identified as _____ interventions.
   (a) Class I
   (b) Class IIa
   (c) Class IIb
   (d) Class III
   (e) Indeterminate

3. Interventions that may be acceptable (supported only by fair-to-good clinical research evidence of effectiveness) but that are considered optional or alternative interventions by the majority of experts are identified as _____ interventions.
   (a) Class I
   (b) Class IIa
   (c) Class IIb
   (d) Class III
   (e) Indeterminate

4. Which of the following statements regarding epinephrine (epi) is false?
   (a) Epi affects both alpha and beta receptors.
   (b) Epi increases myocardial oxygen demand and can increase the size of an infarct.

(c) Epi's alpha effects are stronger than its beta effects.

(d) Epi is a short-acting drug and requires repeat boluses every 3 to 5 minutes to maintain a therapeutic blood level of the drug.

(e) Epi is ineffective when the myocardium is inadequately oxygenated.

5. Epinephrine 1:10,000 (epi 1:10) administration may be indicated in all of the following emergencies except

    (a) ventricular fibrillation.

    (b) asystole.

    (c) pulmonary edema.

    (d) pulseless electrical activity (PEA).

    (e) anaphylactic shock.

6. Which of the following statements regarding epinephrine 1:10,000 is false?

    (a) Epi 1:10 requires protection from light.

    (b) Epi 1:10 is deactivated when combined with alkaline solutions.

    (c) Epi 1:10 is contraindicated in pediatric patients.

    (d) Epi 1:10 is only used in life-threatening emergencies.

    (e) Epi 1:10, even in low doses, can cause myocardial ischemia.

7. Epinephrine 1:10,000 may be administered via which of the following routes?

    (1) SQ   (5) ET

    (2) IM   (6) Rectally

    (3) IV   (7) Transdermally

    (4) IO   (8) Orally (swallowed)

        (a) 1, 2, 3, 4, 5, 6, 7, and 8.

        (b) 1, 2, 3, 4, 5, and 6.

        (c) 2, 3, 4, and 5.

        (d) 3, 4, and 5.

        (e) 3 and 5.

8. In cardiac arrest emergencies, epinephrine 1:10,000 may be administered in doses of

    (a) 1.0 mg IV push (IVP) every 5 to 10 minutes.

    (b) 1.0 mg IVP every 3 to 5 minutes.

    (c) 2.0 to 2.5 mg via ET tube.

    (d) Answers (a) and (c).

    (e) Answers (b) and (c).

9. The most common trade names for the IV form of vasopressin are

    (a) Vancopressin® and Pressyn®.

    (b) Intropin® and Pressyn®.

    (c) Pitressin® and Pressyn®.

    (d) Isoptin® and Pressyn®.

    (e) Stadol® and Pressyn®.

10.   Which of the following statements regarding vasopressin is false?
   (a) Vasopressin is an adrenergic peripheral vasodilator.
   (b) If IV or IO access is unavailable, vasopressin may be administered via ET.
   (c) The recommended IV or IO dose of vasopressin is 40 units.
   (d) Vasopressin causes coronary vasoconstriction.
   (e) Vasopressin causes renal vasoconstriction.

11.   One dose of vasopressin may replace either the first or second dose of epinephrine when
   (a) VF/VT persists after 1 or 2 shocks.
   (b) the rhythm check confirms asystole.
   (c) the rhythm check confirms PEA.
   (d) Answers (a) and (b).
   (e) Answers (a), (b), and (c).

12.   Which of the following statements regarding isoproterenol (Isuprel®) is false?
   (a) Isoproterenol increases cardiac output by increasing the heart rate and strength of contractility.
   (b) Isoproterenol primarily acts only on beta-adrenergic receptors.
   (c) Isoproterenol is still used to increase blood pressure when cardiogenic shock or asystole is refractory to epinephrine and atropine.
   (d) Isuprel® increases myocardial oxygen demand and can increse the size of a myocardial infarct.
   (e) Isoproterenol may be considered for administraton when torsades de pointes is associated with bradycardia and drug-induced QT prolongation.

13.   The most commonly recognized trade name for dopamine is
   (a) Inotropine®.
   (b) Dobutrexine®.
   (c) Intropin®.
   (d) Inopress®.
   (e) Dopastatine®.

14.   Which of the following statements regarding dopamine hydrochloride is false?
   (a) Dopamine is a catecholamine that acts on α, β_1, and dopaminergic adrenergic receptors.
   (b) In therapeutic doses, dopamine provides vasodilation of the renal and mesenteric vasculature, to preserve their functions.
   (c) Infused at a dose rate of less than 20 µg/kg/minute, dopamine primarily affects beta-adrenergic receptors.
   (d) Dopamine's special chemistry allows for concomitant (simultaneous) administration of sodium bicarbonate through an IV in which dopamine is being infused.
   (e) Dopamine dosages should be significantly reduced if the patient has been taking a monoamine-oxidase inhibitor (MAO/MAOI) drug for depression.

15. Dopamine administration is indicated for the treatment of hemodynamically significant hypotension resulting from
    (a) cardiogenic, neurogenic, or hypovolemic shock.
    (b) cardiogenic or neurogenic shock only.
    (c) cardiogenic shock only.
    (d) neurogenic shock only.
    (e) hypovolemic shock only.

16. A dopamine infusion is prepared by adding
    (a) 800 mg of dopamine to 1000 ml of normal saline or D_5W.
    (b) 800 mg of dopamine to 500 ml of D_5W.
    (c) 400 mg of dopamine to 250 ml of D_5W.
    (d) Answer (a) or (b).
    (e) Answer (b) or (c).

17. The recommended initial dopamine infusion rate is
    (a) 1 μg/kg/min.
    (b) 2 to 5 μg/kg/min.
    (c) 10 to 20 μg/kg/min.
    (d) 20 to 25 μg/kg/min.
    (e) 25 to 30 μg/kg/min.

18. Which of the following statements regarding dobutamine is false?
    (a) Elderly patients may require a significantly greater dose of dobutamine in order to achieve a therapeutic response.
    (b) Dobutamine is used only for short-term treatment of congestive heart failure patients who are hypotensive but have no signs of shock.
    (c) The suggested therapeutic dose of dobutamine ranges from 2 to 20 μg/kg per minute.
    (d) Dobutamine produces less increase in heart rate than isoproterenol or dopamine.
    (e) Dobutamine should be administered at a flow rate based upon the patient's response.

19. Which of the following statements regarding amiodarone is false?
    (a) Amiodarone is recommended for ventricular fibrillation refractory to defibrillation.
    (b) Lidocaine is preferred over amiodarone for the treatment of ventricular dysrhythmias.
    (c) Amiodarone is recommended for pulseless ventricular tachycardia refractory to defibrillation.
    (d) Amiodarone is recommended for hemodynamically stable ventricular tachycardia refractory to cardioversion.
    (e) Amiodarone may be used for rate control of rapid atrial dysrhythmias.

20. In cardiac arrest, amiodarone may be initially administered as a loading dose of

(a) 1.0 mg IVP, repeated every 5 to 10 minutes.

(b) 300 mg IVP (consider repeating).

(c) 150 mg slow IVP over 10 minutes (15 mg/min); may repeat every 10 minutes.

(d) Answer (a) or (b).

(e) Answer (a) or (c).

21. In stable wide-complex tachycardia, amiodarone may be initially administered as a loading dose of

(a) 1.0 mg IVP, repeated every 5 to 10 minutes.

(b) 300 mg IVP (consider repeating).

(c) 150 mg slow IVP, over 10 minutes (15 mg/min); may repeat every 10 minutes.

(d) Answer (a) or (c).

(e) None of the above; this is not an indication for amiodarone administration.

22. The most commonly recognized trade name for amiodarone is

(a) Cordarone®.

(b) Inocor®.

(c) Adenocard®.

(d) Aramine®.

(e) Isoptin®.

23. The most common side effect of amiodarone administration is

(a) SVT or PSVT.

(b) widened QRS complexes.

(c) bradycardia.

(d) hypotension from vasodilation.

(e) high-degree heart block.

24. Which of the following statements regarding lidocaine is false?

(a) Lidocaine lowers the threshold of ventricular fibrillation.

(b) Lidocaine should only be considered for administration if amiodarone is unavailable.

(c) Lidocaine may depress the central nervous system.

(d) Successful ectopy suppression requires maintenance of adequate blood levels of lidocaine.

(e) Prophylactic lidocaine administration is no longer recommended for AMI.

25. Lidocaine administration may be considered in all of the following situations except

(a) indeterminate wide complex PSVT.

(b) ventricular fibrillation.

(c) ventricular tachycardia with pulses.

(d) after defibrillation of ventricular fibrillation has resulted in a return of spontaneous circulation without lidocaine having been previously administered.

(e) atrial fibrillation with a ventricular response rate above 100.

26. Lidocaine is contraindicated in all of the following situations, except

(a) second-degree AV block with more than six PVCs per minute.

(b) third-degree AV block with R-on-T PVCs.

(c) bradycardia with couplet PVCs.

(d) bradycardia refractory to atropine and isoproterenol, with R-on-T PVCs.

(e) polymorphic VT with a prolonged baseline QT interval that suggests torsades de pointes.

27. The most common IV dose regimen for lidocaine administration is

(a) 1 mg/kg IV bolus, repeated every 10 minutes until the dysrhythmia is resolved.

(b) 1 mg/kg IV bolus, repeated every 5 minutes until a lidocaine infusion is prepared and administered.

(c) an initial 0.5–0.75 mg/kg IV bolus, followed by 1.0–1.5 mg/kg boluses every 5 to 10 minutes, until a lidocaine infusion is prepared and administered or the maximum bolus dose of 3 mg/kg is reached.

(d) an initial 1.0–1.5 mg/kg IV bolus, repeated every 5 to 10 minutes until a lidocaine infusion is prepared and administered.

(e) an initial 1.0–1.5 mg/kg IV bolus, followed by 0.5–0.75 mg/kg repeat boluses every 5 to 10 minutes until a lidocaine infusion is prepared and administered, or a maximum bolus dose of 3 mg/kg is reached.

28. If the patient has impaired liver function, left ventricular dysfunction, or is over 70 years old, lidocaine should

(a) not be administered.

(b) be administered by maintenance infusion only.

(c) be administered using half of the standard initial (loading) dose and half of the standard repeat bolus regimens or maintenance infusion doses.

(d) be administered with the same standard initial (loading) dose but with reduced repeat bolus regimens or maintenance infusion dose.

(e) be administered with half of the standard initial (loading) dose, followed by standard repeat bolus regimens and maintenance infusion dose.

29. Administration routes for lidocaine during cardiac arrest include

(a) IV only.

(b) IV or IM only.

(c) IV or ET only.

(d) IV, ET, or IM.

(e) IV, IO, or ET.

30. Which of the following statements regarding lidocaine administration during cardiac arrest related to VF or pulseless VT is false?

    (a) Lidocaine has no proven short-term or long-term efficacy in cardiac arrest.

    (b) For ET administration, the initial lidocaine dose should be 2–3 mg/kg, diluted with water instead of 0.9% saline.

    (c) Immediately discontinue lidocaine administration if signs of toxicity develop.

    (d) If the first 3 defibrillations are unsuccessful in converting the rhythm, lidocaine administration should precede epinephrine administration.

    (e) If it is available, amiodarone should be administered before lidocaine.

31. The most commonly recognized trade name for procainamide is

    (a) Bretylol®.

    (b) Pronestyl®.

    (c) Prolixin®.

    (d) Procardia®.

    (e) Betapace®.

32. Which of the following statements regarding procainamide is false?

    (a) Like lidocaine, procainamide is administered only for ventricular dysrhythmias.

    (b) For terminating spontaneously occurring ventricular tachycardia, procainamide administration is considered superior to lidocaine.

    (c) Procainamide may be administered for atrial fibrillation or atrial flutter associated with Wolff-Parkinson-White syndrome.

    (d) Because of the need for slow infusion, efficacy of procainamide in cardiac arrest is limited.

    (e) If blood pressure remains stable and PSVT has been refractory to adenosine and vagal maneuvers, consider procainamide administration.

33. The standard IV dose regimen for procainamide administration is

    (a) 1 mg/kg IV bolus, repeated every 10 minutes until the dysrhythmia is resolved, adverse reactions occur, or a maximum total dose of 20 mg/kg has been administered.

    (b) 1 mg/kg IV bolus, repeated every 5 minutes until a maintenance infusion is prepared and initiated.

    (c) 20 mg/min, via IV infusion, until the dysrhythmia is suppressed, adverse reactions occur, or a maximum total dose of 17 mg/kg has been administered.

    (d) 20 mg/kg IV bolus, repeated every 5 minutes until a maintenance infusion is prepared and initiated.

    (e) single 20 mg/kg IV bolus loading dose, followed by a 5–10 mg/min maintenance infusion.

34. The standard dose rate for a procainamide IV maintenance infusion is

    (a) 1 mg/min.

    (b) 1–4 mg/min.

    (c) 2–6 mg/min.

    (d) 4–8 mg/min.

    (e) 5–10 mg/min.

35.   Which of the following statements regarding procainamide is false?

    (a) Procainamide slows conduction in atrial myocardial tissue.

    (b) Procainamide slows conduction in ventricular myocardial tissue.

    (c) Procainamide can be mixed with either D_5W or normal saline to make a maintenance infusion.

    (d) If the patient has a history of renal failure, the procainamide maintenance infusion should be reduced.

    (e) To be effective in acute cardiac arrest situations, bolus administration of procainamide is required.

36.   Procainamide administration should be discontinued immediately when

    (1) the dysrhythmia has been suppressed.

    (2) the maximum total dose has been administered.

    (3) the QRS diminishes in width by 50%.   (5) hypertension ensues.

    (4) the QRS widens by greater than 50%.   (6) hypotension ensues.

        (a) 1, 2, 3, or 5.

        (b) 2, 3, 5, or 6.

        (c) 1, 2, 4, or 6.

        (d) 3, 4, 5, or 6.

        (e) 1, 3, or 5.

37.   The most commonly recognized trade names for verapamil are

    (a) Isoptin® and Calan®.

    (b) Cardene® and Calan®.

    (c) Isoptin® and Cardene®.

    (d) Isoxsuprine® and Isoptin®.

    (e) Calan® and Cardizem®.

38.   Which of the following statements regarding verapamil is true?

    (a) Verapamil is used to treat narrow-QRS-complex PSVTs but only when adequate blood pressure is present and only after adenosine has failed to terminate the PSVT.

    (b) Verapamil may control ventricular response in patients with uncontrolled narrow-QRS-complex atrial flutter, atrial fibrillation, or multifocal atrial tachycardias.

    (c) Verapamil is used before the administration of adenosine to treat wide-QRS tachycardias of uncertain origin.

    (d) Answers (a) and (b).

    (e) Answers (a), (b), and (c).

39. Which of the following statements regarding verapamil is true?

    (a) Verapamil causes peripheral vasodilation and is contraindicated for administration to hypotensive patients.

    (b) Verapamil is contraindicated for patients on beta-blocking medications or patients with a history of atrial fibrillation and WPW syndrome.

    (c) Verapamil is contraindicated for patients with sick sinus syndrome, or high-degree AV blocks and no pacemaker.

    (d) Answers (a) and (b).

    (e) Answers (a), (b), and (c).

40. Which of the following is/are the accepted IV dose regimen(s) for verapamil administration?

    (a) 2.5–5.0 mg IV, infused over 2 minutes.

    (b) 2.5–5.0 mg IV, infused over 2 minutes; followed by one or more repeat doses of 5–10 mg IV every 15 to 30 minutes (if needed), each infused over 2 minutes, not to exceed a total dose of 20 mg.

    (c) 5 mg bolus every 15 minutes, not to exceed a total dose of 30 mg.

    (d) Answer (a) or (b).

    (e) Answer (a), (b), or (c).

41. Which of the following statements regarding adenosine is false?

    (a) Adenosine is a naturally occurring substance that is present in all body cells.

    (b) The most commonly recognized trade name for adenosine is Sinocard®.

    (c) Adenosine decreases AV conduction of atrial impulses.

    (d) The half-life of adenosine is 5 seconds or less.

    (e) Because the side effects of adenosine are generally self-limited, they are not of as much concern as are the side effects of other drugs.

42. Adenosine administration is indicated for the treatment of

    (a) narrow-complex PSVT, refractory to vagal maneuvers.

    (b) atrial fib/flutter tachycardias associated with Wolff-Parkinson-White (WPW) syndrome, refractory to vagal maneuvers.

    (c) wide-complex tachycardias, refractory to vagal maneuvers.

    (d) Answers (a) and (b).

    (e) Answers (a) and (c).

43. Contraindications for adenosine administration include

    (a) second- or third-degree heart blocks.

    (b) hypersensitivity to adenosine.

    (c) poison- or drug-induced tachycardias.

    (d) Answers (a) and (b).

    (e) Answers (a), (b), and (c).

44.   The side effects of adenosine include all of the following, except
  (a) sudden onset of PVCs, PACs, or other dysrhythmias.
  (b) prolonged periods of nausea.
  (c) sudden complaints of dizziness, chest pain, or chest tightness.
  (d) transient episodes of asystole.
  (e) transient AV blocks or other forms of bradycardia.

45.   Which of the following statements regarding adenosine administration is false?
  (a) Adenosine should always be administered by very rapid IVP (1 to 3 seconds) directly into a major vein or the IV port nearest to the patient's heart.
  (b) Each bolus of adenosine should immediately be followed by a rapid IVP bolus of 20 ml saline.
  (c) Patients taking medications containing theophylline, theobromine, or caffeine should receive only half of the recommended adenosine dose(s).
  (d) Patients taking dipyridamole (Persantine®) or carbamazepine (Tegretol®) should receive only half of the recommended initial adenosine dose.
  (e) If needed, a third bolus of adenosine may be given within 1 to 2 minutes after the second bolus.

46.   The standard initial dose of adenosine is a
  (a) 3 mg bolus.
  (b) 6 mg bolus.
  (c) 9 mg bolus.
  (d) 12 mg bolus.
  (e) 18 mg bolus.

47.   If conversion of the tachycardia is not achieved within 1 to 2 minutes of the initial dose, the standard second dosage of adenosine is a
  (a) 3 mg bolus.
  (b) 6 mg bolus.
  (c) 9 mg bolus.
  (d) 12 mg bolus.
  (e) 18 mg bolus.

48.   Atropine sulfate is
  (a) a sympatholytic drug.
  (b) a parasympatholytic drug.
  (c) an anticholinergic drug.
  (d) Answers (a) and (c).
  (e) Answers (b) and (c).

49. Atropine administration may be indicated for all of the following situations, except

    (a) symptomatic bradycardias with or without ectopy.

    (b) symptomatic infranodal (second- or third-degree) AV blocks with a wide QRS complex.

    (c) symptomatic bronchial asthma and reversible bronchospasm associated with COPD.

    (d) symptomatic bradycardia associated with organophosphate poisoning.

    (e) asystole or bradycardic PEA (EMD).

50. Administration routes for atropine include

    (a) IV only.

    (b) IV or SQ only.

    (c) IV, IO, or SQ only.

    (d) IV or IO only.

    (e) IV, IO, or ET.

51. In a cardiac arrest situation, atropine is administered in IV doses of _____ every 3 to 5 minutes, not to exceed a total dose of _____.

    (a) 0.25 to 0.5 mg / 0.03 mg/kg

    (b) 0.5 to 1.0 mg / 0.03 mg/kg

    (c) 0.5 mg / 3 mg

    (d) 1.0 mg / 3 mg

    (e) 2.0 mg / 4 mg

52. For a symptomatic patient with a pulse (unrelated to organophosphate emergencies), atropine is administered in IV doses of _____ every 3 to 5 minutes, not to exceed _____.

    (a) 0.25 to 0.5 mg / 0.03 mg/kg

    (b) 0.5 to 1.0 mg / 0.03 mg/kg

    (c) 0.5 mg / 3 mg

    (d) 1.0 mg / 3 mg

    (e) 2.0 mg / 4 mg

53. Which of the following statements regarding atropine is false?

    (a) Atropine accelerates sinus and atrial pacemaker activity.

    (b) Large IV doses of atropine (greater than 1 mg) may produce paradoxical bradycardia.

    (c) Atropine increases AV node/junction impulse conduction.

    (d) Larger than normal atropine doses may be required to treat victims of carbamate poisoning.

    (e) Larger than normal atropine doses may be required to treat victims of nerve gas exposure.

54. Which of the following statements regarding sodium bicarbonate is true?
    (a) A single dose of sodium bicarb should be administered immediately before (or after) intubation of a cardiac arrest victim with a known "down time" of more than 5 minutes prior to artificial ventilation initiation; two doses should be administered for a known "down time" of more than 10 minutes prior to artificial ventilation initiation.
    (b) If not needed, sodium bicarb administration will cause metabolic acidosis, even if only a small amount is administered.
    (c) Sodium bicarb administration to patients with known preexisting hyperkalemia, is considered acceptable and useful.
    (d) In cases of hypercarbic acidosis (cardiac arrest with CPR in progress but with repeatedly unsuccessful intubation attempts), administration of a single dose of sodium bicarb has been shown to be highly effective.
    (e) In cases of known preexisting diabetic ketoacidosis, sodium bicarb administration is not considered useful or effective.

55. Sodium bicarbonate administration is considered acceptable and useful for the treatment of
    (a) tricyclic medication overdose.
    (b) prolonged cardiac arrest refractory to hyperventilation.
    (c) persistent or recurrent ventricular fibrillation or pulseless ventricular tachycardia, refractory to defibrillation and lidocaine.
    (d) Answers (a) and (b).
    (e) Answers (a), (b), and (c).

56. Administration routes for sodium bicarbonate include
    (a) IV only.
    (b) IV or SQ only.
    (c) IV or IO only.
    (d) IV or ET only.
    (e) IV, IO, or ET.

57. The standard dose for initial administration of sodium bicarbonate is a
    (a) 0.25 mEq/kg bolus.
    (b) 0.5 mEq/kg bolus.
    (c) 1.0 mEq/kg bolus.
    (d) 2.0 mEq/kg bolus.
    (e) 5.0 mEq/kg bolus.

58. Which of the following statements regarding morphine sulfate is false?

    (a) Morphine decreases peripheral venous capacitance.

    (b) Morphine is a central nervous system depressant.

    (c) Morphine reduces systemic vascular resistance.

    (d) Morphine decreases venous return and diminishes myocardial oxygen demand.

    (e) Morphine alleviates anxiety that might contribute to an increase in infarct size.

59. Morphine sulfate may be indicated for the treatment of all of the following except

    (a) chest pain in the setting of an AMI.

    (b) symptomatic pulmonary edema with associated chest pain.

    (c) symptomatic pulmonary edema without associated chest pain.

    (d) severe pain secondary to blunt chest or abdominal trauma.

    (e) severe pain from isolated extremity trauma or kidney stones.

60. Morphine sulfate is contraindicated for administration to patients who

    (1) are hypotensive.   (4) have head injuries.

    (2) are allergic to barbiturates.   (5) complain of abdominal pain.

    (3) are hypovolemic.   (6) have symptomatic pulmonary edema without associated chest pain.

        (a) 1, 2, 3, 4, and 5.

        (b) 2, 3, 4, and 6.

        (c) 1, 2, 3, 4, 5, and 6.

        (d) 1, 3, 4, and 5.

        (e) 1, 3, 4, 5, and 6.

61. Administration of morphine sulfate may cause

    (a) life-threatening allergic reactions.

    (b) respiratory depression or arrest.

    (c) nausea, vomiting, or abdominal cramping.

    (d) Answers (a) and (b).

    (e) Answers (a), (b), and (c).

62. The most common IV morphine administration regimen for non-cardiac- or pulmonary-related emergencies, is which of the following?

    (a) A single dose of 5–15 mg, depending upon the patient's weight.

    (b) An initial dose of 4 mg, with repeat boluses of 4 mg each, until total pain relief is achieved or respiratory depression occurs.

    (c) An initial dose of 2–10 mg, with repeat boluses of 2 mg each, until pain tolerance is achieved or hypotension or respiratory depression occurs.

    (d) A single dose of 2–4 mg (repeat prehospital doses are contraindicated).

    (e) 2–4 mg slow IVP (over 1 to 5 minutes), titrated to effect; may be repeated every 5 to 30 minutes.

63. Which of the following is the recommended IV morphine administration regimen for cardiac- or pulmonary-related emergencies?

    (a) A single dose of 0.10 mg/kg.

    (b) An initial dose of 4 mg, with repeat boluses of 4 mg each, until total pain relief is achieved or respiratory depression occurs.

    (c) An initial dose of 2–10 mg, with repeat boluses of 2 mg each, until pain tolerance is achieved or hypotension or respiratory depression occurs.

    (d) A single dose of 2–4 mg (repeat prehospital doses are contraindicated).

    (e) An initial dose of 2–4 mg; additional doses of 2–8 mg may be given every 5 to 15 minutes.

64. Administration routes for morphine sulfate include

    (a) IV only.

    (b) IV or IM.

    (c) IV, IM, or IO.

    (d) IV, ET, or IM.

    (e) IV, ET, IM, or IO.

65. Which of the following statements regarding furosemide (Lasix®) is false?

    (a) Lasix inhibits reabsorption of sodium in the kidneys.

    (b) Lasix inhibits reabsorption of chloride in the kidneys.

    (c) Lasix's diuresis effects are not observed until 30 or more minutes after administration.

    (d) Lasix must be protected from light to retain its potency.

    (e) Lasix reduces venous and pulmonary vascular resistance.

66. Indications for furosemide administration include

    (1) acute pulmonary edema in a hypertensive patient.

    (2) acute pulmonary edema in a hypotensive patient.

    (3) emergencies involving hyperkalemia.

    (4) pulmonary edema caused by severe anaphylaxis.

    (5) hypertensive emergencies.

    (6) high altitude pulmonary edema.

    (7) hypertension related to pregnancy (the "first-line" drug of choice).

        (a) 1, 2, 3, 4, 5, 6, and 7.

        (b) 1, 3, 4, 5, 6, and 7.

        (c) 1, 3, 4, 5, and 7.

        (d) 1, 3, 5, and 6.

        (e) 3, 4, 5, 6, and 7.

67. The potential side effects of furosemide administration include all of the following, except

   (a) hyperkalemia.
   (b) allergic reaction in patients allergic to sulfa drugs.
   (c) hypotension.
   (d) dehydration or hypovolemia.
   (e) fetal abnormalities, if administered during pregnancy.

68. Common IV furosemide administration regimens include which of the following?

   (a) 40 mg, if the patient is already on diuretic therapy.
   (b) 20 mg, if the patient is not on diuretic therapy.
   (c) 0.5–1.0 mg/kg administered slowly. If no response within 5 to 15 minutes, a second dose of 1.0–2.0 mg/kg may be administered.
   (d) Answers (a) and (b).
   (e) Answers (a), (b), and (c).

69. Emergency administration routes for furosemide include

   (a) IV only.
   (b) IV or IO.
   (c) IV, IM, or aerosolized (inhalation) administration.
   (d) IV, IM, or ET.
   (e) IV, IM, ET, IO, or aerosolized (inhalation) administration.

70. Which of the following statements regarding nitroglycerin (NTG) administration is false?

   (a) NTG is a smooth-muscle relaxant.
   (b) NTG dilates cerebral arteries, effecting a decrease in intracranial pressure.
   (c) NTG dilates coronary arteries and improves myocardial perfusion.
   (d) NTG products deteriorate (become less effective) when exposed to air or light.
   (e) NTG spray containers should not be shaken prior to administration.

71. Nitroglycerin is indicated for treatment of normotensive or hypertensive patients in all the following emergencies, except

   (a) angina associated with suspected myocardial ischemia.
   (b) chest pain associated with S/Sx of AMI.
   (c) acute pulmonary edema.
   (d) high-altitude cerebral edema (increased intracranial pressure).
   (e) hypertensive emergencies associated with acute coronary syndromes.

72. Contraindications for NTG administration to a patient exhibiting S/Sx of AMI include which of the following?

   (1) Patient ingestion of any antihypertensive medication within the past 24 hours.
   (2) Patient ingestion of a phosphodiesterase inhibitor for erectile dysfunction (Viagra®, Cialis®, Levitra®) within the past 24 hours.

(3) S/Sx of increased ICP.
(4) Severe bradycardia (less than 50 bpm).
(5) Severe tachycardia (greater than 100 bpm).
(6) S/Sx of inferior wall MI with possible right ventricular infarction.

    (a) 1, 2, 3, 4, 5, and 6.
    (b) 2, 3, 4, 5, and 6.
    (c) 1, 3, 4, and 6.
    (d) 1, 2, 4, and 6.
    (e) 2, 3, and 6.

73. Your patient was c/o CP and exhibiting AMI S/Sx. His original systolic B/P was 140 mmHg. You administered an initial dose of NTG. He reports no relief from his CP and has developed a c/o HA. Enough time has passed to allow consideration of a repeat NTG dose, so you recheck his B/P. Which of the following systolic blood pressure measurements indicate a contraindication for repeat administration of NTG to this patient?

    (1) 128   (4) 100
    (2) 118   (5) 90
    (3) 108   (6) 80

    (a) 1, 2, 3, 4, 5, and 6. He developed a headache; this is a contraindication for repeat administration of NTG, regardless of his blood pressure measurement.
    (b) 2, 3, 4, 5, and 6.
    (c) 3, 4, 5, and 6.
    (d) 4, 5 and 6.
    (e) 5 and 6.

74. Your patient was c/o HA, CP, and exhibiting AMI S/Sx. Her original systolic B/P was 190 mmHg. You administered an initial dose of NTG. She reports no relief from her CP, and enough time has passed to allow consideration of a repeat NTG dose, so you recheck her B/P. Which of the following systolic blood pressure measurements indicate a contraindication for repeat administration of NTG to this patient?

    (1) 178   (4) 138
    (2) 168   (5) 128
    (3) 158   (6) 118

    (a) 1, 2, 3, 4, 5, and 6. She was hypertensive with a headache; this was a contraindication for the initial administration of NTG.
    (b) 2, 3, 4, 5, and 6.
    (c) 3, 4, 5, and 6.
    (d) 6.
    (e) None of the above. As long as her blood pressure is above 100 systolic, she may receive another NTG dose.

75. Which of the following statements regarding calcium chloride is false?

(a) Calcium ions reduce myocardial workload by decreasing the strength of myocardial contraction.

(b) Calcium administration to patients on digitalis medications may induce digitalis toxicity.

(c) When calcium is administered via the same IV tubing as sodium bicarb, a precipitate may form within the IV.

(d) Potential side effects of calcium administration include bradycardia, dysrhythmias, and cardiac arrest.

(e) Extravasation of calcium may cause local tissue necrosis.

76. Dosages for calcium chloride administration vary, depending upon the reason it is being given. Which of the following dosage protocols is not acceptable for calcium chloride administration under any circumstance?

(a) 8 to 16 mg/kg (usually 5 to 10 ml) IV/IO, administered slowly and repeated as necessary.

(b) 500 to 1000 mg (5 to 10 ml) IV/IO, administered over 2 to 5 minutes.

(c) Emergency ET administration: dilute half of the standard dose in 10 ml of saline, deliver it rapidly, and immediately perform 2 to 5 minutes of vigorous hyperventilation.

(d) 20 mg/kg IV/IO, administered slowly.

(e) 0.2 ml/kg IV/IO, administered slowly.

77. Which of the following emergencies may indicate the need for calcium chloride administration?

(1) Abdominal pain associated with Portuguese man-of-war stings.

(2) Abdominal pain associated with spider bites.

(3) Overdose of calcium channel blocker medication.

(4) Known hyperkalemia.

(5) Known hypocalcemia.

(6) Beta blocker medication overdose.

(7) Magnesium sulfate overdose.

    (a) 1, 2, 3, 4, 5, 6, and 7.

    (b) 2, 3, 4, 5, 6, and 7.

    (c) 2, 4, 5, 6, and 7.

    (d) 4, 5, 6, and 7.

    (e) 2, 5, and 6.

78. Which of the following statements regarding administration of epinephrine 1:1000 (epi 1:1) is false?

(a) Epi 1:1 may increase myocardial effort and oxygen demand, potentially causing angina or AMI.

(b) Epi 1:1 is contraindicated for use if the patient has cardiovascular disease.

(c) Frequent side effects of epi 1:1 administration are paradoxical bradycardia, complaints of "dizziness" and/or difficulty breathing.

(d) Epi 1:1 must be protected from light to remain effective.

(e) 2005 AHA recommendations for epi 1:1 stipulate that IM administration is preferred over SQ if the patient is in shock (hypotensive) or complains of severe difficulty breathing.

79.   Epinephrine 1:1000 is indicated for the treatment of
(a) allergic reactions.
(b) anaphylactic shock.
(c) bronchial asthma.
(d) Answers (a) and (c).
(e) Answers (a), (b), and (c).

80.   The adult administration dose of epi 1:1 is
(a) 0.1–0.3 mg/kg, IV only.
(b) 0.3–0.5 mg, SQ or IM
(c) 0.5–1.0 mg, SQ or IM.
(d) 1.0–2.0 mg/kg, IV or SQ.
(e) 0.5–1.0 mg, IV or IM.

81.   Which of the following statements regarding aminophylline is false?
(a) The side effects of aminophylline administration include ventricular ectopy, tachycardias, and hypotension.
(b) If the standard dose of aminophylline is administered to patients who are taking beta blockers or erythromycin, aminophylline toxicity may occur.
(c) Rapid aminophylline administration may cause acute hypertension.
(d) For patients taking prescribed theophylline-containing medications, the aminophylline-administration dose should be based upon their theophylline blood level.
(e) For patients taking prescribed medications that contain xanthine, the aminophylline-administration dose should be based upon their xanthine blood level.

82.   Aminophylline administration may be indicated in all of the following emergency situations, except
(a) congestive heart failure with pulmonary edema.
(b) asthma.
(c) chronic bronchitis with bronchospasm.
(d) acute allergic reactions.
(e) emphysema with bronchospasm.

83. In emergency situations involving adults, aminophylline is administered as a
    (a) 250–500 mg slow IV push (over 1 to 2 minutes).
    (b) "piggybacked" IV infusion of 250–500 mg dissolved in 80–90 ml of D_5W, infused over 20 to 30 minutes.
    (c) "piggybacked" IV infusion of 250–500 mg dissolved in 20 ml of D_5W, infused over 20 to 30 minutes.
    (d) Answer (a) or (c), depending upon the nature of the emergency.
    (e) Answer (b) or (c), depending upon the nature of the emergency.

84. All of the following statements regarding racemic epinephrine are true, except
    (a) in the prehospital setting, racemic epi is administered by nebulized inhalation only.
    (b) racemic epi is used to treat croup.
    (c) racemic epi is used to treat epiglottitis.
    (d) racemic epi may cause tachycardia or dysrhythmias.
    (e) within 30 to 60 minutes after racemic epi administration, many patients develop "rebound worsening" of their condition.

85. Which of the following statements regarding albuterol (Proventil®, Ventolin®) is false?
    (a) Albuterol is a sympatholytic medication.
    (b) Albuterol acts only upon β_2-adrenergic receptors.
    (c) Albuterol produces bronchodilation without inducing tachycardia as often as epi or aminophylline administration.
    (d) Albuterol is used to treat asthma or reversible bronchospasm associated with COPD.
    (e) In the prehospital setting, albuterol is administered by inhalation only.

86. Dextrose is classed as a
    (a) sympatholytic.
    (b) vitamin.
    (c) carbohydrate.
    (d) starch.
    (e) sympathomimetic.

87. When unable to obtain reagent-strip confirmation of the patient's blood-sugar level, indications for administration of 50% dextrose in water ($D_{50}W$) include an altered level of consciousness suspected to be caused by
    (a) acute alcohol abuse.
    (b) acute drug overdose.
    (c) hyperglycemic ketoacidosis.
    (d) Answers (a) and (b).
    (e) Answers (a), (b), and (c).

88. When hyperglycemia is confirmed by a reagent strip, administration of $D_{50}W$ is contraindicated in cases involving an altered level of consciousness suspected to be due to
    (a) a seizure disorder.
    (b) a cerebrovascular accident.
    (c) hyperglycemic ketoacidosis.
    (d) All of the above.
    (e) None of the above.

89. $D_{50}W$ is administered in a dosage of
    (a) 25 grams (25 ml of a 50% solution).
    (b) 25 grams (50 ml of a 50% solution).
    (c) 2500 grams (50 ml of a 50% solution).
    (d) 25 mg (50 ml of a 50% solution).
    (e) 250 mg (25 ml of a 25% solution).

90. Administration routes for $D_{50}W$ include
    (a) IV only.
    (b) IV or IO only.
    (c) IV, IO, or ET.
    (d) IV or IM only.
    (e) IV, IM, or IO.

91. Thiamine is classed as a
    (a) sympatholytic.
    (b) vitamin.
    (c) carbohydrate.
    (d) starch.
    (e) sympathomimetic.

92. Thiamine deficiency may
    (a) alter the body's ability to metabolize glucose.
    (b) cause neurologic deficits, especially in alcoholic patients who have been on a "binge."
    (c) contribute to altered levels of consciousness.
    (d) Answers (b) and (c).
    (e) Answers (a), (b), and (c).

93. Administration of thiamine is indicated in situations involving patients exhibiting
    (a) an altered level of consciousness with a history of "brittle diabetes."
    (b) an altered level of consciousness with a history of alcoholism.
    (c) S/Sx of "delirium tremens."
    (d) Answers (b) and (c).
    (e) Answers (a), (b), and (c).

94. Administration routes for thiamine include
     (a) IV only.
     (b) IV or IO only.
     (c) IV, IO, or ET.
     (d) IV or IM only.
     (e) IV, IM, ET, or IO.

95. The standard emergency dosage for thiamine administration is
     (a) 100 mg.
     (b) 100 mEq.
     (c) 50 mg.
     (d) 50 mEq.
     (e) 100 ml of a 50% solution.

96. Which of the following statements regarding the synthetic steroid medication
     dexamethasone (Decadron®, Hexadrol®) is false?
     (a) Dexamethasone has potent diuretic properties, which are believed to be
         responsible for its effectiveness in decreasing cerebral edema.
     (b) Dexamethasone has anti-inflammatory actions.
     (c) Prolonged use of dexamethasone may cause adrenocortical steroid
         suppression or gastrointestinal hemorrhage.
     (d) As much as 100 mg of dexamethasone may be administered in an emergency
         setting.
     (e) In an emergency setting, and with appropriate indications for administration,
         there are no contraindications for administration of dexamethasone.

97. Potential indications for the emergency administration of steroid preparations
     include all of the following, except
     (a) asthma or exacerbated COPD.
     (b) anaphylaxis.
     (c) exacerbated, congestive heart failure or acute, cardiogenic, pulmonary edema.
     (d) spinal cord injury.
     (e) cerebral edema.

98. Which of the following statements regarding methylprednisolone (Solu-Medrol®) is
     false?
     (a) Methylprednisolone has potent anti-inflammatory properties.
     (b) For acute histamine reactions, 80–125 mg IV of methylprednisolone may be
         administered to an adult.

    (c) For adult spinal-cord injury, 30 mg/kg of methylprednisolone is infused IV, over 15 minutes, followed 45 minutes later by a 5.4 mg/kg/hr maintenance infusion.

    (d) Methylprednisolone administration side effects include HA, N/V, hypertension, and hiccups.

    (e) Because methylprednisolone is an intermediate-acting steroid, multiple repeat doses may be administered during the prehospital phase of any emergency indicating consideration of its administration, without concern for adverse side effects.

99. The most familiar trade name for diazepam is

    (a) Valium®.

    (b) Benadryl®.

    (c) Demerol®.

    (d) Vistaril®.

    (e) Haldol®.

100. In the prehospital setting, indications for diazepam administration include all of the following except

    (a) sedation prior to cardioversion.

    (b) treatment of status epilepticus.

    (c) treatment of severe anxiety reactions.

    (d) management of belligerence related to drug or alcohol consumption.

    (e) treatment of pain associated with orthopedic injuries.

101. In the prehospital setting, the most common adult IV dosage of diazepam is

    (a) 2–5 mg, rapid IVP, repeated as needed.

    (b) 5–10 mg, slow IVP, repeated as needed.

    (c) 0.3–0.5 mg bolus, repeated every 5 minutes, to a maximum of 15 mg.

    (d) 0.3–0.5 mg/kg bolus, repeated once only.

    (e) an initial loading dose of 2 mg, followed after 10 minutes by a 5 mg bolus, which may be repeated every 10 minutes as needed.

102. Diazepam may be given via

    (a) IV, IM, or rectal administration only.

    (b) IV, ET, or rectal administration only.

    (c) IV, IM, ET, or rectal administration only.

    (d) IV, IM, ET, IO, or rectal administration.

    (e) IV, IM, IO, or rectal administration only.

103. In the prehospital setting, indications for administration of oxytocin (Pitocin®) include which of the following?
   (1) To induce labor during a prolonged transport time.
   (2) To delay labor during a prolonged transport time.
   (3) To assist in control of severe postpartum hemorrhage.
   (4) To assist in the delivery of another fetus during a multiple-birth situation.
   (5) To assist in the safe delivery of a fetus with a breach- or limb-presentation.

   (a) 1, 2, 3, 4, and 5.
   (b) 2, 3, 4, and 5.
   (c) 3.
   (d) 4 and 5.
   (e) 2, 4, and 5.

104. Which of the following statements regarding magnesium sulfate is false?
   (a) Magnesium sulfate is a CNS stimulant, and its administration may cause tachycardia, tachypnea, or hypertension.
   (b) Magnesium sulfate administration will successfully treat cardiac dysrhythmias related to hypomagnesemia.
   (c) Calcium (chloride or gluconate) administration will antagonize magnesium sulfate effects.
   (d) Even when being administered to a pulseless patient, at least 5 minutes is required to administer magnesium sulfate.
   (e) The side effects of magnesium sulfate include bradycardia, hypotension, respiratory depression, drowsiness, and hypothermia.

105. Prehospital indications for magnesium sulfate administration during cardiovascular emergencies include
   (1) a renal dialysis patient with polymorphic ventricular tachycardia.
   (2) ventricular fibrillation refractory to lidocaine.
   (3) cardiac arrest from suspected hypomagnesemia.
   (4) cardiac arrest from suspected hypocalcemia.
   (5) pulseless torsades de pointes.
   (6) torsades de pointes with a pulse.
   (7) digitalis toxicity with ventricular dysrhythmias.
   (8) narrow-complex tachycardia.

   (a) 2, 3, 5, 6, and 7.
   (b) 1, 2, 3, 5, and 8.
   (c) 2, 3, 5, and 7.
   (d) 1, 2, 3, 4, 5, 7, and 8.
   (e) 1, 2, 3, 4, 5, 6, 7, and 8.

106. Prehospital indications for magnesium sulfate administration include when a hypertensive pregnant patient
   (a) exhibits S/Sx of an imminent seizure (prior to any actual seizure activity).
   (b) has recovered from a single, first-time seizure, and prophylactic administration to prevent further seizures is indicated.
   (c) is in status epilepticus.
   (d) Answers (b) and (c).
   (e) Answers (a), (b), and (c).

107. The recommended dose for initial magnesium sulfate administration to a pulseless cardiovascular emergency patient is
   (a) 2–5 grams diluted in 50–100 ml of D_5W, infused via IV/IO over 25 minutes.
   (b) 1–2 grams diluted in 10 ml of D_5W, infused via IV/IO over 5 to 20 minutes.
   (c) 1–2 grams (undiluted) rapid IV (or IO) push.
   (d) 1–2 grams diluted in 50–100 ml of D_5W, infused via IV/IO over 5 to 60 minutes.
   (e) None of the above. Magnesium sulfate is not administered during pulseless cardiovascular emergencies.

108. The recommended dose for initial magnesium sulfate administration to a cardiovascular emergency patient who has a pulse is
   (a) 2–5 grams diluted in 50–100 ml of D_5W, infused via IV/IO over 25 minutes.
   (b) 1–2 grams diluted in 10 ml of D_5W, infused via IV/IO over 5 to 20 minutes.
   (c) 1–2 grams (undiluted) rapid IV (or IO) push.
   (d) 1–2 grams diluted in 50–100 ml of D_5W, infused via IV/IO over 5 to 60 minutes.
   (e) None of the above. Magnesium sulfate is not administered during cardiovascular emergencies when the patient has a pulse.

109. The dose for initial magnesium sulfate administration to an obstetric patient is
   (a) 2–5 grams, diluted in 50–100 ml of D_5W, infused over 25 minutes, via IV.
   (b) 1–2 grams (2–4 ml of a 50% solution), diluted in 10 ml of D_5W, IVP.
   (c) 1–2 grams (undiluted) rapid IVP.
   (d) 1–2 grams diluted in 50–100 ml of D_5W, infused over 5 to 60 minutes via IV.
   (e) 100 mg, IVP.

110. The most familiar trade name for diphenhydramine is
   (a) Valium®.
   (b) Benadryl®.
   (c) Demerol®.
   (d) Vistaril®.
   (e) Haldol®.

111. Which of the following statements regarding diphenhydramine is false?

    (a) Diphenhydramine is indicated for the treatment of allergic reactions or anaphylactic shock but only after initial administration of epinephrine.

    (b) Diphenhydramine is indicated for the treatment of asthma but only after the initial administration of epinephrine.

    (c) Diphenhydramine is indicated for the treatment of dystonic reactions to phenothiazine medications.

    (d) Diphenhydramine has sedative effects.

    (e) Administration routes for diphenhydramine include IV, IM, and IO.

112. The normal adult dosage of diphenhydramine is

    (a) 1 mg/kg every 2 to 3 minutes, until the desired effect is achieved.

    (b) 10–20 mg diluted in 50 ml D_5W and infused over 15 minutes.

    (c) 25–50 mg bolus.

    (d) Answers (a) and (c).

    (e) Answers (b) and (c).

113. Which of the following statements regarding syrup of ipecac are true?

    (a) Ipecac is particularly effective for the treatment of thorazine overdoses.

    (b) If an overdose patient has a depressed level of consciousness, ipecac may be administered sublingually to effect gastric evacuation without aspiration.

    (c) After the administration of 15–30 ml of ipecac, the patient should remain NPO until arrival at the emergency room.

    (d) All of the above are true.

    (e) None of the above is true.

114. Which of the following statements regarding activated charcoal are false?

    (a) Activated charcoal administration is contraindicated in a patient with a depressed level of consciousness.

    (b) If ipecac has been administered, activated charcoal is contraindicated until after emesis is produced or after 10 minutes have passed without production of emesis.

    (c) Activated charcoal is indicated for treatment of poison ingestions or oral drug ODs that contraindicate antidote administration.

    (d) Unless spontaneous emesis occurs, activated charcoal administration should rapidly be followed by ipecac administration, to purge the charcoal-bound toxins.

    (e) The side effects of activated charcoal administration include nausea and vomiting.

115. Which of the following statements regarding naloxone (Narcan®) is false?

    (a) Naloxone antagonizes the effects of narcotics and opiates.

    (b) The administration of naloxone may precipitate withdrawal signs and symptoms in the narcotic-addicted patient.

    (c) Less naloxone is required to reverse the effects of a synthetic narcotic overdose (such as a Darvon® overdose).

    (d) Naloxone is not effective in the treatment of barbiturate overdose but still may be administered to rule out narcotic ingestion.

    (e) Naloxone is indicated for treatment of unconsciousness of unknown etiology.

116. When narcotic addiction is not suspected, which of the following regimens for naloxone administration is best?

    (a) 1–2 mg doses, repeated every 5 minutes until the desired effect is achieved or a maximum of 10 mg has been administered.

    (b) 2–5 mg doses, repeated every 5 minutes until the desired effect is achieved.

    (c) 1 mg doses, repeated every 5 minutes as needed, until respiratory depression is reversed (a spontaneous respiratory rate of 12–24/min is resumed).

    (d) A single, rapidly administered dose of 10 mg.

    (e) An initial, rapidly administered dose of 10 mg, followed after 5 minutes by one repeat dose of 5 mg if needed.

117. Which of the following regimens for naloxone administration is best when narcotic addiction is suspected?

    (a) 1–2 mg doses, repeated every 5 minutes until the desired effect is achieved or a maximum of 10 mg has been administered.

    (b) 2–5 mg doses, repeated every 5 minutes until the desired effect is achieved.

    (c) 1 mg doses, repeated every 5 minutes as needed, until respiratory depression is reversed (a spontaneous respiratory rate of 12–24/min is resumed).

    (d) A single, rapidly administered dose of 10 mg.

    (e) An initial, rapidly administered dose of 10 mg, followed after 5 minutes by one repeat dose of 5 mg if needed.

118. When a non-narcotic-addicted patient has overdosed on a synthetic narcotic, such as Darvon®, which of the following regimens for naloxone administration is best?

    (a) 1–2 mg doses, repeated every 5 minutes until the desired effect is achieved or a maximum of 10 mg has been administered.

    (b) 2–5 mg doses, repeated every 5 minutes until the desired effect is achieved.

    (c) 1 mg doses, repeated every 5 minutes as needed, until respiratory depression is reversed (a spontaneous respiratory rate of 12–24/min is resumed).

    (d) A single, rapidly administered dose of 10 mg.

    (e) An initial, rapidly administered dose of 10 mg, followed after 5 minutes by one repeat dose of 5 mg if needed.

119. Administration routes for naloxone include

    (a) IV only.

    (b) IV or ET only.

    (c) IV or IM only.

    (d) IV, IO, or ET only.

    (e) IV, IM, ET, or IO.

120. The most familiar trade name for flumazenil is
  - (a) Lorazepam®.
  - (b) Mazenil®.
  - (c) Romazicon®.
  - (d) Atenolol®.
  - (e) Flumax®.

121. Flumazenil is administered to reverse the untoward CNS effects of
  - (a) narcotic overdoses.
  - (b) tricyclic medication overdoses.
  - (c) benzodiazepine overdoses.
  - (d) Answers (a) and (c).
  - (e) Answers (b) and (c).

122. Which of the following situations contraindicate flumazenil administration?
  - (1) A seizure patient who overdoses on her/his prescribed benzodiazepine medication.
  - (2) A mixed benzodiazepine and narcotic overdose (as a diagnostic tool) when the patient is suspected to be narcotic addicted.
  - (3) A tricyclic medication overdose.
  - (4) A mixed cocaine and benzodiazepine overdose.
  - (5) A mixed amphetamine and benzodiazepine overdose.
  - (6) A patient who overdoses on another individual's prescribed benzodiazepine.

  - (a) 5 and 6.
  - (b) 3, 4, and 5.
  - (c) 2, 4, 5, and 6.
  - (d) 1, 2, 3, 4, and 5.
  - (e) 1, 2, 3, 4, 5, and 6.

123. The currently recommended flumazenil administration regimen consists of three doses. Which of the following doses is the first dose to administer?
  - (a) 0.1 mg IV, given over 1 minute.
  - (b) 0.2 mg IV, given over 15 seconds.
  - (c) 0.3 mg IV, given over 30 seconds.
  - (d) 0.5 mg IV, given over 30 seconds.
  - (e) 1.0 mg IV, given over 15 seconds.

124. Which of the following doses is the second flumazenil dose to administer?
  - (a) 0.1 mg IV, given over 1 minute.
  - (b) 0.2 mg IV, given over 15 seconds.
  - (c) 0.3 mg IV, given over 30 seconds.
  - (d) 0.5 mg IV, given over 30 seconds.
  - (e) 1.0 mg IV, given over 15 seconds.

125. Which of the following doses is the third flumazenil dose to administer?

    (a) 0.1 mg IV, given over 1 minute.

    (b) 0.2 mg IV, given over 15 seconds.

    (c) 0.3 mg IV, given over 30 seconds.

    (d) 0.5 mg IV, given over 30 seconds.

    (e) 1.0 mg IV, given over 15 seconds.

126. Which of the following statements about glucagon administration is false?

    (a) Glucagon is especially effective for reversal of acute hypoglycemia in a nondiabetic patient, following excessive physical exertion (such as a marathon event).

    (b) Glucagon is generally only used for the treatment of hypoglycemia when IV access is unobtainable.

    (c) Glucagon may be used for calcium channel blocker overdoses.

    (d) Glucagon may be used for beta blocker overdoses.

    (e) Glucagon is ineffective when insufficient glycogen stores are present in the liver, such as in malnourished individuals.

127. Administration routes for glucagon include

    (a) IV only.

    (b) IM or SQ only.

    (c) IV, IM, or SQ.

    (d) IV, IM, or ET.

    (e) IV, IM, ET, or SQ.

128. Propranolol (Inderal®) is classed as

    (a) a calcium channel blocker.

    (b) an α-blocker.

    (c) an ACE inhibitor.

    (d) a β-blocker.

    (e) a CNS depressant.

129. Indications for propranolol administration include all of the following except

    (a) symptomatic supraventricular tachycardias related to cocaine intoxication, refractory to vagal maneuvers.

    (b) narrow-complex tachycardias originating from a reentry mechanism, refractory to vagal maneuvers and adenosine.

    (c) narrow-complex tachycardias originating from an automatic focus (junctional or ectopic), refractory to vagal maneuvers and adenosine.

    (d) uncontrolled atrial fibrillation refractory to vagal maneuvers and adenosine.

    (e) uncontrolled atrial flutter refractory to vagal maneuvers and adenosine.

130. The recommended dose and administration method for propranolol is

    (a) 1.0 mg slow IV/IO push at 5-minute intervals, not to exceed a total of 5 mg.

    (b) 1.0 mg/kg rapid IVP via port closest to the patient's heart. Do not repeat if ineffective.

    (c) 5 mg slow IV/IO push at 5-minute intervals, not to exceed a total of 15 mg.

    (d) 0.1 mg/kg rapid IVP via the port closest to the patient's heart. May repeat once after 2 minutes.

    (e) 0.1 mg/kg divided into 3 equal doses and administered slow IV push (not to exceed 1 mg/min) at 2- to 3-minute intervals. If needed, repeat the same dose after 2 minutes.

131. Which of the following statements regarding aspirin (ASA) administration is false?

    (a) If chewed, ASA effects occur faster than if the tablet is swallowed.

    (b) ASA interferes with production of new thrombi.

    (c) ASA acts to dissolve existing thrombi.

    (d) ASA inhibits aggregation of platelets.

    (e) ASA administration may be considered for treatment of both coronary ischemia and CVA.

132. In an emergency setting, the recommended dose of ASA is

    (a) 325 mg.

    (b) 160 mg.

    (c) 81 mg.

    (d) Answer (a) or (b).

    (e) Answer (b) or (c).

133. Which of the following is/are contraindications for the emergency administration of aspirin?

    (a) A patient who is allergic to aspirin.

    (b) A patient who cannot swallow.

    (c) A patient who has a recent history of active gastrointestinal bleeding.

    (d) Answers (a) and (c).

    (e) Answers (a), (b), and (c).

134. When an "unstable angina" patient reports no contraindications for aspirin administration,

    (a) 160 mg of aspirin may be orally administered.

    (b) 325 mg of aspirin may be orally administered.

    (c) 300 mg of aspirin may be rectally administered.

    (d) Answer (a), (b), or (c).

    (e) None of the above. Studies have shown emergency aspirin administration to be ineffective in the treatment of unstable angina.

135. When a 911 caller describes a patient complaining of cardiac-related chest pain, emergency dispatchers should determine whether or not the patient has a history of any contraindication to aspirin administration. If none are reported, emergency dispatchers should advise the calling party to have the patient

(a) drink at least 2 quarts (8 cups) of milk, so as to prevent regurgitation by "calming" the patient's stomach, also reducing gastric acid irritation of the myocardium.

(b) chew or otherwise ingest one "adult" aspirin while awaiting the arrival of EMS providers.

(c) remain at "rest," preventing the patient from getting up and walking anywhere.

(d) Answers (a) and (c).

(e) Answers (b) and (c).

The Answer Key to Test Section 4 is on page 416.

Test
Section
Five

5

Test Section 5's subject:

Airway Management and Ventilation

Test Section 5 consists of 102 questions and is allotted 1 hour and 42 minutes for completion.

1.   Structures of the larynx include all of the following, except the
     (a) carina.
     (b) thyroid cartilage.
     (c) cricoid cartilage.
     (d) arytenoid cartilage.
     (e) vocal cords.

2.   _____ is the chemical that decreases the surface tension of the alveoli, making alveolar expansion easier and helping to prevent alveolar collapse.
     (a) Angiotensin
     (b) Albumin
     (c) Globulin
     (d) Surfactant
     (e) Tyrosine

3.   Alveolar collapse is also called
     (a) pleurisy.
     (b) pneumothorax.
     (c) atelectasis.
     (d) parenchyma.
     (e) parietal insufficiency.

4.   The narrowest part of the pediatric airway is the
     (a) carina.
     (b) larynx.
     (c) epiglottis.
     (d) vallecula.
     (e) cricoid cartilage.

5.   The normal resting respiratory rate for an adult is _____ respirations per minute.
     (a) 5–12
     (b) 12–24
     (c) 18–24
     (d) 40–60
     (e) 60–120

6.   The normal resting respiratory rate for a child is _____ respirations per minute.
     (a) 5–12
     (b) 12–24
     (c) 18–24
     (d) 40–60
     (e) 60–120

7.   The normal resting respiratory rate for an infant is _____ respirations per minute.
   (a) 5–12
   (b) 12–24
   (c) 18–24
   (d) 40–60
   (e) 60–120

8.   The approximate percentage of oxygen found in room air is
   (a) 11%.
   (b) 50%.
   (c) 21%.
   (d) 30%.
   (e) 31%.

9.   Which of the following statements regarding the exchange of oxygen and carbon dioxide is true?
   (a) Gas diffuses from areas of higher partial pressure to areas of lower partial pressure.
   (b) Oxygen will diffuse from the pulmonary capillaries into the alveolar spaces.
   (c) Carbon dioxide will diffuse from the alveolar spaces into the pulmonary capillaries.
   (d) All of the above are true.
   (e) None of the above is true.

10.   Oxygen diffuses into blood and primarily combines with
   (a) plasma.
   (b) hemoglobin.
   (c) leukocytes.
   (d) Any of the above.
   (e) None of the above.

11.   Room air contains 79% nitrogen which, when inspired,
   (a) is necessary for maintaining inflation of body cavities.
   (b) serves no physiologic function.
   (c) displaces any oxygen carried by blood.
   (d) Answers (b) and (c).
   (e) Answers (a), (b), and (c).

12.   At sea level, normal arterial PO_2 is
   (a) 8.0–10.0 mmHg (torr).
   (b) 7.35–7.40 mmHg (torr).
   (c) 35–45 mmHg (torr).
   (d) 50–60 mmHg (torr).
   (e) 80–100 mmHg (torr).

13.   At sea level, normal arterial PCO_2 is
  (a) 8.0–10.0 mmHg (torr).
  (b) 7.35–7.40 mmHg (torr).
  (c) 35–45 mmHg (torr).
  (d) 50–60 mmHg (torr).
  (e) 80–100 mmHg (torr).

14.   All of the following factors can adversely affect oxygen concentration in the blood, except
  (a) increased intracranial pressure.
  (b) anemia.
  (c) hemothorax.
  (d) pulmonary embolism.
  (e) hemorrhage.

15.   To initiate respiration, respiratory centers in the brain communicate with the
  (a) intercostal muscles of the chest, via the phrenic nerve.
  (b) diaphragm and intercostal muscles of the chest, via the phrenic and spinal nerves.
  (c) diaphragm and intercostal muscles of the chest, via the phrenic and intercostal nerves.
  (d) diaphragm, via the thoracic and spinal nerves.
  (e) chest wall muscles and diaphragm, via the spinal and phrenic nerves.

16.   Inspiration is achieved by
  (a) enlarging the size of the thoracic cavity.
  (b) creating a negative intrathoracic air pressure, relative to atmospheric air pressure.
  (c) creating a positive intrathoracic air pressure, relative to atmospheric air pressure.
  (d) Answers (a) and (b).
  (e) Answers (a) and (c).

17.   Expiration is achieved by
  (a) decreasing the size of the thoracic cavity.
  (b) creating a negative intrathoracic air pressure, relative to atmospheric air pressure.
  (c) creating a positive intrathoracic air pressure, relative to atmospheric air pressure.
  (d) Answers (a) and (b).
  (e) Answers (a) and (c).

18.   The most common cause of upper airway obstruction is
       (a) aspiration of unchewed food (especially hot dogs or steak).
       (b) occlusion of the posterior pharynx by the tongue.
       (c) consumption of alcohol during meals.
       (d) food consumption while engaged in physical activity.
       (e) laryngeal spasm secondary to foreign body stimulation of the larynx.

19.   Foreign body obstruction of the upper airway frequently occurs secondary to
       (a) children placing toys in their mouths during play.
       (b) adults consuming alcohol with meals.
       (c) trauma producing loose teeth, clotted blood, or vomitus.
       (d) All of the above.
       (e) None of the above.

20.   Aspiration of vomitus may lead to
       (a) increased interstitial fluid and pulmonary edema.
       (b) lethal pulmonary infection.
       (c) lower airway obstruction.
       (d) Answers (a) and (c).
       (e) Answers (a), (b), and (c).

21.   Laryngeal spasm may occur secondary to
       (a) edema of the glottis.
       (b) bronchospasm.
       (c) spinal trauma.
       (d) Answers (a) and (c).
       (e) Answers (a), (b), and (c).

22.   Laryngeal spasm may also occur secondary to
       (a) direct laryngeal trauma.
       (b) anaphylaxis.
       (c) superheated air inhalation.
       (d) Answers (a) and (c).
       (e) Answers (a), (b), and (c).

23.   The most common cause of laryngeal spasm is
       (a) toxic inhalation.
       (b) aggressive intubation efforts.
       (c) bronchospasm.
       (d) epiglottitis.
       (e) smoke inhalation.

24. Which of the following statements about cyanosis is false?

   (a) The bluish tint of cyanosis is produced by the presence of deoxygenated hemoglobin.

   (b) Severe tissue hypoxia is possible, even when cyanosis is not exhibited by the patient.

   (c) Cyanosis occurs almost immediately after a patient's ventilatory status becomes compromised.

   (d) Cyanosis is considered a late sign of respiratory compromise.

   (e) If present, cyanosis can usually be observed at the lips and fingernails.

25. _____ is defined as the absence (or near-absence) of oxygen.

   (a) Hypoxemia

   (b) Hypoxia

   (c) Anoxia

   (d) Hypoventilation

   (e) Anemia

26. _____ is defined as decreased oxygen in the blood.

   (a) Hypoxemia

   (b) Hypoxia

   (c) Anoxia

   (d) Hypoventilation

   (e) Anemia

27. _____ is defined as an oxygen deficiency.

   (a) Hypoxemia

   (b) Hypoxia

   (c) Anoxia

   (d) Hypoventilation

   (e) Anemia

28. Which of the following signs is the earliest indication of an individual experiencing a gradual onset of respiratory distress?

   (a) A respiratory rate that is increased or decreased from normal.

   (b) Cyanotic skin color.

   (c) Difficulty speaking.

   (d) Altered mental status.

   (e) Use of respiratory accessory muscles to assist breathing.

29. During respiratory distress, "accessory" muscles of respiration may be employed. Examples of accessory respiratory muscles include all of the following, except

   (a) intercostal muscles.

   (b) suprasternal muscles.

    (c) abdominal muscles.

    (d) the diaphragm.

    (e) supraclavicular muscles.

30.   _____ respirations are best described as an abnormal respiratory pattern manifested by rapid and deep respirations, often indicative of increased intracranial pressure.

    (a) Agonal

    (b) Central neurogenic hyperventilation

    (c) Biot's or ataxic

    (d) Cheyne-Stokes

    (e) Kussmaul's

31.   _____ respirations are best described as an abnormal respiratory pattern manifested by irregular rates and depths of respirations, with periodic episodes of apnea, often indicative of increased intracranial pressure.

    (a) Agonal

    (b) Central neurogenic hyperventilation

    (c) Biot's or ataxic

    (d) Cheyne-Stokes

    (e) Kussmaul's

32.   _____ respirations are best described as an abnormal respiratory pattern manifested by periods of progressively deeper and faster breathing, alternating with periods of progressively shallow and slower breathing, often indicative of a brain-stem injury.

    (a) Agonal

    (b) Central neurogenic hyperventilation

    (c) Biot's or ataxic

    (d) Cheyne-Stokes

    (e) Kussmaul's

33.   _____ respirations are best described as an abnormal respiratory pattern manifested by either slow or fast, deep, gasping breaths, often accompanying diabetic ketoacidosis.

    (a) Agonal

    (b) Central neurogenic hyperventilation

    (c) Biot's or ataxic

    (d) Cheyne-Stokes

    (e) Kussmaul's

34. _____ respirations are best described as an abnormal respiratory pattern manifested by shallow and slow breaths, and infrequent efforts to breathe, indicative of brain anoxia.

    (a) Agonal

    (b) Central neurogenic hyperventilation

    (c) Biot's or ataxic

    (d) Cheyne-Stokes

    (e) Kussmaul's

35. The preferred location for auscultating breath sounds (the location where they can best be heard) is the

    (a) anterior surface of the chest.

    (b) epigastrium.

    (c) axillary surface of the chest.

    (d) anterior chest, proximal to each side of the sternum.

    (e) posterior surface of the chest.

36. While ventilating an intubated patient with the bag-valve-mask device, you notice that it becomes more and more difficult to compress the bag. This may indicate any of the following, except

    (a) development of pneumothorax or tension pneumothorax.

    (b) an occlusion of the endotracheal tube.

    (c) endobronchial intubation.

    (d) increased intracranial pressure.

    (e) esophageal intubation.

37. Which of the following statements about pulse oximetry ("Pulse-Ox") is false?

    (a) A Pulse-Ox sensor attached to a patient's earlobe can obtain accurate readings.

    (b) A pulse oximeter measures the hemoglobin saturation in peripheral tissues.

    (c) Prehospital Pulse-Ox devices are inaccurate and rarely provide information important to patient assessment or treatment decisions.

    (d) If a patient is displaying any signs of respiratory compromise, 100% oxygen should be administered, even if the patient's Pulse-Ox reading is within normal limits.

    (e) Pulse-Ox readings are sometimes able to indicate oxygenation problems before a patient's blood pressure or pulse rate changes because of them.

38. False Pulse-Ox readings can be caused by all of the following, except

    (a) monitoring an extremity with an absent pulse.

    (b) a severely anemic patient with hypovolemia.

(c) increased intracranial pressure.

(d) anomalous hemoglobin abnormalities.

(e) carbon monoxide poisoning.

39. At sea level, a Pulse-Ox reading of _____ indicates only mild hypoxia, and warrants only administration of supplementary oxygen and further evaluation.

(a) 95%–99%

(b) 91%–94%

(c) 86%–91%

(d) 81%–85%

(e) 75%–80%

40. According to standard (sea-level) Pulse-Ox values, a patient is usually not considered to have severe hypoxia, until his Pulse-Ox reading goes below

(a) 99%.

(b) 94%.

(c) 91%.

(d) 86%.

(e) 80%.

41. Which of the following statements about Sellick's maneuver is false?

(a) Use of Sellick's maneuver is contraindicated in pediatric patients.

(b) Sellick's maneuver may prevent vomiting and gastric distention.

(c) Sellick's maneuver may assist intubation by improving visualization of the larynx.

(d) Only gentle pressure should be used when employing Sellick's maneuver.

(e) Because regurgitation frequently follows the use of Sellick's maneuver, once employed, Sellick's maneuver should be maintained until the patient is protected from aspiration of vomitus.

42. Which of the following statements regarding an oropharyngeal airway is false?

(a) The oropharyngeal airway is intended to prevent the tongue from occluding the posterior oropharynx.

(b) Once the patient is orally intubated, the oropharyngeal airway may be used as a "bite block" to prevent occlusion of the endotracheal tube.

(c) The oropharyngeal airway will assist in preventing aspiration of vomitus.

(d) The oropharyngeal airway may obstruct the airway.

(e) Use of a properly sized oropharyngeal airway in patients with a gag reflex may stimulate regurgitation, followed by aspiration of vomitus.

43. Which of the following statements regarding a nasopharyngeal airway is false?

  (a) The presence of oral cavity trauma does not contraindicate the use of a nasopharyngeal airway.

  (b) The nasopharyngeal airway may cause severe epistaxis.

  (c) The nasopharyngeal airway may be used when the patient's teeth are clenched closed.

  (d) The nasopharyngeal airway is intended to bypass the tongue and facilitate a patient's airway.

  (e) Use of a properly sized nasopharyngeal airway in patients with a gag reflex may stimulate regurgitation, followed by aspiration of vomitus.

44. The inflatable cuff at the distal end of the endotracheal tube holds approximately _____ ml of air.

  (a) 0–5

  (b) 5–10

  (c) 20–25

  (d) 25–30

  (e) 30–35

45. Which of the following medications may be administered via the endotracheal tube?

  (1) naloxone   (5) $D_{50}W$

  (2) epinephrine 1:10,000   (6) epinephrine 1:1000

  (3) atropine sulfate   (7) diazepam

  (4) lidocaine   (8) vasopressin

  (a) 1, 2, 3, 4, and 8.

  (b) 1, 2, 3, and 4.

  (c) 1, 2, 3, 4, and 7.

  (d) 1, 2, 3, 4, 7, and 8.

  (e) 1, 2, 3, 4, 5, 6, 7, and 8.

46. The straight laryngoscope blade is also called a Wisconsin, Flagg, or _____ blade.

  (a) Macintosh

  (b) Maclean

  (c) Miller

  (d) Monroe

  (e) Magill

47. The curved laryngoscope blade is also called a _____ blade.

  (a) Macintosh

  (b) Maclean

(c) Miller

(d) Monroe

(e) Magill

48. The forceps commonly used to remove visualized obstructive matter from the airway are called _____ forceps.

   (a) Macintosh

   (b) Maclean

   (c) Miller

   (d) Monroe

   (e) Magill

49. The straight laryngoscope blade is designed to facilitate viewing of the vocal cords by

   (a) preventing the tongue from obscuring the view.

   (b) elevating the tongue and (indirectly) the epiglottis.

   (c) directly elevating the epiglottis.

   (d) Answers (a) and (b).

   (e) Answers (a) and (c).

50. The curved laryngoscope blade is used to facilitate viewing of the vocal cords by

   (a) preventing the tongue from obscuring the view.

   (b) elevating the tongue and (indirectly) the epiglottis.

   (c) directly elevating the epiglottis.

   (d) Answers (a) and (b).

   (e) Answers (a) and (c).

51. The curved laryngoscope blade is designed to be placed

   (a) into the vallecula.

   (b) under the epiglottis.

   (c) into the larynx.

   (d) under the larynx.

   (e) below the soft palate.

52. The straight laryngoscope blade is designed to be placed

   (a) into the vallecula.

   (b) under the epiglottis.

   (c) into the larynx.

   (d) under the larynx.

   (e) below the soft palate.

53. Which of the following statements about laryngoscope blade preferences is false?

   (a) The curved laryngoscope blade is preferred for infant intubation.

   (b) For adult intubation, the personal preference of the individual intubating the patient is the best laryngoscope blade to use.

   (c) The curved blade is considered less likely to traumatize the adult larynx or stimulate a gag reflex.

   (d) Patient anatomy variety requires that care providers be skilled in using either style of laryngoscope blade.

   (e) The straight laryngoscope blade is preferred when the patient has a very large tongue.

54. Children younger than 8 years old

   (a) should not be intubated in the prehospital environment.

   (b) should only be intubated with an uncuffed endotracheal tube.

   (c) are at greater risk for suffering tachycardias due to the vagal response triggered by intubation.

   (d) Answers (a) and (c).

   (e) Answers (b) and (c).

55. When using a flexible or malleable stylet within an endotracheal tube to provide better control during intubation, the distal stylet end should be recessed from the distal end of the ET tube by at least

   (a) 6 cm, or $2\frac{1}{4}$ inches.

   (b) 5 cm, or 2 inches.

   (c) 4 cm, or $1\frac{1}{2}$ inches.

   (d) 2 cm, or $\frac{3}{4}$ inch.

   (e) 1 cm, or $\frac{1}{4}$ inch.

56. Hyperventilation with 100% oxygen should

   (a) precede any intubation attempt (or any repeat attempt) for at least 5–10 seconds.

   (b) only be performed upon a COPD patient after intubation has been accomplished.

   (c) never be performed upon head-injured patients (respiratory alkalosis may be precipitated).

   (d) Answers (a) and (b).

   (e) Answers (b) and (c).

57. According to the 2005 AHA ACLS recommendations, endotracheal intubation

   (a) is no longer considered the optimal airway management method if the intubating paramedic is inexperienced.

   (b) causes hypoxemia whenever intubation attempts are prolonged, or whenever tube misplacement (or displacement) goes unnoticed.

(c) can cause unacceptable interruptions of CPR when performed by paramedics with little intubation experience.

(d) Answers (b) and (c).

(e) Answers (a), (b), and (c).

58. Which of the following statements regarding esophageal intubation is false?

(a) Uncorrected esophageal intubation may result in severe hypoxia, brain damage, or brain death.

(b) Aspiration of vomitus is encouraged by esophageal intubation.

(c) Mist condensation observed within the endotracheal tube indicates esophageal intubation.

(d) Gurgling sounds auscultated at the epigastrium indicate esophageal intubation.

(e) If esophageal intubation is even slightly suspected, immediately extubate the patient.

59. Endobronchial intubation is most frequently indicated by

(a) absent breath sounds on ventilation, when auscultating the right chest.

(b) absent breath sounds on ventilation, when auscultating the left chest.

(c) bilaterally absent breath sounds on ventilation, when auscultating the chest.

(d) gurgling sounds on ventilation, when auscultating the epigastrium.

(e) observance of abdominal distention after 4 or 5 ventilations.

60. When endobronchial intubation is suspected, the paramedic should

(a) hyperinflate the ET's distal cuff to provide an improved seal, and increase the volume of ventilation.

(b) leave the tube in place but completely deflate the ET's distal cuff to allow for the passage of air around the tube, and increase the volume of ventilation.

(c) immediately extubate the patient.

(d) deflate the ET's distal cuff and withdraw the ET tube until breath sounds are bilaterally present and equal.

(e) leave the ET's distal cuff inflated (to act as an "anchor" preventing extubation), and withdraw the ET tube until breath sounds are bilaterally present and equal.

61. Which of the following is the most reliable indication that endotracheal intubation has been successful?

(a) Observation of anterior chest wall rising and falling.

(b) Auscultation of equal bilateral breath sounds.

(c) The absence of gastric noises auscultated on ventilation.

(d) Confirmation provided by an end-tidal CO_2 detector or esophageal detector device.

(e) Direct visualization of the ET tube passing between the vocal cords.

62.   Which of the following statements regarding orotracheal intubation is false?

   (a) Whether the provider is right- or left-handed, the laryngoscope is held in the left hand.
   (b) Whether the provider is right- or left-handed, the laryngoscope should first be inserted into the right side of the patient's mouth.
   (c) Dentures or partial dental plates should remain in place to maintain the shape of the oropharynx and improve visualization of the vocal cords.
   (d) Lift the laryngoscope handle upward and forward to avoid using the patient's upper teeth as a fulcrum.
   (e) Whether the provider is right- or left-handed, the ET tube is held in the right hand.

63.   Which of the following statements regarding blind nasotracheal intubation is false?

   (a) Suspicion of spinal injury is a strong indication for blind nasotracheal intubation.
   (b) Blind nasotracheal intubation may be performed on any apneic patient.
   (c) A laryngoscope is not required to perform blind nasotracheal intubation.
   (d) Nasotracheal intubation presents a greater risk of pulmonary infection than orotracheal intubation.
   (e) Blind nasotracheal intubation is potentially more traumatic than orotracheal intubation.

64.   Nasotracheal intubation is indicated in all of the following situations, except

   (a) an unconscious patient with nasal and basilar skull fractures requiring hyperventilation.
   (b) a conscious patient nearing respiratory arrest secondary to asthma, anaphylactic shock, or inhalation injury.
   (c) a conscious patient with severe oral trauma and suspected spine injury.
   (d) an unconscious patient with a fractured jaw.
   (e) an unconscious patient with clenched teeth or a profound gag reflex.

65.   To successfully accomplish blind nasotracheal intubation, passage of the ET tube through the glottic opening should be precisely timed to coincide with _____, when the glottis is open.

   (a) the moment between the patient's inspiratory/expiratory efforts
   (b) the patient's next expiration
   (c) the patient's next inspiration
   (d) the next time the patient swallows
   (e) the next time the patient coughs

66.   When the anterior neck is palpated from the top downwards, the cricoid cartilage is

   (a) the first prominent structure felt.
   (b) the second prominent structure felt.

(c) the third prominent structure felt.

(d) between the first and second prominent structures.

(e) between the second and third prominent structures.

67. When the anterior neck is palpated from the top downwards, the thyroid cartilage is

    (a) the first prominent structure felt.

    (b) the second prominent structure felt.

    (c) the third prominent structure felt.

    (d) between the first and second prominent structures.

    (e) between the second and third prominent structures.

68. When the anterior neck is palpated from the top downwards, the cricothyroid membrane is

    (a) the first prominent structure felt.

    (b) the second prominent structure felt.

    (c) the third prominent structure felt.

    (d) between the first and second prominent structures.

    (e) between the second and third prominent structures.

69. Which of the following statements regarding needle cricothyrotomy and transtracheal jet insufflation (ventilation) is true?

    (a) The needle and catheter are inserted into the cricoid cartilage.

    (b) Upon removal of the needle, the secured catheter allows for both ventilation and suctioning of secretions.

    (c) Attaching the catheter to a 15 mm adapter allows for bag-valve ventilation.

    (d) Needle cricothyrotomy is contraindicated when a complete upper airway obstruction exists.

    (e) Deliver only 10 slow ventilations per minute, to ensure that overinflation does not occur.

70. Pleural-tissue injury, caused by overinflation or excessive high-pressure ventilations, is often referred to as

    (a) subcutaneous emphysema.

    (b) barrel chest.

    (c) insufflation.

    (d) traumatic asphyxia.

    (e) barotrauma.

71. Which of the following statements regarding open cricothyrotomy is true?

    (a) A surgical incision is made in the cricoid cartilage.

    (b) Insertion of a cuffed ET tube or tracheostomy tube allows for BVM ventilation.

    (c) For children under the age of 12, a large enough incision must be made to accommodate a 5.0 ET tube (the smallest cuffed ET tube).

    (d) The first incision (the skin incision) should be horizontal.

    (e) The second incision (through the cricoid cartilage) should be horizontal.

72. Which of the following statements regarding suction is true?
    (a) A whistle-tip suction device may be used for up to 30 seconds each time, as its smaller bore depletes less oxygen than the tonsil-tip suction device.
    (b) Tonsil-tip suction devices remove larger particles but less fluid volume than whistle-tip suction devices.
    (c) A Yankauer suction device can be used to suction the nares, oropharynx, larynx, and esophagus.
    (d) A whistle-tip suction device can be used to suction the nares, oropharynx, nasopharynx, or endotracheal tube.
    (e) Suction is applied as the device is inserted and discontinued as the device is withdrawn.

73. Complications of suctioning include all of the following, except
    (a) hypoxia.
    (b) hypotension due to the vagal response that may be triggered by suctioning.
    (c) tachycardias due to the vagal response that may be triggered by suctioning.
    (d) cardiac dysrhythmias from decreased myocardial oxygen supply.
    (e) increased intracranial pressure due to cough responses triggered by suctioning.

74. The liters-per-minute (LPM) flow rate range for optimal nasal cannula oxygen administration is
    (a) 4 LPM.
    (b) 6 LPM.
    (c) 8 LPM.
    (d) 10 LPM.
    (e) 12 LPM.

75. If set at 6 LPM, a nasal cannula will deliver an oxygen concentration of
    (a) only 25%, because that LPM rate is less than optimal.
    (b) 40%.
    (c) as much as 60%, hence the reason that LPM rate is considered optimal.
    (d) as much as 80%, hence the reason that LPM rate is considered optimal.
    (e) only 20%, because that LPM rate is excessive for a nasal cannula.

76. The LPM flow rate range for oxygen administration via a simple face mask is
    (a) 15 LPM.
    (b) 1–6 LPM.
    (c) 8–15 LPM.
    (d) 6–10 LPM.
    (e) 10–15 LPM.

77.   If set at 10 LPM, a simple face mask can deliver an oxygen concentration of
        (a) only 26–54%, because that LPM rate is inadequate for a simple face mask.
        (b) 36–54%.
        (c) 40–60%.
        (d) 60–80%.
        (e) 80–95%.

78.   The LPM flow rate range for oxygen administration via a nonrebreather mask is
        (a) 15 LPM.
        (b) 1–6 LPM.
        (c) 4–15 LPM.
        (d) 6–10 LPM.
        (e) 8–15 LPM.

79.   If set at 15 LPM, a nonrebreather mask will deliver an oxygen concentration of
        (a) 100%.
        (b) 36–54%.
        (c) 40–60%.
        (d) 60–80%.
        (e) 80–95%.

80.   The Venturi mask is an oxygen-delivery mask designed especially for
        (a) pediatric patients.
        (b) geriatric patients.
        (c) COPD patients.
        (d) cancer patients.
        (e) AIDS patients.

81.   The Venturi mask delivers four different oxygen concentrations, depending upon
        the components assembled and the liter flow of oxygen. The four concentrations are
        (a) 4%, 6%, 20%, or 40%.
        (b) 20%, 30%, 40%, or 50%.
        (c) 24%, 28%, 35%, or 40%.
        (d) 28%, 44%, 58%, or 80%.
        (e) 40%, 48%, 58%, or 84%.

82.   The limitations of mouth-to-mouth, mouth-to-nose, and mouth-to-stoma
        ventilations include all of the following, except
        (a) inadequate ventilatory volumes are delivered.
        (b) only 17% oxygen is delivered.
        (c) lack of protection against disease transmission.
        (d) lack of protection against patient aspiration.
        (e) inability to hyperventilate the patient without causing rescuer
             hyperventilation.

83. Which of the following statements regarding mouth-to-mask ventilations is false?

(a) With an oxygen inlet valve and 10 LPM of oxygen instilled, mouth-to-mask ventilations will provide the patient with approximately 50% oxygen.

(b) Mouth-to-mask ventilations provide inadequate tidal volumes.

(c) With a one-way valve device, the risk of contact with patient secretions and expired air is minimized.

(d) Safe hyperventilation is not possible with mouth-to-mask ventilations.

(e) Hyperinflation of the patient's lungs is a possible complication of mouth-to-mask ventilations.

84. Without supplemental oxygen delivery, the bag-valve-mask (BVM) device

(a) should not be used.

(b) will deliver up to 21% oxygen to the patient.

(c) will deliver up to 50% oxygen to the patient.

(d) will deliver 60–70% oxygen to the patient.

(e) will deliver 90–95% oxygen to the patient.

85. With oxygen run at 15 LPM but without a reservoir attached, the BVM will deliver ____ oxygen to the patient.

(a) 21–50%

(b) 50–60%

(c) 60–70%

(d) 80–90%

(e) 90–95%

86. With a reservoir attached, and oxygen run at 15 LPM, the BVM will deliver ____ oxygen to the patient.

(a) 21–50%

(b) 50–60%

(c) 60–70%

(d) 80–90%

(e) 90–95%

87. A BVM reservoir device is the

(a) portion of the BVM that is compressed to deliver ventilations to the patient.

(b) bag that comes off of the end of the BVM compression section.

(c) corrugated tubing that extends from the end of the BVM compression section.

(d) Answer (a) or (b), depending upon the commercial model.

(e) Answer (b) or (c), depending upon the commercial model.

88. Which of the following statements regarding demand-valve ventilation is true?

(a) The demand valve can be connected to a mask, esophageal intubation device, or endotracheal tube.

(b) The demand valve provides 100% oxygenation when operated at 40 LPM.

(c) The demand valve can ventilate past a minor airway obstruction.

(d) All of the above are true.

(e) None of the above is true.

89. Which of the following statements regarding demand valve ventilation is false?

    (a) Lung compliance is not detectable when employing demand valve ventilation.

    (b) Because of the risk of lung injury from hyperexpansion, demand-valve ventilation is contraindicated in patients less than 16 years old.

    (c) The potential for pulmonary rupture (causing pneumothorax) is increased with use of demand-valve ventilation on intubated patients, or patients with chest trauma.

    (d) Gastric distention is a frequent side effect when demand-valve ventilation is used to ventilate nonintubated patients.

    (e) Due to the relatively low oxygen flow rates required by modern demand-valve devices, demand-valve ventilations can be provided over a prolonged period of time.

90. Gastric distention

    (a) occurs when ventilation is performed without adequate airway positioning.

    (b) may occur in the intubated patient if ventilatory volumes are excessive.

    (c) will significantly increase the risk of aspiration and decrease the ventilatory capacity of the lungs.

    (d) Answers (a) and (c).

    (e) Answers (a), (b), and (c).

91. Capnometry is

    (a) accomplished by using an end-tidal CO_2 ($ETCO_2$) monitoring device.

    (b) the measurement of carbon dioxide present in expired air.

    (c) the graphic recording of expired carbon dioxide over a period of time.

    (d) Answers (a) and (b).

    (e) Answers (a), (b), and (c).

92. Which of the following statements regarding capnometry is false?

    (a) $ETCO_2$ monitoring is a safe and effective noninvasive indicator of cardiac output during CPR.

    (b) $ETCO_2$ monitoring may be an early indicator of return of spontaneous circulation in intubated patients during CPR.

    (c) If the ET tube was clearly seen to pass through the vocal cords, $ETCO_2$ monitor confirmation of tube placement is unnecessary, and hyperventilation should not be interrupted to attach one.

    (d) Detection of exhaled CO_2 is not an infallible means of confirming ET tube placement, particularly during cardiac arrest.

    (e) If ventilation is reasonably consistent during CPR, changes in end-tidal CO_2 concentration reflect changes in cardiac output.

93. Which of the following statements regarding $ETCO_2$ monitoring is false?

  (a) When exhaled CO_2 is detected during CPR, it is usually a reliable indicator of ET tube placement in the trachea.

  (b) A false-positive CO_2 reading may be detected if the patient ingested a large quantity of carbonated beverage prior to cardiac arrest and the ET tube is in the esophagus.

  (c) If a colorimetric $ETCO_2$ detector is contaminated with gastric contents, the device may display a constant color rather than a breath-to-breath color change.

  (d) If a colorimetric $ETCO_2$ detector is contaminated with acidic drugs (eg, ET-administered epinephrine), the device may display a constant color rather than a breath-to-breath color change.

  (e) IV bolus administration of epinephrine during CPR will not effect elimination and detection of $ETCO_2$.

94. If the patient is experiencing a rising body temperature, the end-tidal CO_2 monitoring device may show

  (a) a slow, gradual increase in $ETCO_2$.

  (b) a sudden increase in $ETCO_2$.

  (c) a slow, gradual decrease in $ETCO_2$.

  (d) a rapid, gradual decrease in $ETCO_2$.

  (e) a sudden drop of $ETCO_2$ to zero.

95. If an obstructed lung area suddenly becomes unobstructed, the end-tidal CO_2 monitoring device may show

  (a) a slow, gradual increase in $ETCO_2$.

  (b) a sudden increase in $ETCO_2$.

  (c) a slow, gradual decrease in $ETCO_2$.

  (d) a rapid, gradual decrease in $ETCO_2$.

  (e) a sudden drop of $ETCO_2$ to (or close to) zero.

96. If the patient is moved and the ET tube enters the esophagus, the end-tidal CO_2 monitoring device should show

  (a) a slow, gradual increase in $ETCO_2$.

  (b) a sudden increase in $ETCO_2$.

  (c) a slow, gradual decrease in $ETCO_2$.

  (d) a rapid, gradual decrease in $ETCO_2$.

  (e) a sudden drop of $ETCO_2$ to zero.

97. Excessive hyperventilation can cause the end-tidal CO_2 monitoring device to show

  (a) a slow, gradual increase in $ETCO_2$.

  (b) a sudden increase in $ETCO_2$.

  (c) a slow, gradual decrease in $ETCO_2$.

  (d) a rapid, gradual decrease in $ETCO_2$.

  (e) a sudden drop of $ETCO_2$ to (or close to) zero.

98. A sudden loss of blood pressure can cause the end-tidal CO_2 monitoring device to show
     (a) a slow, gradual increase in $ETCO_2$.
     (b) a sudden increase in $ETCO_2$.
     (c) a slow, gradual decrease in $ETCO_2$.
     (d) a rapid, gradual decrease in $ETCO_2$.
     (e) a sudden drop of $ETCO_2$ to zero.

99. If the patient is experiencing a decreasing body temperature, the end-tidal CO_2 monitoring device may show
     (a) a slow, gradual increase in $ETCO_2$.
     (b) a sudden increase in $ETCO_2$.
     (c) a slow, gradual decrease in $ETCO_2$.
     (d) a rapid, gradual decrease in $ETCO_2$.
     (e) a sudden drop of $ETCO_2$ to zero.

100. If a partial airway obstruction develops, the end-tidal CO_2 monitoring device may show
     (a) a slow, gradual increase in $ETCO_2$.
     (b) a sudden increase in $ETCO_2$.
     (c) a slow, gradual decrease in $ETCO_2$.
     (d) a rapid, gradual decrease in $ETCO_2$.
     (e) a sudden drop of $ETCO_2$ (close to zero).

101. If a pulmonary embolism obstruction occurs, the end-tidal CO_2 monitoring device may show
     (a) a slow, gradual increase in $ETCO_2$.
     (b) a sudden increase in $ETCO_2$.
     (c) a slow, gradual decrease in $ETCO_2$.
     (d) a rapid, gradual decrease in $ETCO_2$.
     (e) a sudden drop of $ETCO_2$ (close to zero).

102. If the patient's blood pressure suddenly begins to rise, the end-tidal CO_2 monitoring device may show
     (a) a slow, gradual increase in $ETCO_2$.
     (b) a sudden increase in $ETCO_2$.
     (c) a slow, gradual decrease in $ETCO_2$.
     (d) a rapid, gradual decrease in $ETCO_2$.
     (e) a sudden drop of $ETCO_2$ to (or close to) zero.

The Answer Key to Test Section 5 is on page 423.

Test
Section
Six

6

Test Section 6's subjects:

- Soft-Tissue Trauma
- Burns
- Head and Face Trauma
- Spine Trauma
- Thoracic Trauma
- Abdominal Trauma
- Musculoskeletal Trauma

Test Section 6 consists of 165 questions, and is allotted 2 hours and 45 minutes for completion.

1.   Trauma scenarios (mechanisms of injury) that indicate the need for immediate and rapid transport to a trauma facility include all of the following, except
      (a) any fall of twice the victim's height (10–12 feet).
      (b) any event involving a pedestrian or bicyclist and an automobile.
      (c) any motorcycle impact of greater than 20 mph.
      (d) any patient ejected from a motor vehicle.
      (e) any MVA patient (regardless of complaints or lack thereof) who shared a vehicle compartment with an individual who was pronounced dead at the scene.

2.   _____ is a medical phrase describing what is commonly called "Battle's sign."
      (a) Periorbital ecchymosis
      (b) Anterotemporal ecchymosis
      (c) Retroauricular ecchymosis
      (d) Posteromandibular ecchymosis
      (e) Supraclavicular ecchymosis

3.   _____ is a medical phrase describing what is commonly called "raccoon eyes."
      (a) Bilateral periorbital ecchymosis
      (b) Bilateral anterotemporal ecchymosis
      (c) Bilateral retroauricular ecchymosis
      (d) Bilateral posteromandibular ecchymosis
      (e) Bilateral supraclavicular ecchymosis

4.   Which of the following statements regarding signs of basilar skull fracture is true?
      (a) Raccoon eyes develop within 10 minutes of a basilar skull fracture.
      (b) Battle's sign indicates that basilar skull fracture has just occurred.
      (c) Cerebrospinal fluid (CSF) is the only substance that will separate from blood when placed on paper or gauze, producing a "target" or "halo" sign.
      (d) Checking for the "target" or "halo" sign is most reliable when using blood from the nose.
      (e) Leakage of blood and/or CSF may diminish the increase of intracranial pressure (ICP), and act to limit brain damage from ICP.

5.   Spinal fracture occurs most commonly in the
      (a) cervical and thoracic spine.
      (b) thoracic and sacral spine.
      (c) cervical and lumbar spine.
      (d) lumbar and sacral spine.
      (e) thoracic and lumbar spine.

6. When a traumatic event produces compression stress that is transmitted along the length of the spine, injury caused by this type of force is often referred to as compression injury or
   (a) axial loading injury.
   (b) distraction injury.
   (c) spinal stair-step injury.
   (d) hyperextension injury.
   (e) flexion-rotation injury.

7. A hanging mechanism of injury can produce
   (a) an axial loading injury.
   (b) a distraction injury.
   (c) a spinal stair-step injury.
   (d) a hyperextension injury.
   (e) a flexion-rotation injury.

8. Reduced ocular range of motion from an entrapped extraocular muscle may indicate a fracture of the
   (a) occipital bone.
   (b) zygomatic bone.
   (c) mandible.
   (d) Answer (a) or (b).
   (e) Answer (b) or (c).

9. A contrecoup brain injury is best defined as an injury occurring
   (a) at the site of a skull impact.
   (b) as a result of intracerebral hemorrhage, hours after a skull impact.
   (c) midway between an entry and exit wound of the skull.
   (d) at an exit wound of the skull.
   (e) at a site opposite that of an original impact.

10. The meninges are best described as three concurrent tissue membranes that surround and protect
   (a) the brain.
   (b) the spinal cord.
   (c) the peripheral nerves.
   (d) the brain and spinal cord.
   (e) the brain, spinal cord, and the peripheral nerves.

11. The outermost meningeal tissue layer is called the _____, which can be literally translated as "tough mother."
- (a) during mater
- (b) Macho Madre
- (c) dura mater
- (d) Gran Madre
- (e) touros matter

12. The innermost meningeal tissue layer is called the _____, which can be literally translated as "tender mother."
- (a) pia mater
- (b) Petite Madre
- (c) petit matter
- (d) Poco Madre
- (e) tendros matter

13. The middle meningeal layer, a layer of connective tissue between the outermost and innermost meninges, is called the _____, a name that reflects the "spider-web" appearance of its fibers.
- (a) spinne mater
- (b) arachnoid membrane
- (c) arachnid mater
- (d) Charlotte's layer
- (e) web matter

14. Bleeding that occurs within the meninges, between the outermost and innermost meningeal layers, produces
- (a) an epidural hematoma.
- (b) a subdural hematoma.
- (c) intracerebral hemorrhage.
- (d) spider angioedema.
- (e) ocular petechiae.

15. A 16-year-old male is struck in the head with a baseball while watching the game from the stands. His hysterical girlfriend reports that he was immediately "knocked out" and remained unconscious "until just a minute ago!" On your arrival, he is alert and oriented to person, place, and time. Based only upon this information, you immediately suspect that he has suffered
- (1) a cerebral concussion.
- (2) a cerebral contusion.
- (3) an epidural hematoma.
- (4) a subdural hematoma.
- (5) intracerebral hemorrhage.
- (6) spider angioedema.
- (7) a transient ischemic attack.
- (8) a spinal injury.

     (a) 1, 2, 3, 4, 5, 6, 7, and 8.
     (b) 1, 3, 6, and 8.

    (c) 1, 3, and 8.

    (d) 1 and 8.

    (e) 1.

16. As you continue your assessment of this 16-year-old male's injury, he begins asking you, "Who are you again?", "Why are you here?", and similar questions. This is an example of a patient exhibiting signs of

    (a) retrograde amnesia.

    (b) anterograde amnesia.

    (c) idiosyncratic amnesia.

    (d) angioedema amnesia.

    (e) illicit drug use or abuse.

17. Next, this 16-year-old male suddenly turns to his girlfriend and asks, "Did I bring you here?", "Did you drive?", and so on. With these questions he is exhibiting

    (a) new signs of retrograde amnesia.

    (b) new signs of anterograde amnesia.

    (c) more signs of idiosyncratic amnesia.

    (d) more signs of angioedema amnesia.

    (e) more signs of illicit drug use or abuse.

18. As you continue your treatment and ongoing assessment of this 16-year-old male, he begins to become drowsy and disoriented. His speech becomes slurred and nonsensical at times, as though he might be hallucinating. Based upon all the preceding information (from this set of four questions), you now suspect that this 16-year-old male is suffering from

    (1) a cerebral concussion.   (6) spider angioedema.

    (2) a cerebral contusion.   (7) a transient ischemic attack.

    (3) an epidural hematoma.   (8) a spinal injury.

    (4) a subdural hematoma.   (9) a drug abuse or overdose incident.

    (5) intracerebral hemorrhage.

        (a) 1, 2, 3, 4, 5, 6, 7, 8, and 9.

        (b) 1, 3, 6, 8, and 9.

        (c) 1, 3, 8, and 9.

        (d) 1, 3, and 8.

        (e) 1 and 9.

19. Cerebral ischemia may produce a response called "Cushing's reflex." This response is classically characterized by evidence of

    (a) decreasing blood pressure, increasing pulse rate, increasing respirations.

    (b) increasing blood pressure, increasing pulse rate, increasing respirations.

    (c) increasing blood pressure, decreasing pulse rate, and erratic respirations.

    (d) decreasing blood pressure, decreasing pulse rate, and decreasing respirations.

    (e) decreasing blood pressure, with flexion or extension of the extremities ("posturing").

20. When provided with a pain stimulus, your head-injured patient flexes and adducts both arms. This response classically characterizes
     (a) a withdrawal response to pain.
     (b) decerebrate posturing.
     (c) the ability to localize pain and coordinate a response.
     (d) decorticate posturing.
     (e) Answer (a) or (c).

21. When provided with a pain stimulus, your head-injured patient extends and abducts both arms. This response characterizes
     (a) a withdrawal response to pain.
     (b) decerebrate posturing.
     (c) the ability to localize pain and coordinate a response.
     (d) decorticate posturing.
     (e) Answer (a) or (c).

22. The painful-stimuli responses described in the two preceding questions indicate the presence of
     (a) normal, coordinated and localized, muscular responses to painful stimulation.
     (b) reflex anomalies often seen with narcotic drug overdoses.
     (c) a cervical spinal cord lesion (C-1 to C-3) or injury, resulting in reflexive muscle movement of the extremities.
     (d) a significant brain stem injury, from which few patients recover.
     (e) minor brain injuries, and typically represent only transient response anomalies.

23. Immediate onset of stroke-like signs and symptoms most frequently accompanies
     (a) an epidural hematoma.
     (b) a subdural hematoma.
     (c) intracerebral hemorrhage.
     (d) a spider angioedema.
     (e) ocular petechiae.

24. Several hours or days may pass before signs and symptoms of _____ are shown.
     (a) an epidural hematoma
     (b) a subdural hematoma
     (c) intracerebral hemorrhage
     (d) a spider angioedema
     (e) ocular petechiae

25. Which of the following statements regarding increased intracranial pressure (ICP) is false?

    (a) Edema or hemorrhage within the cranium will increase ICP.

    (b) Increased ICP results in diminished cerebral blood flow and shunting of cerebrospinal fluid to the spinal column.

    (c) Uncompensated ICP causes hypoxia and hypercarbia, which produces cerebral vascular constriction.

    (d) Increased ICP triggers a rise in systemic blood pressure, in an attempt to improve cerebral circulation.

    (e) Increased ICP and hypercarbia triggers hyperventilation.

26. All of the following signs and symptoms may be caused by increased ICP, except

    (1) Cushing's reflex (or triad).   (5) vomiting.
    (2) Kussmaul's respirations.   (6) pupils slow to react, bilaterally.
    (3) Cheyne-Stokes respirations.   (7) a unilaterally dilated, nonreactive pupil.
    (4) Biot's respirations.   (8) seizure.

    (a) 4, 6, and 8.

    (b) 4 and 7.

    (c) 2 and 6.

    (d) 2.

    (e) 6.

27. The inner ear is susceptible to injury from

    (a) blast trauma.

    (b) diving trauma.

    (c) basilar skull fracture.

    (d) Answers (a) and (b).

    (e) Answers (a), (b), and (c).

28. Dermatomes are best defined as specific body surface areas innervated by pairs of

    (a) cranial sensory nerves.

    (b) cranial motor nerves.

    (c) non-myelinated nerves.

    (d) spinal motor nerves.

    (e) spinal sensory nerves.

29. The very base of a patient's neck (the "collar region") is an area innervated by nerves originating from

    (a) C-3.

    (b) C-4.

    (c) C-7.

    (d) T-1.

    (e) the cranium (one pair of the "cranial nerves").

30. The nipple-line of a patient's body is innervated by nerves originating from
  (a) T-1.
  (b) T-3.
  (c) T-4.
  (d) T-10.
  (e) the cranium (one pair of the "cranial nerves").

31. The umbilical area of a patient's body is innervated by nerves originating from
  (a) T-1.
  (b) T-3.
  (c) T-4.
  (d) T-10.
  (e) the cranium (one pair of the "cranial nerves").

32. The phrenic nerve originates from
  (a) peripheral nerve roots at C-1.
  (b) peripheral nerve roots C-3 through C-5.
  (c) peripheral nerve roots C-6 through C-7.
  (d) peripheral nerve roots T-4 through T-10.
  (e) the cranium (one pair of the "cranial nerves").

33. Neurogenic shock results from
  (a) vasodilation below the site of the injury.
  (b) vasoconstriction below the site of the injury.
  (c) autonomic nervous system interruption.
  (d) Answers (a) and (c).
  (e) Answers (b) and (c).

34. Spine injury may result in
  (a) apnea or diaphragmatic breathing.
  (b) priapism.
  (c) "hold-up" positioning of the arms above the head.
  (d) Answers (a) and (b).
  (e) Answers (a), (b), and (c).

35. Which of the following medications may be administered to treat signs or symptoms caused by spinal injury?
  (1) methylprednisolone   (4) atropine
  (2) dexamethasone   (5) 250 cc bolus of LR or NS
  (3) dopamine

    (a) 1, 2, 3, 4, and 5.
    (b) 1, 2, 3, and 5.
    (c) 1, 2, and 3.
    (d) 1 and 2.
    (e) 2.

36. Complaint of a "dark curtain" suddenly obstructing a portion of the patient's field of vision classically accompanies an ocular emergency involving
   (a) hyphema.
   (b) acute retinal artery occlusion.
   (c) retinal detachment.
   (d) subconjunctival hemorrhage.
   (e) corneal abrasion.

37. A patient complaining of extreme anterior eye pain and exclaiming, "Something is in my eye!" classically accompanies an ocular emergency involving
   (a) hyphema.
   (b) acute retinal artery occlusion.
   (c) retinal detachment.
   (d) subconjunctival hemorrhage.
   (e) corneal abrasion.

38. Sudden onset of painless vision loss, in only one eye, is caused by
   (a) hyphema.
   (b) acute retinal artery occlusion.
   (c) retinal detachment.
   (d) subconjunctival hemorrhage.
   (e) corneal abrasions.

39. The appearance of blood within the iris and pupil is called
   (a) hyphema.
   (b) acute retinal artery occlusion.
   (c) retinal detachment.
   (d) subconjunctival hemorrhage.
   (e) corneal abrasions.

40. A strong sneeze or violent vomiting may produce
   (a) hyphema.
   (b) acute retinal artery occlusion.
   (c) retinal detachment.
   (d) subconjunctival hemorrhage.
   (e) corneal abrasions.

41. Bleeding within the anterior chamber of the eye produces
   (a) hyphema.
   (b) acute retinal artery occlusion.
   (c) retinal detachment.
   (d) subconjunctival hemorrhage.
   (e) corneal abrasions.

42.   Loss of vitreous or aqueous humor (liquid) from the eye
       (a) will cause only transient loss of sight, because the lost humor will naturally
            regenerate.
       (b) can be replaced with synthetic humor to preserve the esthetic appearance of
            the eye, but will not preserve the function of sight.
       (c) can be replaced with synthetic humor to preserve both the esthetic
            appearance of the eye and the function of sight.
       (d) will threaten the patient's sight in the affected eye, possibly resulting in
            permanent loss of sight in that eye.
       (e) will cause only transient loss of depth perception, because the lost humor
            will naturally regenerate.

43.   Endotracheal intubation may stimulate a parasympathetic response, which could
       result in
       (a) increased ICP.
       (b) bradycardias or dysrhythmias.
       (c) tachycardias or dysrhythmias.
       (d) Answers (a) and (b).
       (e) Answers (a) and (c).

44.   The presence of tachycardia and hypotension in a patient with a closed
       head injury
       (a) confirms the absence of increased ICP.
       (b) confirms the presence of increased ICP.
       (c) does not rule out the presence of increased ICP and suggests the presence of
            additional (hidden) injuries elsewhere.
       (d) rules out the presence of increased ICP and suggests the presence of
            additional (hidden) injuries elsewhere.
       (e) confirms the presence of increased ICP but does not rule out the presence of
            additional (hidden) injuries elsewhere.

45.   Which of the following treatments is not indicated for the care of a head-injured
       patient?
       (a) Spinal immobilization
       (b) Airway maintenance and management
       (c) Prolonged hyperventilation with high-flow oxygen after intubation
       (d) Hemorrhage control
       (e) Suctioning of fluids or vomitus from the airway

46.   Which of the following medications can reduce ICP?
       (a) Morphine sulfate
       (b) Furosemide

(c) Mannitol

(d) Answers (b) and (c).

(e) Answers (a), (b), and (c).

47. The Glasgow Coma Scale evaluates a patient's level of consciousness based upon

(a) respiratory rate, systolic blood pressure, and capillary refill.

(b) the history mnemonic, AMPLE.

(c) level of consciousness, pulse rate, respiratory rate, blood pressure, and motor responses.

(d) the patient's ability to localize and identify auditory stimuli.

(e) eye-opening responses, verbal responses, and motor responses to a variety of stimuli.

48. The highest possible Glasgow Coma Scale score is

(a) 10.

(b) 15.

(c) 25.

(d) 50.

(e) 100.

49. The lowest possible Glasgow Coma Scale score is

(a) 0.

(b) 1.

(c) 3.

(d) 15.

(e) 50.

50. Your patient has an open neck wound. Your management considerations should include

(a) use of an occlusive dressing.

(b) anticipation of airway compromise.

(c) strong suspicion of spinal injury.

(d) Answers (a) and (b).

(e) Answers (a), (b), and (c).

51. To reduce the potential for an air embolus occurring when a patient has an open neck wound, you should place the patient

(a) on her/his left side.

(b) on her/his right side.

(c) in the Trendelenburg position.

(d) in the reverse-Trendelenburg position.

(e) in the semi-Fowler's position.

52. Which of the following statements regarding the management of a head-injured patient and use of the pneumatic antishock garment (PASG) is true?

    (a) Application of PASG is contraindicated for patients with head injuries.

    (b) PASG should be applied but not inflated until the patient displays signs and symptoms of increased ICP.

    (c) PASG may be applied but not inflated until the patient displays signs and symptoms of neurogenic or hypovolemic shock.

    (d) Application and inflation of PASG is recommended to assist in obtaining IV access in the presence of increased ICP.

    (e) Application and inflation of PASG is recommended to assist in decreasing ICP.

53. The IV fluid of choice for management of a head-injured patient is

    (a) NS or LR only, because additional (hidden) injuries must be anticipated.

    (b) D_5W only, because seizures can be anticipated and because Dilantin® is incompatible with NS or LR.

    (c) D_5W only, because fluid administration is contraindicated in the presence of increased ICP.

    (d) any colloid solution.

    (e) any crystalloid solution.

54. _____ is the condition in which a patient normally has unequal pupil sizes.

    (a) Dysconjugate gaze

    (b) Doll's eye gaze

    (c) Ophthalmia

    (d) Nystagmus

    (e) Aniscoria

55. _____ describes the eyes of a patient who appears to be looking in different directions simultaneously (failure of the pupils to move together).

    (a) Dysconjugate gaze

    (b) Doll's eye gaze

    (c) Ophthalmia

    (d) Nystagmus

    (e) Aniscoria

56. An expanding lesion placing increased pressure on the third cranial nerve will produce

    (a) reactivity and size changes of the pupil on the affected side.

    (b) reactivity and size changes of the pupil on the opposite side of the lesion.

    (c) bilateral pupillary reactivity and size changes.

    (d) Answer (a) or (c).

    (e) Answer (b) or (c).

57. _____ may cause a patient's pupils to be bilaterally dilated and delayed in reactivity to light.

  (a) A congenital defect common to approximately 20% of the population
  (b) Cerebral hypoxia
  (c) Some depressant drugs
  (d) Answer (b) or (c).
  (e) Answer (a), (b), or (c).

58. Which of the following statements regarding mannitol is false?

  (a) Mannitol is a large glucose molecule that acts as an osmolar diuretic and is very effective for reducing ICP associated with cerebral edema.
  (b) Mannitol should be administered via very slow IVP at the IV port closest to the patient's heart.
  (c) If intracranial hemorrhage is present, mannitol administration may increase the rate of hemorrhage and cause more serious brain damage.
  (d) Mannitol administration may cause CHF or sodium depletion.
  (e) Mannitol administration is contraindicated if the patient has acute pulmonary edema or has recently developed CHF.

59. Which of the following statements regarding furosemide (Lasix®) is false?

  (a) Because Lasix is a systemic diuretic, it may cause dehydration or electrolyte disturbances.
  (b) Lasix's vasodilating effect may lower the patient's blood pressure, thus increasing cerebral hypoxia and hypercarbia.
  (c) Lasix is contraindicated for administration to patients allergic to the "sulfa" class of drugs.
  (d) Lasix is often administered in combination with mannitol, to increase the rate of cerebral diuresis.
  (e) Lasix is a more effective cerebral diuretic than mannitol.

60. Which of the following statements regarding seizures and head injury is true?

  (a) Cerebral hypoxia may cause seizure activity.
  (b) Head trauma may cause seizure activity.
  (c) Because diazepam administration is contraindicated in the presence of head injury, trauma-related seizures are treated with hyperventilation and fluid administration only.
  (d) Answers (a) and (b).
  (e) Answers (a), (b), and (c).

61. Lung-tissue contusion may result in

  (a) localized atelectasis.
  (b) localized pulmonary edema.
  (c) loss of 1000 to 1500 ml of blood.
  (d) Answers (a) and (b).
  (e) Answers (a), (b), and (c).

62. The medical term for "coughing up blood" is

    (a) hematemesis.

    (b) hemolysis.

    (c) hemocytophagia.

    (d) hemoptysis.

    (e) hemolytic sputum.

63. Which of the following statements about rib fracture is false?

    (a) The pain of a rib fracture can be so great that respiratory excursion may be limited, and hypoventilation may occur.

    (b) Pediatric blunt chest trauma patients are less likely to have a rib fracture than adults but more likely to have underlying organ trauma.

    (c) Geriatric blunt chest trauma patients are more likely than younger patients to have both a rib fracture and underlying organ trauma.

    (d) Anterior blunt chest trauma may cause a rib fracture at the posterior angle of one or more ribs.

    (e) Because ribs 10 to 12 are not attached to the sternum, they are the ribs most frequently fractured.

64. A flail chest segment is best defined as

    (a) one rib, fractured in two or more places.

    (b) two or more adjacent ribs, each fractured in one or more places.

    (c) three or more adjacent ribs, each fractured in two or more places.

    (d) Answer (a) or (b).

    (e) Answer (a), (b), or (c).

65. Which of the following statements regarding paradoxical movement of a flail-chest segment is true?

    (a) Even the "smallest" flail-segment section frequently exhibits paradoxical movement.

    (b) Positive pressure ventilation reverses the mechanism that produces paradoxical movement.

    (c) Paradoxical movement is observed as a section of the chest bulging out during inspiration and sinking into the chest during expiration.

    (d) Paradoxical movement is observed as a side of the chest that does not move with each respiration, whereas the opposite side of the chest moves normally.

    (e) As long as the patient's diaphragm remains functional, paradoxical movement of a flail-chest segment results in only superficial reduction of ventilatory volume.

66. Pulsus paradoxus is best defined as a

    (a) pulse that speeds up during each expiration.

    (b) pulse that is absent, in spite of the presence of an ECG rhythm.

(c) pulse that is too fast to count.

(d) greater than 10 mmHg drop in systolic blood pressure during expiration.

(e) greater than 10 mmHg drop in systolic blood pressure during inspiration.

67. Electrical alternans is best defined as when the
  (a) ECG rhythm speeds up every three or four cardiac cycles.
  (b) ECG rhythm slows down every three or four cardiac cycles.
  (c) amplitude of the P, QRS, and T waves alternates every three or four cardiac cycles.
  (d) amplitude of the P, QRS, and T waves alternates with every other cardiac cycle.
  (e) amplitude of the P, QRS, and T waves is opposite from normal.

68. Traumatic asphyxia is best defined as a
  (a) severe crushing injury of the upper airway, producing sudden and complete airway obstruction, resulting in asphyxia.
  (b) severe compression injury of the chest, reversing the flow of venous blood (moving it away from the heart), often accompanied by inadequate chest excursion, and rapidly resulting in asphyxia.
  (c) total airway obstruction, produced by any foreign body inspired during traumatic injury, resulting in asphyxia.
  (d) Answer (a) or (b).
  (e) Answer (a) or (c).

69. The most "classic" combination of signs indicative of traumatic asphyxia include which of the following?
  (1) A horizontal "line" identifying the site of impact, produced by the grossly obvious presence of cyanosis extending only below the site of impact (the patient remains pale and dry above this "line").
  (2) Deep red, purple, or blue blotchy discoloration of the head and neck.
  (3) A face that appears swollen.
  (4) Bulging eyes.
  (5) Conjunctival hemorrhages.
  (6) JVD.

  (a) 1, 2, 3, 4, 5, and 6.
  (b) 1, 3, 4, 5, and 6.
  (c) 2, 3, 4, 5, and 6.
  (d) 1, 5, and 6.
  (e) 1.

70. A pneumothorax may be caused by
  (a) blunt chest trauma.
  (b) penetrating trauma.
  (c) aggressive coughing or heavy lifting (in the absence of trauma).
  (d) Answers (a) and (b).
  (e) Answers (a), (b), and (c).

71. The pleural injury commonly referred to as the "paper bag syndrome" is most frequently produced by
    (a) penetrating trauma to a hyperinflated thorax.
    (b) deceleration trauma from a fall (while the thorax is hyperinflated).
    (c) blunt trauma to a hyperinflated thorax.
    (d) Answer (a) or (b).
    (e) Answer (a) or (c).

72. All of the following mechanisms may result in a tension pneumothorax except
    (a) an open pneumothorax that develops an intrathoracic pressure greater than that of atmospheric pressure.
    (b) a simple pneumothorax that develops an intrathoracic pressure greater than that of atmospheric pressure.
    (c) needle decompression of the chest with implementation of a one-way valve, allowing air to escape the pleural space.
    (d) the application of an occlusive dressing when an internal injury is continuing to leak air into the pleural space.
    (e) an injury that creates a one-way valve, allowing air to continue entering the pleural space but not to escape.

73. The earliest sign of a tension pneumothorax is
    (a) shifting of the trachea toward the injured side.
    (b) shifting of the trachea away from the injured side.
    (c) severe dyspnea and tachypnea.
    (d) Answer (a) or (c).
    (e) Answer (b) or (c).

74. A late-developing sign of tension pneumothorax is
    (a) shifting of the trachea toward the injured side.
    (b) shifting of the trachea away from the injured side.
    (c) severe dyspnea and tachypnea.
    (d) Answer (a) or (c).
    (e) Answer (b) or (c).

75. Possible signs and symptoms of a tension pneumothorax include all of the following except
    (a) agitation, increasing dyspnea, increasing resistance to ventilation.
    (b) subcutaneous emphysema.
    (c) jugular venous distention.
    (d) hyperresonant percussion response on the side opposite the injury.
    (e) tachypnea, tachycardia, and hypotension.

76. Possible signs and symptoms of a hemothorax include all of the following except
    (a) diminished or absent breath sounds on the entire side of the injury site.
    (b) flat jugular veins.
    (c) tachypnea, tachycardia, and hypotension.
    (d) cyanosis.
    (e) diaphoresis.

77. Which of the following statements regarding myocardial contusion is false?
    (a) Severe, blunt, anterior chest trauma is the most frequent cause of myocardial contusion.
    (b) In the absence of preexisting cardiac disease, a myocardial contusion produces chest pain similar to that of acute myocardial infarction but, unlike AMI, is rarely accompanied by dysrhythmias or ectopy.
    (c) Myocardial contusion may reduce the strength of cardiac contraction and diminish cardiac output.
    (d) In the absence of preexisting cardiac disease, a myocardial contusion heals with less scarring than does an acute myocardial infarction.
    (e) Myocardial contusion may result in myocardial hematoma, hemoperitoneum, or myocardial tissue necrosis.

78. Traumatic compression of the heart between the sternum and the thoracic spine is most likely to produce myocardial contusion to the
    (a) right atrium and right ventricle.
    (b) right and left atria.
    (c) left atrium and left ventricle.
    (d) right and left ventricles.
    (e) aorta.

79. The heart is surrounded by a protective sac composed of two layers. The inner layer, which lines the external cardiac muscle surface, is called the
    (a) epicardium, or visceral pericardium.
    (b) parietal pericardium.
    (c) endocardium.
    (d) myocardium.
    (e) epicardium, or parietal pericardium.

80. The outer layer of the sac surrounding the heart is called the
    (a) epicardium, or visceral pericardium.
    (b) parietal pericardium.
    (c) endocardium.
    (d) myocardium.
    (e) epicardium, or parietal pericardium.

81. The area between the layers of the heart's protective sac
(a) contains a cushion of air.
(b) contains a cushion of cartilage.
(c) contains a small amount of lubricating fluid.
(d) is completely empty unless penetrating trauma fills it with blood.
(e) is completely empty unless penetrating trauma fills it with air.

82. Pericardial tamponade occurs when cardiac filling is restricted by
(a) a penetration of both layers of the pericardial sac, with blood flowing into the lungs.
(b) a penetration of both layers of the pericardial sac, with blood flowing into the mediastinum.
(c) a penetration of the outer layer of the pericardial sac, with blood flowing into the mediastinum.
(d) blood or other types of fluid accumulating within the mediastinal space.
(e) blood or other types of fluid accumulating within the pericardial space.

83. Possible signs and symptoms of pericardial tamponade include all of the following except
(a) tachycardia and tachypnea.
(b) distended jugular veins, which become less distended (or "flat") during inspiration.
(c) diminished breath sounds in the left chest.
(d) signs of pulsus paradoxus or electrical alternans.
(e) decreasing pulse pressure.

84. Management of the patient with pericardial tamponade includes
(a) administration of high-flow oxygen.
(b) rapid transportation to the closest appropriate emergency department.
(c) prehospital performance of pericardiocentesis, using the subclavicular approach.
(d) Answers (a) and (b).
(e) Answers (a), (b), and (c).

85. Signs and symptoms that may accompany a thoracic dissecting aortic aneurysm include all of the following except
(a) complaint of a "tearing" or "burning" pain in the periumbilical area, described as radiating "backward" to the spine.
(b) a history of deceleration trauma.
(c) a diminished or absent radial pulse in one wrist but present and normal in the other wrist.
(d) hypertension.
(e) hypotension.

86. Nitrous oxide administration is indicated for all of the following emergencies except
    (a) an alert and oriented patient complaining of pain from an isolated extremity fracture.
    (b) an atraumatic patient complaining of severe abdominal pain.
    (c) an atraumatic patient complaining of chest pain.
    (d) an atraumatic patient who is hyperventilating.
    (e) a burn patient who does not have additional blunt or penetrating trauma.

87. Nitrous oxide administration is contraindicated in all of the following emergencies except
    (a) an alert and oriented atraumatic patient who denies all complaints of pain but is experiencing a severe anxiety reaction.
    (b) an atraumatic patient with an altered LOC who complains of chest pain.
    (c) a patient with an altered LOC (apparently from alcohol intoxication) complaining of severe pain from an isolated extremity fracture.
    (d) a chest trauma patient complaining of chest pain.
    (e) a COPD patient complaining of pain from an isolated extremity fracture.

88. Nitrous oxide is administered
    (a) for one minute, via a nonrebreather mask securely affixed to the patient's face, and then discontinued for the remainder of transport.
    (b) only during the last minute of transport time, via a nonrebreather mask securely affixed to the patient's face.
    (c) continuously, via a nonrebreather mask securely affixed to the patient's face, and discontinued only when the patient develops an altered level of consciousness, or upon arrival at the emergency department.
    (d) continuously, via a face mask held in place by the patient, and discontinued only when the pain is significantly reduced or the patient drops the mask.
    (e) only via bag-valve-mask (or bag-valve-ET) ventilations, if the patient is unconscious.

89. When applying an occlusive dressing to a penetrating wound of the thorax, the dressing should be sealed (taped) on
    (a) three sides only, so that air can escape during expiration.
    (b) three sides only, so that air can escape during inspiration.
    (c) two sides only, so that air can escape during expiration.
    (d) two sides only, so that air can escape during inspiration.
    (e) all four sides so that no air exchange is allowed via the wound.

90. Prehospital needle decompression of a tension pneumothorax should be performed in the _____ intercostal space, at the _____ line.
    (a) fifth or sixth / midclavicular
    (b) second or third / midaxillary
    (c) fifth / midclavicular
    (d) second / midclavicular
    (e) second / midaxillary

91. Which of the following statements regarding needle decompression of a tension pneumothorax is false?

    (a) If a pneumothorax is not present, needle decompression may create one.

    (b) The needle should be inserted along the lower edge of the rib that forms the upper border of the appropriate intercostal space.

    (c) If the patient remains symptomatic, a second decompression may be necessary, and should be performed proximal to the original site.

    (d) If patient improvement is transient, and signs or symptoms reappear, perform another decompression proximal to the original site.

    (e) In the absence of a flutter valve, needle decompression converts a tension pneumothorax into a simple pneumothorax and improves the patient's condition.

92. Treatment of a patient with a cardiac contusion

    (a) includes administration of ACLS medications, according to standard ACLS protocols.

    (b) is limited to supportive therapy, with oxygen being the only "medication" administered to the trauma patient.

    (c) includes needle decompression to diminish chest pain secondary to increased mediastinal pressure.

    (d) Answers (a) and (c).

    (e) Answers (b) and (c).

93. Management of an object impaled in the chest includes

    (a) occlusive dressing about the entrance and exit sites of the object.

    (b) stabilization with bulky dressings.

    (c) removal of the object if its location interferes with required performance of CPR.

    (d) Answers (a) and (b).

    (e) Answers (a), (b), and (c).

94. Which of the following structures are not completely covered by the peritoneum, and thus are located in the retroperitoneal space?

(1) liver	(6) urinary bladder
(2) duodenum	(7) gall bladder
(3) pancreas	(8) abdominal aorta
(4) kidneys	(9) inferior vena cava
(5) ureters	(10) spleen

    (a) 1, 4, 5, and 7.

    (b) 4, 5, 7, and 10.

    (c) 2, 4, 5, and 10.

    (d) 4, 5, 8, 9, and 10.

    (e) 2, 3, 4, 5, 6, 8, and 9.

95. Trauma to the lateral left upper abdominal quadrant should cause suspicion of injury to the

    (a) liver and left kidney.

    (b) spleen and left kidney.

    (c) ascending colon, spleen, and left kidney.

    (d) descending colon, liver, and left kidney.

    (e) appendix, liver, and left kidney.

96. The abdominal aorta bifurcates at its distal end, becoming the left and right

    (a) intra-abdominal arteries.

    (b) femoral arteries.

    (c) common iliac arteries.

    (d) great saphenous veins.

    (e) popliteal arteries.

97. Peritonitis is most rapidly produced by

    (a) bleeding in the abdomen.

    (b) bowel contents spilled within the abdomen.

    (c) digestive fluids loose within the abdomen.

    (d) air entering the abdominal cavity.

    (e) chemical ingestions.

98. Rebound tenderness is best defined as the patient's complaint of pain

    (a) during application of direct pressure to the abdomen.

    (b) upon the release of direct pressure applied to the abdomen.

    (c) during application of direct pressure to the flanks.

    (d) Answers (a) and (c).

    (e) Answers (b) and (c).

99. Abdominal guarding is best defined as abdominal muscle spasm or contraction that occurs in response to

    (a) pressure applied to the abdomen.

    (b) movement of abdominal contents.

    (c) pressure applied to the flanks.

    (d) Answers (a) and (b).

    (e) Answers (a), (b), and (c).

100. Which of the following statements regarding IV management for patients with a serious mechanism of abdominal injury is false?

    (a) If the patient is hypotensive, fluid resuscitation (at least one IV) should be initiated before transportation.

    (b) Only large-bore catheters and trauma tubing should be used.

    (c) If tachycardia does not respond to a 250 ml bolus of fluid, another 250 ml bolus may be given.

    (d) IV flow rates should be titrated to maintain a systolic blood pressure of 90 mmHg.

    (e) The need for aggressive fluid resuscitation should be anticipated, before shock signs and symptoms develop.

101. A commonly accepted limit for the amount of IV fluid that may be infused during the prehospital phase of treatment is

    (a) 500 ml.

    (b) 1000 ml.

    (c) 2000 ml.

    (d) 3000 ml.

    (e) 4000 ml.

102. Treatment of an eviscerated abdomen includes

    (a) high-flow oxygen and large-bore IVs.

    (b) application of moist dressings, covered with an occlusive dressing.

    (c) application of PASG and gentle inflation of the abdominal section only.

    (d) Answers (a) and (b).

    (e) Answers (a), (b), and (c).

103. Management of an object impaled in the LUQ of the abdomen includes

    (a) high-flow oxygen and large-bore IVs.

    (b) occlusive dressing about the entrance site of the object, and stabilization with bulky dressings.

    (c) removal of the object only if life-threatening hypotension requires inflation of the abdominal section of the PASG.

    (d) Answers (a) and (b).

    (e) Answers (a), (b), and (c).

104. A strain is a muscular injury characterized by

    (a) edema and pooling of blood occurring within the closed fascia of a muscle.

    (b) muscle fibers that have been overstretched.

    (c) one or more of a joint capsule's ligaments being torn.

    (d) Answer (a) or (b).

    (e) Answer (b) or (c).

105. Compartment syndrome is a muscular injury characterized by
  (a) edema and pooling of blood occurring within the closed fascia of a muscle.
  (b) muscle fibers that have been overstretched.
  (c) one or more of a joint capsule's ligaments being torn.
  (d) Answer (a) or (b).
  (e) Answer (b) or (c).

106. A sprain is a joint injury characterized by
  (a) edema and pooling of blood occurring within the closed fascia of a joint.
  (b) muscle fibers that have been overstretched.
  (c) one or more of a joint capsule's ligaments being torn.
  (d) Answer (a) or (b).
  (e) Answer (b) or (c).

107. Specialized bands of tissue that connect muscle to bone are called
  (a) synovials.
  (b) bursae.
  (c) fascicles (or fasciculi).
  (d) ligaments.
  (e) tendons.

108. Specialized bands of tissue that connect bone to bone are called
  (a) synovials.
  (b) bursae.
  (c) fascicles (or fasciculi).
  (d) ligaments.
  (e) tendons.

109. _____ may cause the affected leg to appear shortened.
  (a) An anterior hip dislocation
  (b) A posterior hip dislocation
  (c) A hip fracture
  (d) Answer (b) or (c).
  (e) Answer (a), (b), or (c).

110. _____ usually presents with lateral rotation of the affected leg and a palpable bulge in the inguinal area.
  (a) An anterior hip dislocation
  (b) A posterior hip dislocation
  (c) A hip fracture
  (d) Answer (b) or (c).
  (e) Answer (a), (b), or (c).

111. _____ usually causes the affected leg to rotate inward, toward the body, with the knee typically flexed.

    (a) An anterior hip dislocation

    (b) A posterior hip dislocation

    (c) A hip fracture

    (d) Answer (b) or (c).

    (e) Answer (a), (b), or (c).

112. Your patient's arm is extended above her head, and she is unable to move it at the shoulder. She most likely has _____ of her shoulder.

    (a) an anterior dislocation

    (b) a posterior dislocation

    (c) an inferior dislocation

    (d) a superior dislocation

    (e) a transverse fracture

113. Your patient's shoulder looks "hollow" or "squared-off." He's holding his arm close to, and in front of, his chest. He most likely has _____ of his shoulder.

    (a) an anterior dislocation

    (b) a posterior dislocation

    (c) an inferior dislocation

    (d) a superior dislocation

    (e) a transverse fracture

114. Your patient's arm is medially rotated, and he holds his elbow and forearm away from his body. He most likely has _____ of his shoulder.

    (a) an anterior dislocation

    (b) a posterior dislocation

    (c) an inferior dislocation

    (d) a superior dislocation

    (e) a transverse fracture

115. The pulse that can be found in the posterior area of the knee is called the _____ pulse.

    (a) popliteal

    (b) dorsalis pedis

    (c) posterior fibial

    (d) medial malleolus

    (e) posterior tibial

116. The pulse that can be found on the medial side of the ankle, just posterior and inferior to the "ankle bone," is called the _____ pulse.
  (a) popliteal
  (b) dorsalis pedis
  (c) posterior fibial
  (d) medial malleolus
  (e) posterior tibial

117. The pulse that is found on the top of the foot is called the _____ pulse.
  (a) popliteal
  (b) dorsalis pedis
  (c) posterior fibial
  (d) medial malleolus
  (e) posterior tibial

118. Which of the following statements regarding musculoskeletal injury management is false?
  (a) Use gentle traction to straighten an angulated fracture before splinting, unless a significant increase in pain occurs, or resistance to movement is encountered.
  (b) Splint dislocations or deformities near a joint without changing the position in which they are found.
  (c) Dislocations or deformities near a joint may be manipulated if distal circulation is compromised; however, manipulation must stop if a significant increase in pain occurs, or resistance to movement is encountered.
  (d) Any open fracture with protruding bone ends must be gently tractioned until the bone ends are drawn back into place.
  (e) Immobilization is not complete unless it includes immobilization of the joint above the injury and the joint below the injury.

119. When using a bandage to affix an extremity to a padded splint, begin wrapping the bandage around the limb and splint
  (a) at the midpoint of the splint, wrapping first toward the body and then away from the body.
  (b) at the injury site, wrapping first away from the body and then toward the body.
  (c) at the most distal point of the splint, wrapping toward the body.
  (d) at the most proximal point of the splint, wrapping away from the body.
  (e) wherever you are most accustomed to beginning. No "medical preference" has been established for the manner in which a limb is wrapped.

120. Which of the following statements regarding pelvic fracture management is false?
- (a) Pelvic fractures present a potential life-threat, because more than 2 liters of "hidden" blood loss may occur.
- (b) To avoid increasing intra-pelvic hemorrhage, inflate only the abdominal section of the PASG to stabilize a pelvic fracture. Leg sections should remain uninflated unless one or both are required to manage associated lower extremity injury.
- (c) Even if initial vital signs are stable, two large-bore IVs should be initiated, using trauma tubing. Run both IVs at TKO, unless hypotension ensues.
- (d) 1000 ml boluses of LR or NS may be required to maintain a systolic blood pressure of 90–100 mmHg.
- (e) All pelvic fracture patients are considered candidates for rapid transport to a trauma center.

121. A traction splint is appropriate for stabilization and immobilization of
- (a) an isolated midshaft femoral fracture, only.
- (b) an isolated midshaft tibial fracture, only.
- (c) an isolated patellar fracture, only.
- (d) a midshaft femoral fracture or a tibial or fibular fracture, but only when joint injury (knee or ankle) is absent.
- (e) any femoral fracture, regardless of additional associated injuries in the same limb.

122. Which of the following statements regarding functions of the skin is false?
- (a) Skin protects the body from penetration of bacteria and other microorganisms.
- (b) Sweat evaporation on the skin surface serves to warm the body in cold weather, whereas blood-vessel constriction provides cooling in hot weather.
- (c) Skin acts to prevent loss of body fluids.
- (d) The skin is an organ of sensation, allowing us to perceive touch, pressure, pain, heat, and cold.
- (e) Skin protects underlying structures from minor trauma.

123. Sweat glands, sebaceous glands, and hair follicles originate in the skin layer called the _____.
- (a) subcutaneous tissue
- (b) subdermis
- (c) hyperdermis
- (d) dermis
- (e) epidermis

124. The skin layer called the _____ is composed largely of fat.
- (a) subcutaneous tissue
- (b) subdermis

    (c) hyperdermis

    (d) dermis

    (e) epidermis

125. Pigmentation cells that provide protection from ultraviolet radiation are contained within the skin layer called the _____.

    (a) subcutaneous tissue

    (b) subdermis

    (c) hyperdermis

    (d) dermis

    (e) epidermis

126. The skin layer called the _____ serves as the body's layer of insulation.

    (a) subcutaneous tissue

    (b) subdermis

    (c) hyperdermis

    (d) dermis

    (e) epidermis

127. Sebum is a substance that

    (a) coats the epidermis, and helps the skin prevent fluid from entering or leaving the body.

    (b) moisturizes and lubricates the epidermis, keeping it soft and pliant.

    (c) is secreted by sebaceous glands that originate in the epidermis.

    (d) Answers (a) and (b).

    (e) Answers (a), (b), and (c).

128. Sweat is secreted by the

    (a) sebaceous glands.

    (b) pituitary gland.

    (c) adrenal gland.

    (d) arrector pili glands.

    (e) sudoriferous glands.

129. The dark, blue-black color produced by extravasation of deoxygenated blood is called

    (a) erythema.

    (b) ecchymosis.

    (c) jaundice.

    (d) mottling.

    (e) cyanosis.

130. Reddened inflammation of the skin is also called
- (a) erythema.
- (b) ecchymosis.
- (c) jaundice.
- (d) mottling.
- (e) cyanosis.

131. A closed skin injury caused by crushing damage of only the smallest blood vessels and accompanied by only superficial edema is called
- (a) a hematoma.
- (b) a laceration.
- (c) an abrasion.
- (d) an incision.
- (e) a contusion.

132. Commonly called a "goose egg" by nonmedical personnel, the medical term for a pocket of blood formed within the tissue beneath an injury site is
- (a) a hematoma.
- (b) a laceration.
- (c) an abrasion.
- (d) an incision.
- (e) a contusion.

133. An injury characterized by the scraping-removal of epidermal skin layers (sometimes removing portions of the dermis), is called
- (a) a hematoma.
- (b) a laceration.
- (c) an abrasion.
- (d) an incision.
- (e) a contusion.

134. An open, soft-tissue wound with smooth, linear edges is called
- (a) a hematoma.
- (b) a laceration.
- (c) an abrasion.
- (d) an incision.
- (e) a contusion.

135. An open, soft-tissue wound with ragged, uneven edges is called
- (a) a hematoma.
- (b) a laceration.
- (c) an abrasion.
- (d) an incision.
- (e) a contusion.

136. An avulsion is best defined as
   (a) the complete removal of a section of skin from the body.
   (b) the incomplete removal of a section of skin from the body, producing an attached flap.
   (c) the complete removal of a digit, a limb portion, or an entire limb from the body.
   (d) Answer (a) or (b).
   (e) Answer (b) or (c).

137. An amputation is best defined as
   (a) the complete removal of a section of skin from the body.
   (b) the incomplete removal of a section of skin from the body, producing an attached flap.
   (c) the complete removal of a digit, a limb portion, or an entire limb from the body.
   (d) Answer (a) or (b).
   (e) Answer (b) or (c).

138. A _____ is also called a partial-thickness burn.
   (a) first-degree burn
   (b) second-degree burn
   (c) third-degree burn
   (d) Answer (a) or (b).
   (e) Answer (b) or (c).

139. A _____ is also called a full-thickness burn.
   (a) first-degree burn
   (b) second-degree burn
   (c) third-degree burn
   (d) Answer (a) or (b).
   (e) Answer (b) or (c).

140. According to the rule of nines, a burn covering an adult's posterior head and neck represents a _____ body surface area burn.
   (a) 4.5%
   (b) 9%
   (c) 10%
   (d) 13.5%
   (e) 18%

141. According to the rule of nines, a burn covering an adult's upper and lower back, including the buttocks, represents a _____ body surface area burn.
   (a) 4.5%
   (b) 9%
   (c) 10%
   (d) 13.5%
   (e) 18%

142. According to the rule of nines, a burn covering an adult's anterior lower leg from knee to toes represents a _____ body surface area burn.

    (a) 4.5%

    (b) 9%

    (c) 10%

    (d) 13.5%

    (e) 18%

143. According to the rule of nines, a burn covering the anterior and posterior of an adult's arm from shoulder to fingertips represents a _____ body surface area burn.

    (a) 4.5%

    (b) 9%

    (c) 10%

    (d) 13.5%

    (e) 18%

144. According to the rule of nines, a burn covering an adult's anterior abdomen, including the genitalia, represents a _____ body surface area burn.

    (a) 4.5%

    (b) 9%

    (c) 10%

    (d) 13.5%

    (e) 18%

145. Using the Palmar Surface Area technique (also called the "rule of palms"), an area equivalent to the palmar surface of the patient's hand is approximately _____ of the patient's body surface area.

    (a) 5%

    (b) 4.5%

    (c) 3%

    (d) 2%

    (e) 1%

146. When assessing (measuring) the palmar surface of the patient's hand, measure the area of the patient's

    (a) palm and fingers.

    (b) palm, fingers, and thumb (all digits extended).

    (c) palm, fingers, and thumb while making a "fist."

    (d) palm only; do not include the fingers or thumb area.

    (e) fingers only; do not actually include the area of the palm.

147. According to the rule of nines, a burn covering a child's posterior head and neck represents a _____ body surface area burn.

    (a) 4.5%

    (b) 9%

    (c) 10%

    (d) 13.5%

    (e) 18%

148. According to the rule of nines, a burn covering a child's upper and lower back, including the buttocks, represents a _____ body surface area burn.

    (a) 4.5%

    (b) 9%

    (c) 10%

    (d) 13.5%

    (e) 18%

149. According to the rule of nines, a burn covering the anterior and posterior of a child's arm from shoulder to fingertips represents a _____ body surface area burn.

    (a) 4.5%

    (b) 9%

    (c) 10%

    (d) 13.5%

    (e) 18%

150. According to the rule of nines, a burn covering a child's anterior abdomen, including the genitalia, represents a _____ body surface area burn.

    (a) 4.5%

    (b) 9%

    (c) 10%

    (d) 13.5%

    (e) 18%

151. The term eschar refers to

    (a) the scabs that form on second-degree burns.

    (b) the blisters that form on first-degree burns.

    (c) the red discoloration of burned tissue.

    (d) the white discoloration of burned tissue.

    (e) the inelastic, hard and leathery, necrotic tissue that forms in an area of third-degree burns.

152. Systemic complications commonly associated with serious thermal burns include all of the following except

    (a) hyperthermia.

    (b) hypothermia.

    (c) hypovolemia.

    (d) organ failure.

    (e) infection.

153. Circumferential third-degree burns of an extremity are especially serious because
- (a) the extremity will require amputation.
- (b) the patient will be more susceptible to hyperthermia.
- (c) the patient will be more susceptible to systemic infection.
- (d) even if the extremity heals, it will lose all functional ability.
- (e) blood flow to the tissue underlying and distal to the burn may become seriously restricted.

154. When someone is located near an explosion, the most frequent and most life-threatening injuries that occur are those involving
- (a) blast trauma to the ears.
- (b) blast trauma to solid abdominal organs.
- (c) burns of up to 35% of the body surface area.
- (d) penetrating wounds from debris.
- (e) blast trauma to the lungs.

155. Actual thermal injury to the lower airways is most likely to occur due to
- (a) smoke inhalation.
- (b) air pressure changes caused by blast injury.
- (c) toxic fume inhalation.
- (d) inhalation of extremely hot, dry air.
- (e) superheated steam inhalation.

156. Which of the following statements regarding electrical burn injuries is false?
- (a) Electrical current contact causes internal, unseen thermal burns, extending from the entry to the exit site.
- (b) Electrical current contact may immobilize respiratory muscles and cause acute respiratory arrest.
- (c) Even without direct physical contact with the electrical source, close proximity to high-voltage electrical current may ignite articles of clothing or produce flash burns, and result in external thermal burns.
- (d) The passage of electrical current through the body may induce ventricular fibrillation and cause acute cardiac arrest.
- (e) Internal injury caused by electrical current contact is usually greater along muscles and bones, because these tissues conduct electricity more readily than blood vessels or myelinated nerves.

157. All of the following burn patients complain of equally severe pain. Mandatory transportation to a burn center is not required for which of these patients?
- (a) A 20-year-old female with a 50% BSA sunburn.
- (b) A 35-year-old male with a 20% BSA partial-thickness burn.
- (c) A 15-year-old male with a 10% BSA partial-thickness burn, including one foot.
- (d) A 20-year-old male with superficial erythema of his face, after superheated steam exposure.
- (e) A 30-year-old female with partial-thickness thermal burns limited to the outer half of her right hand.

158. Using either the Parkland or Brooke methods of calculating IV fluid-resuscitation flow rates for a burn victim, after determining the patient's total IV fluid infusion needs for the first 24 hours post-burn, you must run your IV at a rate based upon the fact that 50% of this 24-hour total fluid volume must be infused within the first ___ hours, post-burn.

    (a) 4

    (b) 6

    (c) 8

    (d) 10

    (e) 12

159. Chemical burns caused by acid-containing substances usually destroy cells by

    (a) forming a thick, insoluble mass (a "coagulum") at the point of contact, causing a process known as coagulation necrosis, which actually limits the depth of burn.

    (b) thermally burning away cell membranes, causing the cell contents to leak out and rapidly producing serious hypovolemia.

    (c) ionizing the cell membranes, and creating progressively deeper burns as the process is allowed to continue.

    (d) deionizing the cell membranes, creating progressively deeper burns as the process is allowed to continue.

    (e) progressively penetrating surface and underlying tissues, causing liquefying necrosis to all tissues contacted, progressively causing deeper and deeper tissue injury if not arrested.

160. Chemical burns caused by alkali-containing substances usually destroy cells by

    (a) forming a thick, insoluble mass (a "coagulum"), at the point of contact, causing a process known as coagulation necrosis, which actually limits the depth of burn.

    (b) thermally burning away cell membranes, causing the cell contents to leak out and rapidly producing serious hypovolemia.

    (c) ionizing the cell membranes, creating progressively deeper burns as the process is allowed to continue.

    (d) deionizing the cell membranes, creating progressively deeper burns as the process is allowed to continue.

    (e) progressively penetrating surface and underlying tissues, causing liquefying necrosis to all tissues contacted, progressively causing deeper and deeper tissue injury if not arrested.

161. Treatment of skin exposure to _____ first requires removal via gentle brushing, followed by copious water lavage.

    (a) dry lime

    (b) phenol

    (c) sodium metal

    (d) oleoresin capsicum ("OC" or "pepper spray")

    (e) tear gas ("CS")

162. Treatment of skin exposure to _____ first requires a thorough alcohol lavage, followed by a copious water lavage.

    (a) dry lime

    (b) phenol

    (c) sodium metal

    (d) All of the above.

    (e) None of the above.

163. Treatment of skin exposure to _____ first requires removal via gentle brushing, followed by covering the wound with oil.

    (a) dry lime

    (b) phenol

    (c) sodium metal

    (d) oleoresin capsicum ("OC" or "pepper spray")

    (e) tear gas ("CS")

164. Place the following types of radiation in order of their power of penetration (injury-severity risk). Start with the least-penetrating radiation type (very weak penetration power—radiation stopped by paper, clothing, or the epidermis), and end with the strongest-penetrating radiation type.

    (1) neutron radiation   (3) gamma radiation

    (2) beta radiation   (4) alpha radiation

        (a) 1, 2, 3, 4.

        (b) 4, 3, 1, 2.

        (c) 4, 2, 3, 1.

        (d) 4, 2, 1, 3.

        (e) 1, 4, 2, 3.

165. Which of the following statements regarding radiation-injury management is false?

    (a) If it is necessary to remove the patient from the contamination source, only the oldest responding care providers should do it.

    (b) Pregnant responders should never participate in the extrication or decontamination of radiation-injury patients.

    (c) Even after performance of all decontamination measures, the human body remains a potential source of ionizing-radiation contamination. Thus, all radiation-injury patients must remain isolated from other individuals.

    (d) Decontamination of radiation-injury patients is accomplished by removing all clothing and thoroughly washing with soap and water.

    (e) All removed clothing and all run-off water from decontamination must be safely contained and subsequently transported to a site where it may be securely disposed of.

The Answer Key to Test Section 6 is on page 427.

Test
Section
Seven

7

Test Section 7's subjects:

- Electrophysiology of the Heart
- Basic Dysrhythmia Identification (Using Lead II)
- Cardiovascular Disorders and Management

Test Section 7 consists of 180 questions, and is allotted 3 hours for completion.

1.   The outermost layer of heart muscle (the outer surface of the heart) is called the
      (a) endocardium.
      (b) epicardium.
      (c) parietal pericardium.
      (d) myocardium.
      (e) intercardium.

2.   The thick middle layer of heart muscle is called the
      (a) endocardium.
      (b) epicardium.
      (c) parietal pericardium.
      (d) myocardium.
      (e) intercardium.

3.   The smooth, innermost layer (inner surface) of heart muscle is called the
      (a) endocardium.
      (b) epicardium.
      (c) parietal pericardium.
      (d) myocardium.
      (e) intercardium.

4.   Which of the following statements regarding myocardial muscle and chambers is false?
      (a) Heart muscle is composed of specialized muscle cells found only in the heart.
      (b) Heart muscle is striated, like skeletal muscle.
      (c) Heart muscle has electrical properties similar to those of smooth muscle.
      (d) The ventricular chamber walls are more muscular than those of the atrial chambers.
      (e) The ventricular chambers are located at the base of the heart, the atrial chambers at the heart's apex.

5.   Beginning with the return of blood to the heart from the peripheral circulation (via the inferior/superior vena cava), place the following in order, describing the normal flow of blood.
      (1) inferior/superior vena cava
      (2) left atrium
      (3) right atrium
      (4) left ventricle
      (5) right ventricle
      (6) aorta
      (7) pulmonary arteries
      (8) bicuspid (mitral) valve
      (9) tricuspid valve
      (10) pulmonic valve
      (11) aortic valve
      (12) pulmonary veins
      (13) pulmonary capillaries

     (a) 1, 3, 8, 5, 11, 12, 13, 7, 2, 9, 4, 10, 6.
     (b) 1, 3, 8, 5, 10, 7, 13, 12, 2, 9, 4, 11, 6.
     (c) 1, 3, 8, 5, 10, 12, 13, 7, 2, 9, 4, 11, 6.
     (d) 1, 3, 9, 5, 10, 7, 13, 12, 2, 8, 4, 11, 6.
     (e) 1, 3, 9, 5, 10, 12, 13, 7, 2, 8, 4, 11, 6.

6. Heart muscle receives its blood supply exclusively via the coronary arteries. The coronary arteries originate in the
     (a) coronary sinus of the right atrium.
     (b) pulmonary artery just above the leaflets of the pulmonic valve.
     (c) pulmonary vein just beyond the leaflets of the pulmonic valve.
     (d) aorta just above the leaflets of the aortic valve.
     (e) aorta just above the leaflets of the mitral valve.

7. Fluid, gas, and nutrient exchange occurs between body tissue and
     (a) capillaries.
     (b) veins.
     (c) arteries.
     (d) Answers (a) and (b).
     (e) Answers (a), (b), and (c).

8. One-way valves aid the direction of blood flow within
     (a) capillaries.
     (b) veins.
     (c) arteries.
     (d) All of the above.
     (e) None of the above.

9. The ventricular relaxation phase of a cardiac cycle is called
     (a) systole.
     (b) diastole.
     (c) asystole.
     (d) Answer (a) or (b).
     (e) Answer (b) or (c).

10. Normally, the right and left atria contract together, _____, during the cardiac cycle called atrial _____.
     (a) just before the ventricles contract / diastole
     (b) just after the ventricles contract / diastole
     (c) just before the ventricles contract / systole
     (d) just after the ventricles contract / systole
     (e) at the exact same time as the ventricles / diastole

11. Normally, the right and left ventricles contract together, _____, during the cardiac cycle called ventricular _____.
     (a) just before the atria contract / diastole
     (b) just after the atria contract / diastole
     (c) just before the atria contract / systole
     (d) just after the atria contract / systole
     (e) at the exact same time as the ventricles / diastole

12. The bicuspid (mitral) and tricuspid valves are open during
     (a) systole.
     (b) diastole.
     (c) asystole.
     (d) Answer (a) or (b).
     (e) Answer (b) or (c).

13. The pulmonic and aortic valves are open during
     (a) systole.
     (b) diastole.
     (c) asystole.
     (d) Answer (a) or (b).
     (e) Answer (b) or (c).

14. The coronary arteries are able to fill with blood only during
     (a) systole.
     (b) diastole.
     (c) asystole.
     (d) Answer (a) or (b).
     (e) Answer (b) or (c).

15. The average adult stroke volume is
     (a) 120 milliliters.
     (b) 15 milliliters.
     (c) 7.45 milliliters.
     (d) 40 milliliters.
     (e) 70 milliliters.

16. Preload significantly influences stroke volume, and is defined as
     (a) the volume of blood delivered to the heart.
     (b) the pressure within the filled ventricles, at the end of diastole.
     (c) the resistance against which the ventricles contract.
     (d) Answers (a) and (b).
     (e) Answer (a) or (c).

17.   Afterload significantly influences stroke volume, and is defined as
      (a) the volume of blood delivered to the heart.
      (b) the pressure within the filled ventricles at the end of diastole.
      (c) the resistance against which the ventricles contract.
      (d) Answers (a) and (b).
      (e) Answer (a) or (c).

18.   Which of the following statements regarding nervous systems and their control of
      the heart is false?
      (a) The autonomic nervous system influences the rate, conductivity, and
           contractility of the heart.
      (b) The sympathetic nervous system influences the rate, conductivity, and
           contractility of the heart.
      (c) Parasympathetic control of the heart occurs via the phrenic nerve (the tenth
           cranial nerve).
      (d) The chemical neurotransmitter for the parasympathetic nervous system is
           acetylcholine.
      (e) The parasympathetic and sympathetic nervous systems directly oppose each
           other in their effects on the heart.

19.   The chemical neurotransmitter norepinephrine has both alpha and beta effects.
      Alpha effects on the heart include
      (a) increased heart rate.
      (b) increased myocardial contractility.
      (c) increased automaticity and conductivity.
      (d) All of the above.
      (e) None of the above.

20.   Beta effects on the heart include
      (a) increased heart rate.
      (b) increased myocardial contractility.
      (c) increased automaticity and conductivity.
      (d) All of the above.
      (e) None of the above.

21.   Chronotropy and chronotropic are terms that refer to the
      (a) heart's size.
      (b) rate of heart contraction.
      (c) strength of cardiac muscular contraction.
      (d) speed of nervous impulse conduction through the heart.
      (e) amount of blood ejected from the heart.

22. Dromotropy and dromotropic are terms that refer to the
    (a) heart's size.
    (b) rate of heart contraction.
    (c) strength of cardiac muscular contraction.
    (d) speed of nervous impulse conduction through the heart.
    (e) amount of blood ejected from the heart.

23. Inotropy and inotropic are terms that refer to the
    (a) heart's size.
    (b) rate of heart contraction.
    (c) strength of cardiac muscular contraction.
    (d) speed of nervous impulse conduction through the heart.
    (e) amount of blood ejected from the heart.

24. Electrolytes play an important role in cardiac electrical and mechanical functions. Sodium's influence primarily affects
    (a) the depolarization phase of myocardial cells.
    (b) the force of myocardial contraction.
    (c) the repolarization phase of myocardial cells.
    (d) Answers (a) and (b).
    (e) Answers (b) and (c).

25. Potassium's influence primarily affects
    (a) the depolarization phase of myocardial cells.
    (b) the force of myocardial contraction.
    (c) the repolarization phase of myocardial cells.
    (d) Answers (a) and (b).
    (e) Answers (b) and (c).

26. Calcium's influence primarily affects
    (a) the depolarization phase of myocardial cells.
    (b) the force of myocardial contraction.
    (c) the repolarization phase of myocardial cells.
    (d) Answers (a) and (b).
    (e) Answers (b) and (c).

27. A specialized property of the heart's pacemaker cells is their ability to depolarize and generate an electrical impulse without stimulation from another source. This ability is called
    (a) excitability.
    (b) reciprocity.
    (c) automaticity.
    (d) conjunctivity.
    (e) conductivity.

28.   A property of all myocardial cells is the ability to respond to an electrical stimulus. This ability is called
   (a) excitability.
   (b) reciprocity.
   (c) automaticity.
   (d) conjunctivity.
   (e) conductivity.

29.   The ability to propagate an impulse from cell to cell is another property of all myocardial cells. This ability is called
   (a) excitability.
   (b) reciprocity.
   (c) automaticity.
   (d) conjunctivity.
   (e) conductivity.

30.   Beginning with the most common pacemaker site, place the following structures in order, indicating the normal route of electrical impulse conduction through the heart.

   (1) AV junction/node   (4) Interatrial and internodal tracts
   (2) Bundle branches   (5) Purkinje fibers
   (3) Bundle of His   (6) SA node

   (a) 6, 5, 4, 1, 2, 3.
   (b) 6, 5, 1, 4, 3, 2.
   (c) 6, 4, 1, 3, 2, 5.
   (d) 4, 6, 1, 3, 2, 5.
   (e) 4, 6, 1, 5, 3, 2.

31.   The normal intrinsic rate of ventricular self-excitation (the normal spontaneous discharge rate of Purkinje fibers) is
   (a) 10–20 beats per minute.
   (b) 15–40 beats per minute.
   (c) 40–60 beats per minute.
   (d) 60–100 beats per minute.
   (e) 80–100 beats per minute.

32.   The atrioventricular node's normal intrinsic rate of self-excitation is
   (a) 10–20 beats per minute.
   (b) 15–40 beats per minute.
   (c) 40–60 beats per minute.
   (d) 60–100 beats per minute.
   (e) 80–100 beats per minute.

33. The sinoatrial node's normal intrinsic rate of self-excitation is
  (a) 10–20 beats per minute.
  (b) 15–40 beats per minute.
  (c) 40–60 beats per minute.
  (d) 60–100 beats per minute.
  (e) 80–100 beats per minute.

34. The light- and heavy-lined squares of standard ECG paper horizontally represent a specific interval of time. One small, light-lined, square of standard ECG paper horizontally represents
  (a) 0.02 seconds of time.
  (b) 0.04 seconds of time.
  (c) 0.10 seconds of time.
  (d) 0.15 seconds of time.
  (e) 0.20 seconds of time.

35. One large, heavy-lined, square of standard ECG paper horizontally represents
  (a) 0.02 seconds of time.
  (b) 0.04 seconds of time.
  (c) 0.10 seconds of time.
  (d) 0.15 seconds of time.
  (e) 0.20 seconds of time.

36. The light- and heavy-lined squares of standard ECG paper vertically represent the voltage amplitude of deflections inscribed by the ECG stylus. When the ECG monitor is properly calibrated, 1 millivolt of energy creates a vertical wave deflection that extends to the height of
  (a) one small square.
  (b) two small squares.
  (c) one large square.
  (d) two large squares.
  (e) three large squares.

37. Normally, the ECG's representation of atrial depolarization is
  (a) indicated by the P wave.
  (b) indicated by the T wave.
  (c) indicated by the QRS complex.
  (d) indicated by the P-R interval.
  (e) hidden within the QRS.

38. Normally, the ECG's representation of atrial repolarization is
  (a) indicated by the P wave.
  (b) indicated by the T wave.

    (c) indicated by the QRS complex.

    (d) indicated by the P-R interval.

    (e) hidden within the QRS.

39.   Normally, the ECG's representation of ventricular depolarization is

    (a) indicated by the P wave.

    (b) indicated by the T wave.

    (c) indicated by the QRS complex.

    (d) indicated by the P-R interval.

    (e) hidden within the QRS.

40.   Normally, the ECG's representation of ventricular repolarization is

    (a) indicated by the P wave.

    (b) indicated by the T wave.

    (c) indicated by the QRS complex.

    (d) indicated by the P-R interval.

    (e) hidden within the QRS.

41.   Normally, the ECG's representation of the time required for an atrial impulse to reach the ventricles is

    (a) indicated by the P wave.

    (b) indicated by the T wave.

    (c) indicated by the QRS complex.

    (d) indicated by the P-R interval.

    (e) hidden within the QRS.

42.   During repolarization of ventricular cells, the period of time when no amount of electrical stimulation can induce early depolarization of any ventricular cells is called the

    (a) absolute repolarized state.

    (b) absolute unpolarized state.

    (c) relative repolarizing period.

    (d) secondary repolarizing period.

    (e) absolute refractory period.

43.   The ECG's representation of the period of time when no amount of electrical stimulation can induce early ventricular depolarization is

    (a) the first half of the QRS complex.

    (b) the first half of the T wave.

    (c) the second half of the T wave.

    (d) the second half of the QRS complex.

    (e) any part of the T wave.

44. During ventricular repolarization, there comes a time when enough ventricular cells have repolarized that a sufficiently strong stimulus may produce premature ventricular depolarization. This period of time is called the
  (a) relative repolarized state.
  (b) relative unpolarized state.
  (c) absolute repolarizing period.
  (d) secondary repolarizing period.
  (e) relative refractory period.

45. The ECG's representation of the period of time when a sufficiently strong stimulus may induce premature ventricular depolarization is
  (a) the first half of the QRS complex.
  (b) the first half of the T wave.
  (c) the second half of the T wave.
  (d) the second half of the QRS complex.
  (e) any part of the T wave.

46. The normal P-R interval duration is _____ seconds.
  (a) 0.04–0.08
  (b) 0.04–0.12
  (c) 0.08–0.20
  (d) 0.12–0.20
  (e) 0.14–0.22

47. The normal QRS complex duration is _____ seconds.
  (a) 0.04–0.08
  (b) 0.04–0.12
  (c) 0.08–0.20
  (d) 0.12–0.20
  (e) 0.14–0.22

48. Your patient has an irregular pulse. On the ECG monitor you note regularly irregular groups of QRS complexes. There are more P waves present than QRSs. The P-R intervals become progressively longer, until a P wave is not followed by a QRS. The R-R intervals progressively shorten, until a long R-R interval occurs (containing the nonconducted P wave). Your patient has a
  (a) first-degree AV block.
  (b) second-degree AV block, type I ("Mobitz I" or "Wenckebach").
  (c) second-degree AV block, type II ("Mobitz II" or "infranodal").
  (d) third-degree AV block ("complete" heart block).
  (e) None of the above.

49. Your patient has a bradycardic pulse. On the ECG monitor you note regular R-R intervals, PRIs of 0.14 seconds, and QRSs of 0.14 seconds. There are twice as many P waves as QRSs and there is no ectopy present. Your patient has a
    (a) first-degree AV block.
    (b) second-degree AV block, type I (Mobitz I or Wenckebach).
    (c) second-degree AV block, type II (Mobitz II or infranodal).
    (d) third-degree AV block ("complete" heart block).
    (e) None of the above.

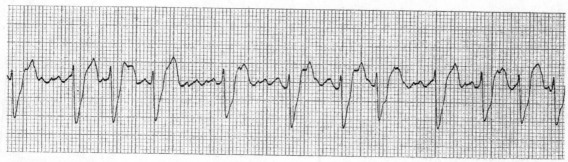

FIGURE 7–1

50. Which of the following descriptions best identifies the ECG strip in Figure 7–1?
    (a) Sinus dysrhythmia with aberrantly conducted PACs.
    (b) Sinus dysrhythmia with frequent unifocal PVCs.
    (c) Atrial flutter with variable ventricular response.
    (d) Uncontrolled Atrial fibrillation with a bundle branch block (BBB).
    (e) Controlled Atrial fibrillation with aberrantly conducted PACs.

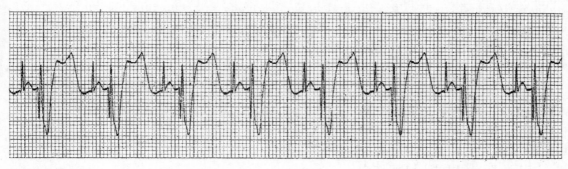

FIGURE 7–2

51. Which of the following descriptions best identifies the ECG strip in Figure 7–2?
    (a) Sinus rhythm with bigeminal, unifocal PVCs.
    (b) Atrial tachycardia with bigeminal, unifocal PVCs.
    (c) AV sequential (dual-chambered) pacemaker rhythm.
    (d) Pacemaker rhythm with interpolated sinus tachycardia.
    (e) Junctional rhythm with bigeminal, unifocal PVCs.

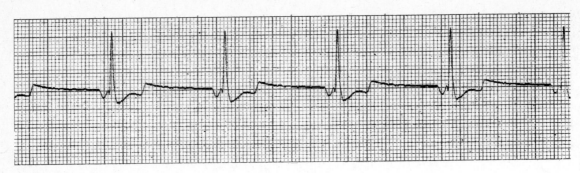

FIGURE 7–3

52. Which of the following descriptions best identifies the ECG strip in Figure 7–3?
(a) A junctional rhythm.
(b) Sinus bradycardia.
(c) Accelerated idioventricular rhythm.
(d) Wandering atrial pacemaker.
(e) Atrial bradycardia.

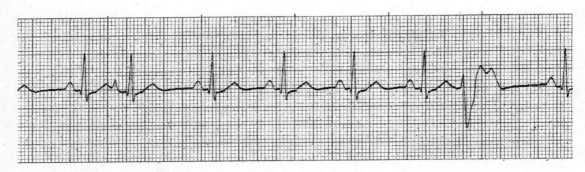

FIGURE 7–4

53. Which of the following descriptions best identifies the ECG strip in Figure 7–4?
(a) Sinus rhythm with one PVC.
(b) Sinus rhythm with multifocal PVCs.
(c) Sinus rhythm with one PAC.
(d) Sinus rhythm with multifocal PACs.
(e) Sinus rhythm with one PAC and one PVC.

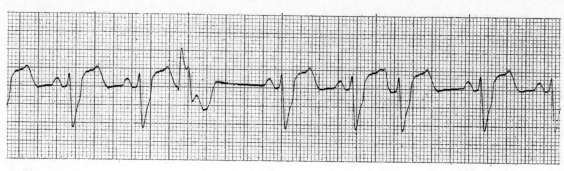

FIGURE 7–5

54. Which of the following descriptions best identifies the ECG strip in Figure 7–5?
  (a) Sinus dysrhythmia with one PVC.
  (b) Sinus rhythm with two, multifocal, PVCs.
  (c) Sinus dysrhythmia with one PAC.
  (d) Wenckebach, second-degree AV block with one PVC.
  (e) Sinus rhythm with one PAC and one PVC.

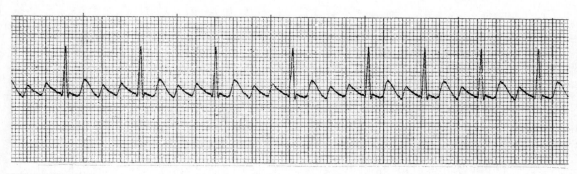

FIGURE 7–6

55. Which of the following descriptions best identifies the ECG strip in Figure 7–6?
  (a) Atrial flutter with variable (4:1 and 3:1) ventricular response.
  (b) Atrial flutter with variable (3:1 and 2:1) ventricular response.
  (c) Atrial fibrillation.
  (d) Atrial fibrillation with uncontrolled ventricular response.
  (e) Classic second-degree AV block with variable ventricular response.

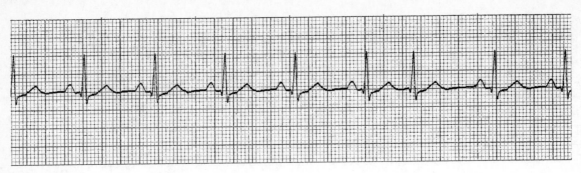

FIGURE 7–7

56.   Which of the following descriptions best identifies the ECG strip in Figure 7–7?
         (a) Normal sinus rhythm.
         (b) Sinus rhythm with one PAC.
         (c) Sinus rhythm with one PJC.
         (d) Sinus rhythm with one PVC.
         (e) Sinus dysrhythmia.

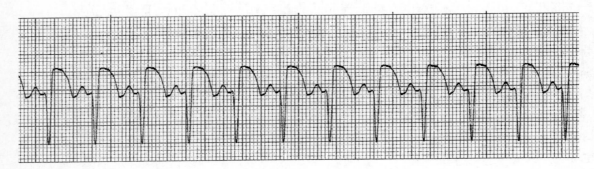

FIGURE 7–8

57.   Which of the following descriptions best identifies the ECG strip in Figure 7–8?
         (a) Sinus tachycardia, QRS within normal limits, elevated S-T segment.
         (b) Sinus tachycardia with a BBB, no ectopy.
         (c) Junctional tachycardia, QRS within normal limits, elevated S-T segment.
         (d) Ventricular tachycardia.
         (e) Runaway pacemaker rhythm.

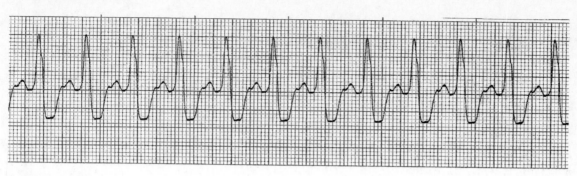

FIGURE 7–9

58.   Which of the following descriptions best identifies the ECG strip in Figure 7–9?
    (a) Sinus tachycardia with a BBB, inverted T wave, no ectopy.
    (b) Atrial tachycardia, QRS within normal limits, depressed S-T wave.
    (c) Junctional tachycardia with a BBB, inverted T wave.
    (d) Ventricular tachycardia.
    (e) Runaway pacemaker rhythm.

In each of the following scenarios, you are out of radio contact, without access to a phone, and have standing orders to follow 2005 AHA ACLS protocols when unable to communicate with your base physician.

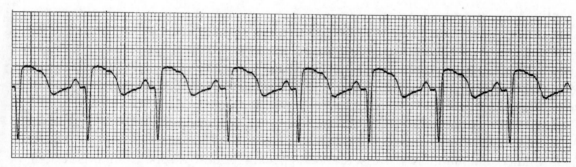

FIGURE 7–10

59.   Figure 7–10 is the ECG strip of a 45 y/o female, AAOx3, with cool, pale, and
    diaphoretic skin. Her chief complaint is substernal chest pain, radiating to her left
    shoulder, for the past two hours. P: 80; B/P: 110/70; R: 24. Which of the following
    descriptions best identifies the ECG strip in Figure 7–10?
    (a) Sinus rhythm, QRS WNL, S-T elevation indicative of AMI, no ectopy.
    (b) Underlying atrial tachycardia, with 2:1 second-degree AV block, type II.
    (c) Sinus rhythm with a BBB, no ectopy.
    (d) Atrial rhythm with a BBB, no ectopy.
    (e) Slow V-tach.

60. *(Preceding scenario continued:)* You administer oxygen and start an IV lifeline. Which of the following medications would you elect to administer?

(1) sublingual nitroglycerine

(2) 160–325 mg aspirin, chewed

(3) IV morphine sulfate (if pain unrelieved by NTG)

(4) standard, initial lidocaine bolus

(5) maintenance infusion of lidocaine

(6) prophylactic lidocaine bolus

    (a) 1 and 3.

    (b) 1, 2, 3, 4, and 5.

    (c) 1, 2, 3, 5, and 6.

    (d) 1, 2, 3, and 4.

    (e) 1, 2, and 3.

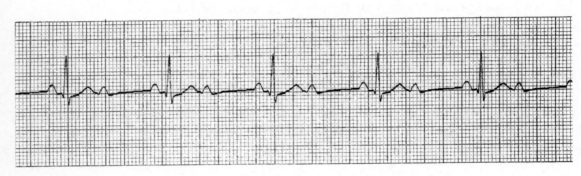

FIGURE 7–11

61. Figure 7–11 is the ECG strip of a 65 y/o male, AAOx3, with warm, pale, and dry skin. His only complaint is of feeling "weaker than usual today." P: 50; B/P: 100/70; R: 20. You administer oxygen and start an IV lifeline. Which of the following would you elect to perform?

    (a) Administration of atropine 0.5 mg, repeated every 3 to 5 minutes as needed, to a maximum dose of 3 mg.

    (b) Immediate initiation of transcutaneous pacing (prior to atropine administration).

    (c) Initiation of transcutaneous pacing only after the maximum amount of atropine has failed to increase the heart rate.

    (d) Answers (a) and (c).

    (e) None of the above; transportation only.

62. Figure 7–12's ECG strip is that of a 70 y/o female; awake, slightly disoriented, anxious and fearful; skin cool, pale, and diaphoretic. She c/o chest pain, nausea, and SOB. Her husband denies any recent trauma. Her breath sounds are bilaterally clear and equal. P: weak and thready at 170; B/P: 80/P; R: 40 and shallow. You administer oxygen and start an IV lifeline. Which of the following would you elect to perform?

    (a) Administration of diazepam, 10 mg IV, immediately followed by synchronized cardioversion.

    (b) Administration of a 1000 ml NS fluid challenge prior to medication or cardioversion.

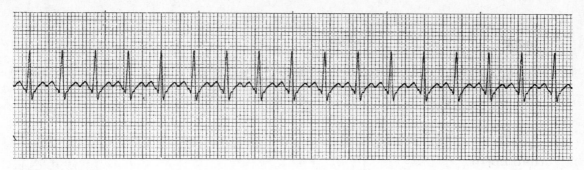

FIGURE 7–12

(c) Administration of an IV dopamine infusion, titrated to a B/P of 100 systolic.
(d) Trial performance of vagal maneuvers and/or carotid sinus massage prior to medication or cardioversion.
(e) None of the above; transportation only.

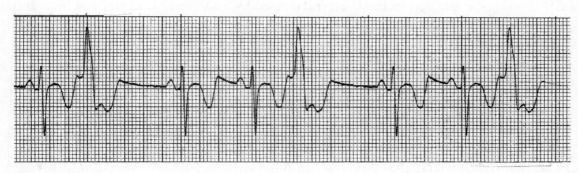

FIGURE 7–13

63. Figure 7–13's ECG strip is that of a 47 y/o male, AAOx3, with warm, pale, and dry skin. He reports weakness and nausea (without vomiting) for approximately two hours. He denies SOB and chest pain. P: 50 and irregular; B/P: 100/P; R: 24. You administer oxygen and start an IV lifeline. Which of the following medications would you elect to administer?
   (a) 160–325 mg aspirin, PO.
   (b) 0.5 mg IV atropine, repeated every 3 to 5 minutes, as needed, to a maximum of 3 mg.
   (c) 1.0–1.5 mg/kg IV lidocaine, followed every 5 to 10 minutes by a 0.5–0.75 mg/kg bolus, to a maximum dose of 3 mg/kg.
   (d) Answers (a) and (b).
   (e) Answers (a) and (c).

64. Pre-excitation syndromes are most often characterized by a P-R interval that is shorter than normal, indicating that the
    (a) ventricles are depolarizing at the same time as the atria.
    (b) ventricles are depolarizing earlier than the atria.
    (c) ventricles are generating a separate impulse, producing earlier depolarization.
    (d) sinus or atrial impulses are bypassing the AV node/junction and depolarizing the ventricles earlier than normal.
    (e) sinus or atrial impulses are traveling through the His-Purkinje system faster than normal.

65. In addition to a shorter-than-normal PRI, Wolff-Parkinson-White syndrome is characterized by a QRS of greater-than-normal duration, which is caused by the inclusion of
    (a) a bundle branch block.
    (b) a beta wave at the end of the QRS complex.
    (c) a beta wave at the beginning of the QRS complex.
    (d) a delta wave at the end of the QRS complex.
    (e) a delta wave at the beginning of the QRS complex.

66. Lown-Ganong-Levine syndrome is characterized by a shorter-than-normal PRI, and a QRS complex that is
    (a) always inverted.
    (b) of greater-than-normal duration.
    (c) within normal duration limits, unless the patient also has a bundle branch block.
    (d) of shorter-than-normal duration because of a beta-equivalent wave.
    (e) of shorter-than-normal duration because of a delta-equivalent wave.

67. The ECG of hypothermic patients often exhibits a slow, positive deflection at the end of the QRS complex (in lead II). This wave is called the
    (a) Jackson wave.
    (b) J wave.
    (c) Osborn wave.
    (d) Answer (a) or (b) may be used to refer to the same wave.
    (e) Answer (b) or (c) may be used to refer to the same wave.

68. Your 36 y/o male patient has a history of renal dialysis three times a week, and c/o extreme weakness. After questioning, he admits missing his dialysis appointment yesterday. You suspect he is suffering from _____, especially because his ECG exhibits _____.
    (a) hypokalemia / a U wave following each T wave
    (b) hyperkalemia / a U wave preceding each T wave
    (c) hypokalemia / a U wave preceding each T wave
    (d) hyperkalemia / very tall and peaked (pointy) T waves
    (e) hypocalcemia / very tall and peaked (pointy) T waves

69. Your 56 y/o female patient had an atraumatic syncopal episode. She is on Lasix®
for hypertension, and has not been taking her "vitamin K" because she "can't stand
the smell or taste of it!" She reports that her antidepressant medication makes her
mouth dry, so she drinks excessive amounts of water and urinates a lot. You
suspect she is suffering from _____, especially because her ECG exhibits
_____.

    (a) hypokalemia / a U wave following each T wave
    (b) hyperkalemia / a U wave preceding each T wave
    (c) hypokalemia / a U wave preceding each T wave
    (d) hyperkalemia / very tall and peaked (pointy) T waves
    (e) hypoglycemia / very tall and peaked (pointy) T waves

70. When a supraventricular impulse reaches the ventricles and depolarizes them
before they are completely repolarized, the PRI is often followed by a QRS of
greater-than-normal duration. Which of the following best describes this
phenomenon?

    (a) PVCs.
    (b) AV dissociation.
    (c) Aberrant conduction.
    (d) Fusion beats.
    (e) Ventricular escape.

71. Your patient's ECG shows pacemaker spikes, each followed by a QRS of greater-than-
normal duration. However, occasionally a pacemaker spike is followed by a wide QRS
that has a very different morphology than the others. This is most likely caused by

    (a) the pacemaker impulse coming earlier than before (a pacer PVC).
    (b) periodic AV dissociation between the pacemaker and the ventricles.
    (c) the aberrant conduction of an occasionally lower-than-normal pacemaker
        voltage output.
    (d) the patient's natural pacemaker spontaneously firing at the same time, or
        immediately after, a pacemaker impulse, both impulses meeting the ventricles
        at relatively the same time, and creating a fusion beat.
    (e) ventricular escape, because the pacemaker was late in firing.

72. Atrial fibrillation often produces impulses that reach the ventricles before they are
fully repolarized. Because the _____ bundle branch is the one that most often
requires more time to repolarize, the wide and bizarre QRS complex created by a
supraventricular impulse prematurely depolarizing the ventricles often appears to
have "rabbit ears" (i.e., two R waves, a pattern called R-S-R prime, and written
RSR') in the V_1 (MCL$_1$) lead.

    (a) left anterior
    (b) right anterior
    (c) left posterior
    (d) left main
    (e) right

73. Which of the following statements regarding differentiation between supraventricular tachycardia and ventricular tachycardia is true?

    (a) AV dissociation is always indicative of V-tach.

    (b) Even if the patient is awake, if there is no pulse, the rhythm is always V-tach.

    (c) If the patient is awake and reasonably well oriented, the rhythm is always SVT.

    (d) If a pulse is present, the rhythm is always SVT.

    (e) A QRS duration of greater than 0.14 sec is usually indicative of SVT.

74. Which of the following statements regarding the treatment of supraventricular and ventricular tachycardias is true?

    (a) Adenosine can be equally effective for both SVT and V-tach (with a pulse).

    (b) Carotid sinus massage can be equally effective for both SVT and V-tach (with or without a pulse).

    (c) Cardioversion can be equally effective for both SVT and V-tach (with a pulse).

    (d) Answers (a) and (b).

    (e) Answers (b) and (c).

75. Your patient denies c/o chest pain; denies c/o shoulder, arm, neck, or jaw discomfort; but c/o substernal chest "pressure." This patient's complaint

    (a) is less serious than a patient with "stable" angina.

    (b) may be as serious as that of a patient with "unstable" angina.

    (c) may be as serious as that of a patient with stable angina but is not as serious as unstable or preinfarction angina.

    (d) is one rarely related to cardiac ischemia or infarction.

    (e) strongly indicates esophageal reflux disease, not cardiac ischemia.

76. Onset of chest pain during physical exertion that is quickly relieved by rest

    (a) is typically referred to as stable angina.

    (b) may be stable or unstable angina.

    (c) is typically referred to as unstable, or preinfarction, angina.

    (d) is rarely related to cardiac ischemia.

    (e) strongly indicates esophageal reflux disease, not cardiac ischemia.

77. Onset of chest pain during rest

    (a) is typically considered to be stable angina.

    (b) may be stable or unstable angina.

    (c) is typically considered to be unstable, or preinfarction, angina.

    (d) is rarely related to cardiac ischemia.

    (e) strongly indicates esophageal reflux disease, not cardiac ischemia.

78. A myocardial infarction (MI) has occurred as soon as an area of myocardial tissue has

    (a) caused a complaint of chest pain.

    (b) become ischemic.

    (c) become necrotic.

    (d) Answer (a) or (b).

    (e) Answer (a) or (c).

79. In the adult, acute myocardial infarction (AMI) is most often caused by complications related to (or associated with)

    (a) cardiac dysrhythmias.

    (b) coronary vasospasm.

    (c) acute volume overload.

    (d) acute respiratory failure.

    (e) atherosclerotic heart disease (ASHD).

80. The most common complication or side effect of an AMI is

    (a) cardiac dysrhythmias.

    (b) coronary vasospasm.

    (c) acute volume overload.

    (d) acute respiratory failure.

    (e) atherosclerotic heart disease (ASHD).

81. The most common cause of death from AMI is

    (a) cardiac dysrhythmias.

    (b) coronary vasospasm.

    (c) acute volume overload.

    (d) acute respiratory failure.

    (e) atherosclerotic heart disease (ASHD).

82. The location and size of an infarct is dependent on the site of coronary vessel obstruction. The majority of infarcts, however, occur in the

    (a) right ventricle.

    (b) right atrium.

    (c) left atrium.

    (d) left ventricle.

    (e) aorta.

83. An MI involving only a partial thickness of the heart muscle wall is called

    (a) a subendocardial, or non-Q-wave, infarction.

    (b) a subendocardial, or pathological Q-wave, infarction.

    (c) a second-degree infarction.

    (d) a transmural, or non-Q-wave, infarction.

    (e) a transmural, or pathological Q-wave, infarction.

84.   An MI involving the full thickness of the heart muscle wall is called
      (a) a subendocardial, or non-Q-wave, infarction.
      (b) a subendocardial, or pathological Q-wave, infarction.
      (c) a third-degree infarction.
      (d) a transmural, or non-Q-wave, infarction.
      (e) a transmural, or pathological Q-wave, infarction.

85.   Which of the following statements regarding pathologies associated with left
      ventricular failure is false?
      (a) As the left ventricle fails, left atrial pressure rises, transmitting "back
           pressure" to pulmonary veins and capillaries.
      (b) When pulmonary capillary pressure becomes too high, capillaries burst,
           producing areas of hemothorax (which sound like areas of pulmonary
           edema).
      (c) When pulmonary capillary pressure becomes too high, blood plasma is forced
           into the alveoli, creating pulmonary edema.
      (d) Progressive alveolar fluid congestion will lead to death from hypoxia unless
           intervention occurs.
      (e) Because AMI is a common cause of left-ventricular failure, all patients with
           atraumatic pulmonary edema must be presumed to be having an AMI.

86.   Bronchoconstriction associated with pulmonary edema produces
      (a) rales.
      (b) rhonchi.
      (c) wheezing.
      (d) stridor.
      (e) snoring.

87.   Jugular vein distention
      (a) results almost immediately after acute left-ventricular failure occurs.
      (b) is associated only with incidents of isolated right-ventricular failure (unrelated
           to left-ventricular failure).
      (c) is associated only with pericardial tamponade and tension pneumothorax.
      (d) may be present if back pressure from left-ventricular failure progresses all the
           way into the right heart and the venous system.
      (e) occurs only when pulmonary embolism has occluded the pulmonary
           vasculature.

88.   The most common cause of right-ventricular failure is
      (a) left-ventricular failure.
      (b) chronic hypertension.
      (c) COPD.
      (d) pulmonary embolus.
      (e) infarct of right atrium or ventricle.

89. All of the following signs or symptoms may be caused by isolated right-ventricular failure, except
    (a) tender upper right abdominal quadrant from liver engorgement.
    (b) extremity edema.
    (c) fluid accumulation in serous cavities (ascites).
    (d) pulmonary edema.
    (e) JVD.

90. Which of the following statements regarding prehospital management of a patient with congestive heart failure (CHF) is false?
    (a) If the CHF patient is awake and can assist, employ a demand valve to provide positive-pressure ventilations.
    (b) Consider MS and/or NTG administration.
    (c) If the patient is hypotensive, a trial bolus of 300 ml of NS or LR must be administered prior to dopamine administration, to rule out hypovolemia caused by third-spacing of fluids.
    (d) Consider Lasix® and/or Intropin® administration.
    (e) Aggressively treat underlying dysrhythmias (especially if they appear to be contributing to the CHF), according to protocol.

91. Cardiogenic shock can be most completely defined as
    (a) shock that persists after correction of dysrhythmias.
    (b) shock that persists after correction of hypovolemia.
    (c) the most severe form of pump failure.
    (d) Answer (a) or (b).
    (e) Answers (a), (b) and (c).

92. Which of the following statements regarding cardiogenic shock is false?
    (a) Cardiogenic shock occurs when left-ventricular function is so compromised that the heart cannot meet the metabolic needs of the body.
    (b) Cardiogenic shock is often due to an infarction involving 40% or more of the left ventricle.
    (c) Adolescent patients may suffer cardiogenic shock secondary to tension pneumothorax or cardiac tamponade.
    (d) "Recreational" drug use may cause acute cardiogenic shock.
    (e) Unless the patient has a history of atypical organ disease or chronic alcoholism, a diagnosis of cardiogenic shock requires that the patient be over 30 years of age.

93. Management of a patient in cardiogenic shock includes all of the following, except
    (a) nonemergent transport to avoid increasing anxiety and infarct size.
    (b) consideration of vasopressor administration.
    (c) supine positioning if hypotensive.
    (d) consideration of MS and/or NTG administration.
    (e) aggressive treatment of bradycardias or tachycardias.

94.   The definition of sudden death requires that death occur
   (a) without any warning signs or symptoms.
   (b) secondary to a cause that could have been corrected, had appropriate treatment been administered prior to death.
   (c) within one hour of the onset of signs and symptoms.
   (d) within the second hour after the onset of signs and symptoms.
   (e) anytime within the first 24 hours following the onset of signs and symptoms.

95.   Which of the following statements regarding the management of cardiopulmonary arrest is false?
   (a) Ensuring that effective BLS measures are continued is more important than accomplishing endotracheal intubation, IV access, or ACLS medication administration.
   (b) If the patient is initially found in V-fib, CPR performance always precedes endotracheal intubation or IV access.
   (c) IV access should be obtained as close to the central circulation as possible (EJ or AC access).
   (d) Because medication administration is of greater priority than endotracheal intubation, advanced airway placement should be delayed until after IV access is obtained and "first line" ACLS medications have been administered.
   (e) Following defibrillation, do not delay resumption of ventilation and chest compressions to check the ECG rhythm or to check for a pulse.

96.   Which of the following etiologies may cause pulseless electrical activity (PEA)?

   (1) Hypovolemia   (7) "Tablets" (drug overdose)
   (2) Hypoxia   (8) Tamponade, cardiac
   (3) Hydrogen ion (acidosis)   (9) Tension pneumothorax
   (4) Hyper-/Hypokalemia   (10) Thrombosis, coronary
   (5) Hypoglycemia   (11) Thrombosis, pulmonary
   (6) Hypothermia   (12) Trauma

   (a) 1, 2, 3, 4, 5, 6, 7, 8, 9, 10, 11, and 12.
   (b) 1, 3, 4, 6, 7, 8, 9, and 10.
   (c) 1, 3, 5, 6, 7, 8, and 12.
   (d) 1, 2, 3, 5, 6, 7, and 12.
   (e) 1, 3, 6, 7, and 12.

97.   Your patient is a 39 y/o male who undergoes kidney dialysis three times each week. He is apneic and pulseless. CPR is being performed. His ECG shows a sinus tachycardia with tall, peaked T waves. The underlying rhythm's QRS complexes are WNL. There are occasional PVCs (≈ 6/minute). You confirm that the patient has no pulse. Based upon this information, which of the following is the most likely cause of this patient's PEA?
   (a) Hypervolemia.
   (b) Hyperkalemia.

(c) Hypoxia.

(d) Hypoglycemia.

(e) Hypokalemia.

98. Based solely upon the information provided in the previous question, it would be considered beneficial to administer which of the following medications?

   (a) Lidocaine.

   (b) Atropine.

   (c) Isuprel.

   (d) $D_{50}W$.

   (e) Sodium bicarbonate.

99. The most common site for an abdominal aneurysm is

   (a) the ascending aorta.

   (b) the distal curve of the aortic arch.

   (c) the descending aorta, at or immediately above the point where it passes through the diaphragm into the abdomen.

   (d) the abdominal aorta, below the renal arteries and just above the common iliac bifurcation.

   (e) the abdominal aorta, below the iliac arteries and just above the renal artery bifurcation.

100. The most common site for a thoracic aneurysm is

   (a) the ascending aorta.

   (b) the distal curve of the aortic arch.

   (c) the descending aorta, at or immediately above the point where it passes through the diaphragm into the abdomen.

   (d) the abdominal aorta, below the renal arteries and just above the common iliac bifurcation.

   (e) the abdominal aorta, below the iliac arteries and just above the renal artery bifurcation.

101. In addition to complaints of abdominal pain, common signs and symptoms of dissecting abdominal aortic aneurysm include all of the following, except

   (a) hypotension.

   (b) bilateral cramping of the lower extremities.

   (c) the urge to defecate.

   (d) back or flank pain.

   (e) decreased or unequal femoral pulses.

102. All of the following may predispose an individual to developing an aneurysm, except

   (a) syphilis.

   (b) neuroleptic malignant syndrome.

   (c) Marfan's syndrome.

   (d) atherosclerosis.

   (e) chronic hypertension.

103. Which of the following statements regarding dissecting aortic aneurysms is false?

    (a) Dissecting aortic aneurysms may rupture spontaneously, without a precipitating strain or exertion.

    (b) Dissecting aortic aneurysms are most commonly caused by atherosclerosis and hypertension.

    (c) Once started, a dissecting thoracic aneurysm may extend to involve the aortic valve, the carotid and subclavian arteries, the abdominal aorta and its branches.

    (d) The majority of dissecting aortic aneurysms occur in the abdominal aorta.

    (e) Pregnant patients may develop dissecting aortic aneurysm.

104. Hypertensive emergency (also called hypertensive crisis) is best characterized by

    (a) any systolic blood pressure greater than 130 mmHg.

    (b) any diastolic blood pressure greater than 130 mmHg.

    (c) any diastolic blood pressure greater than 100 mmHg.

    (d) any systolic blood pressure less than 100 mmHg.

    (e) any diastolic blood pressure greater than 90 mmHg.

105. Common signs and symptoms of hypertensive emergency include any of the following, except

    (a) restlessness or confusion.

    (b) blurred vision or headache.

    (c) vomiting without c/o nausea.

    (d) seizure or stroke.

    (e) an attitude of "elation" inappropriate to the situation.

106. Which of the following statements regarding defibrillation is false?

    (a) Successful conversion after defibrillation is less likely in the presence of hypoxia, acidosis, hypothermia, electrolyte imbalance, or drug toxicity.

    (b) Repeated defibrillation causes increased transthoracic electrical current resistance, requiring a greater energy-administration setting each time more than one shock is required.

    (c) Creams or pastes used for defibrillation must be those made specifically for defibrillation (ECG monitoring creams or pastes must not be used).

    (d) If any type of defibrillation conduction medium leaks across the surface of the patient's chest, it may provide an external pathway for electricity to travel, and also may present a fire (and/or burn) hazard.

    (e) Placement of defibrillator paddles on or within 1 inch of the generator ("battery") of an implanted automatic defibrillator or pacemaker can damage or disable the generator.

107. Which of the following statements regarding defibrillator paddle placement is false?

    (a) One paddle is positioned just below the right clavicle, lateral to the upper sternum.

    (b) One paddle is positioned lateral to the left nipple, in the anterior axillary line.

(c) Although not a preferred positioning, it is acceptable for one paddle to be placed lateral to the left breast (anterior axillary line) and the other one placed on the right or left upper back.

(d) Even if a defibrillator paddle is marked "sternum," it should never be placed on the sternum.

(e) If the "positive" paddle is placed where the "negative" paddle should be placed, and vice versa, the direction of defibrillation energy delivery will be reversed, causing ineffective defibrillation, and increasing myocardial damage.

108. Which of the following statements regarding the types of defibrillators available to prehospital care providers is false?

(a) Monophasic waveform defibrillators deliver current in only one direction (polarity) of current flow.

(b) Biphasic waveform defibrillators deliver a current that first travels in a positive direction, then reverses to flow in a negative direction.

(c) Biphasic waveform shocks require higher energy level settings than do monophasic waveform shocks, but are more successful at terminating ventricular fibrillation.

(d) The first-introduced defibrillators were monophasic waveform defibrillators.

(e) Few monophasic waveform defibrillators are currently being manufactured, but many are still in use.

109. Your 55 y/o patient is in ventricular fibrillation. The first time you defibrillate her with a **monophasic defibrillator** her rhythm converts to a supraventricular tachycardia with faint carotid pulses. After 4 minutes, her rhythm suddenly deteriorates to ventricular fibrillation again and she is pulseless. The energy setting for your next defibrillation should be

(a) 360 joules.

(b) 300 joules.

(c) 200 joules.

(d) 100 joules.

(e) 1 joule per kilogram.

110. Your 55 y/o patient is in ventricular fibrillation. The first time you defibrillate her with a **biphasic defibrillator** her rhythm converts to a supraventricular tachycardia with faint carotid pulses. After 4 minutes, her rhythm suddenly deteriorates to ventricular fibrillation again and she is pulseless. The energy setting for your next defibrillation should be

(a) 360 joules.

(b) 300 joules.

(c) 200 joules.

(d) 100 joules.

(e) 1 joule per kilogram.

Questions number 111 and 112 are far more time consuming than "standard" written examination questions. If you are timing your test-taking performance, note the time NOW and STOP timing yourself. Resume your timing after you've answered question 112.

111. According to the 911 caller, a 55 y/o male was playing cards at his local bridge center when he suddenly began complaining of an awful "squeezing" sensation in his chest, SOB, and nausea. When you arrive on scene, bystanders report that "He was sitting in that chair and just went out—not five seconds ago! We carefully put him on the floor, and we've been fanning him to give him air!" You immediately discover that the patient is apneic and pulseless. (No one initiated CPR prior to your arrival.) There is no visually apparent or reported trauma associated with his emergency. **You are equipped with an ECG monitor that has a monophasic defibrillator.** From the following list of ACTION OPTIONS, select those actions you would perform, and then list them in order of their preferred performance (first-to-last), ending with the DELIVERY OF A THIRD SHOCK. You may use any option more than once, and you may not need to use all the options. **Note: Every time you "Analyze the patient's cardiac rhythm" he is in ventricular fibrillation.** Given that fact, once you have listed all the action options you would perform, in the order that you would perform them, select the ANSWER option that most closely represents your planned course of treatment.

    (1) Analyze the patient's cardiac rhythm.
    (2) Defibrillate the patient with 360 joules.
    (3) Defibrillate the patient with 300 joules.
    (4) Defibrillate the patient with 200 joules.
    (5) Perform 30 chest compressions and 2 ventilations.
    (6) Perform 3 cycles of 30 chest compressions and 2 ventilations.
    (7) Perform 5 cycles of 30 chest compressions and 2 ventilations.
    (8) Obtain IV or IO access.

      (a) 1, 4, 1, 3, 1, 2.
      (b) 7, 1, 2, 7, 8, 1, 2, 7, 1, 2.
      (c) 1, 4, 6, 1, 3, 6, 1, 2.
      (d) 7, 1, 4, 7, 8, 1, 3, 7, 1, 2.
      (e) 1, 4, 5, 1, 3, 5, 8, 1, 2.

112. You are faced with exactly the same scenario and action options as described in question 111. However, this time you are equipped with an ECG monitor that has a **biphasic defibrillator.** As in question 111, every time you **"Analyze the patient's cardiac rhythm" he is in ventricular fibrillation.** Given that fact, once you have selected all the action options you would perform and listed them in order of their preferred performance (first-to-last) through to the DELIVERY OF A THIRD SHOCK, select the ANSWER option that most closely represents your planned course of treatment.

      (a) 1, 4, 1, 3, 1, 2.
      (b) 7, 1, 2, 7, 8, 1, 2, 7, 1, 2.
      (c) 1, 4, 6, 1, 3, 6, 1, 2.
      (d) 7, 1, 4, 7, 8, 1, 3, 7, 1, 2.
      (e) 1, 4, 5, 1, 3, 5, 8, 1, 2.

Resume timing your test-taking performance NOW.

113. Which of the following statements regarding defibrillation is false?

(a) Due to the "false" asystole phenomenon, if ACLS medications fail to convert asystole to a rhythm, the patient should be defibrillated at least once, with either a mono- or biphasic defibrillator, set at 360 joules.

(b) Never interrupt CPR to defibrillate asystole.

(c) The best way to minimize risk of causing fire or burns during defibrillation is to use self-adhesive defibrillation pads.

(d) Fires have been started when ventilator tubing was disconnected from the ET tube and left adjacent to the patient's head during defibrillation.

(e) Oxygen should never be allowed to flow across the patient's chest during defibrillation.

114. The synchronizing function of a defibrillator is specifically programmed to recognize various portions of the patient's ECG rhythm and deliver an electrical countershock discharge during

(a) any upward P-wave deflection.

(b) any downward Q-wave deflection.

(c) any upward R-wave deflection.

(d) the S-T segment of the patient's ECG.

(e) any upward T-wave deflection.

115. Emergency synchronized cardioversion is contraindicated for the treatment of all of the following, except

(a) junctional tachycardia with a pulse.

(b) multifocal atrial tachycardia with a pulse.

(c) monomorphic V-tach without a pulse.

(d) monomorphic V-tach with a pulse.

(e) an unconscious patient with a pulseless SVT.

116. Which of the following S/Sx indicate that a conscious tachycardic patient is "unstable"?

(1) Confused, but alert.
(2) Complains of chest pain.
(3) Hypotension.
(4) Hypertension.
(5) Cold, pale, sweaty skin.
(6) Cold, pale, dry skin.
(7) Heart rate less than 150/min.
(8) Heart rate greater than 150/min.

(a) 1, 2, 3, 4, 5, 6, 7, or 8.

(b) 1, 2, 3, 5, or 8.

(c) 3, 6, 7, or 8.

(d) 3, 5, 6, or 8.

(e) 1, 2, 3, 4, 7, or 8.

117. Which of the following statements regarding the treatment of a conscious tachycardic patient who is determined to be "stable" is false?

(a) Synchronized cardioversion should immediately be performed, in order to prevent a "stable" patient from becoming "unstable."

(b) IV access should be established.

(c) If a regular, narrow QRS tachycardic rhythm is not converted by vagal maneuvers, adenosine should be administered.

(d) If a regular, wide QRS tachycardic rhythm is observed, amiodarone should be administered.

(e) If a regular, wide QRS tachycardic rhythm is refractory to amiodarone, synchronized cardioversion should be performed.

118. When a patient has a pulse, emergency synchronized cardioversion is recommended for which of the following tachycardias?

(1) Unstable SVT due to a reentry mechanism (LGL and WPW).

(2) Unstable atrial fibrillation with a regular ventricular response of less than 150 bpm.

(3) Unstable atrial fibrillation with a regular ventricular response of greater than 150 bpm.

(4) Unstable atrial flutter with a regular ventricular response of less than 150 bpm.

(5) Unstable atrial flutter with a regular ventricular response of greater than 150 bpm.

(6) Unstable junctional tachycardia with a regular ventricular response of less than 150 bpm.

(7) Unstable junctional tachycardia with a regular ventricular response of greater than 150 bpm.

(8) Unstable monomorphic ventricular tachycardia.

(9) Unstable polymorphic (irregular) ventricular tachycardia.

(10) A "stable" patient with any tachycardia generating a regular or irregular pulse of greater than 150 bpm.

    (a) 1, 2, 3, 4, 5, 6, 7, 8, 9, and 10.

    (b) 1, 2, 3, 4, 5, 6, 7, 8, and 10.

    (c) 2, 4, 6, 8, 9, and 10.

    (d) 1, 2, 3, 4, 5, and 8.

    (e) 3, 5, 7, 8, and 9.

119. Which of the following statements regarding sedation administration to a conscious but unstable patient prior to performing synchronized cardioversion is false?

(a) If IV access cannot be rapidly established, perform cardioversion without sedation.

(b) 5–15 mg of diazepam can be administered IV to sedate a conscious patient prior to cardioversion.

    (c) 2–5 mg of midazolam (Versed®) can be administered IV to sedate a conscious patient prior to cardioversion.

    (d) Do not delay cardioversion to administer sedation via IM injection.

    (e) Cardioverting a conscious patient (even if unstable) is extremely painful. If medical advisor contact is required to obtain an order to sedate a conscious patient, to prevent accusations of unnecessary cruelty, cardioversion should be delayed until after the order is obtained and the patient is sedated.

It is highly unlikely that any tachycardia would remain unchanged after delivery of 2 countershocks. However, for the next three questions, assume that that is the situation.

120. The energy setting progression (from initial to final countershock) recommended for synchronized cardioversion of an adult patient with PSVT or atrial flutter is

    (a) 100 J, 200 J, 300 J, 360 J.

    (b) 25 J, 50 J, 75 J, 100 J.

    (c) 50 J, 100 J, 200 J, 300 J.

    (d) 50 J (all subsequent shocks should remain at the same energy setting).

    (e) 100 J (all subsequent shocks should remain at the same energy setting).

121. The energy setting progression (from initial to final countershock) recommended for synchronized cardioversion of an adult patient with atrial fibrillation is

    (a) 100 J, 200 J, 300 J, 360 J.

    (b) 25 J, 50 J, 75 J, 100 J.

    (c) 50 J, 100 J, 200 J, 300 J.

    (d) 50 J (all subsequent shocks should remain at the same energy setting).

    (e) 100 J (all subsequent shocks should remain at the same energy setting).

122. The energy setting progression (from initial to final countershock) recommended for synchronized cardioversion of an adult patient with ventricular tachycardia is

    (a) 100 J, 200 J, 300 J, 360 J.

    (b) 25 J, 50 J, 75 J, 100 J.

    (c) 50 J, 100 J, 200 J, 300 J.

    (d) 100 J (all subsequent shocks should remain at the same energy setting).

    (e) 200 J (all subsequent shocks should remain at the same energy setting).

123. Valsalva maneuvers include all of the following, except directing the patient to

    (a) "bear down" as if to have a bowel movement.

    (b) "hiss like a cat," forcing air past a partially closed glottis.

    (c) stand up and forcefully sit back down ("plop!"), two to three times.

    (d) take a deep breath, and then forcefully expel it between tightly pursed lips.

    (e) take a deep breath, and then forcefully expel it between rounded, closed lips (as if "blowing out a candle").

124. Carotid sinus massage stimulates

    (a) carotid body baroreceptors, increasing vagal tone and decreasing the heart rate.

    (b) carotid body baroreceptors, decreasing vagal tone and decreasing the heart rate.

    (c) carotid artery glands, causing rate-deceasing hormones to be released into the blood stream and travel to the heart muscle.

    (d) the blood/brain barrier, decreasing vagal tone and decreasing the heart rate.

    (e) the sympathetic nervous system, resulting in a slowing of the heart rate.

125. Which of the following statements regarding carotid-artery massage is false?

    (a) If carotid bruits are present on only one side, do not perform carotid-artery massage on either side (even on the "clear-sounding" side).

    (b) If the carotid pulses are unequal, massage the side with the stronger pulse.

    (c) As soon as the heart rate begins to slow, even if after only a few seconds of massage, immediately discontinue carotid-artery massage (do not continue massage until the "target heart rate" is reached).

    (d) Carotid-artery massage is contraindicated if the patient has a history of CVA.

    (e) Carotid-artery massage should continue for no longer than 15–20 seconds per session.

126. Performance of transcutaneous cardiac pacing (TCP) is recommended for symptomatic patients with an ECG showing

    (1) sinus bradycardia, refractory to atropine.

    (2) bradycardic Type II, second-degree AV block.

    (3) complete (third-degree) heart block.

    (4) atrial fibrillation with slow ventricular response.

    (5) asystole (when onset is "witnessed" and pacing can quickly be implemented).

    (6) asystole (only after all first- and second-line ACLS resuscitation efforts have failed).

    (7) any bradycardia, when the patient is severely hypothermic, and drug administration is contraindicated.

        (a) 1, 2, 3, 4, 6, and 7.

        (b) 1, 2, 3, and 4.

        (c) 1, 2, 3, and 6.

        (d) 1, 2, 3, 6, and 7.

        (e) 1, 2, 3, 4, and 5.

127. To perform standard TCP, set the desired heart rate to approximately

(a) 40 bpm (minimum setting); once capture is achieved, increase the rate to 60 bpm and keep it there.

(b) 60 bpm; once capture is achieved, increase the rate as needed, up to 80 bpm.

(c) 80 bpm, and remain at that setting before and after capture is achieved.

(d) 110 bpm; once capture is achieved, decrease the rate to 80 bpm.

(e) 170 bpm (maximum setting); once capture is achieved, decrease the rate to 100 bpm (the rate required to sustain the body during emergency).

128. Your ECG monitor/defibrillator/pacemaker unit has a mA (milliamperes) output range of 0 to 200 mA. Before turning on the pacemaker, set the mA output at

(a) the minimum setting ("0").

(b) a low setting ("50").

(c) a medium setting ("100").

(d) a high setting ("150").

(e) the maximum setting ("200").

129. After turning on the pacer, the mA output should be

(a) maintained at the setting established prior to turning it on.

(b) increased by increments of 50 mA at a time until the maximum setting is reached.

(c) decreased by increments of 50 mA at a time until ventricular capture is accomplished.

(d) slowly increased until ventricular capture is accomplished.

(e) slowly decreased until ventricular capture is accomplished.

130. Which of the following statements about TCP is false?

(a) When a significantly symptomatic patient has a block at or below the His-Purkinje level, do not delay TCP to administer atropine.

(b) Whenever possible, a conscious patient should be sedated prior to, or immediately after, initiation of TCP.

(c) TCP electrode placement is entirely dependent upon the type of pacing electrodes being used; read and follow the packaging instructions.

(d) If two paramedics are available, one should administer atropine to a symptomatic bradycardia patient while the other prepares to initiate TCP (in the event that atropine fails to correct the bradycardia).

(e) "Overdrive" TCP is highly recommended by the 2005 AHA guidelines (a Class Ia treatment) for treatment of torsades de pointes refractory to drug therapy.

Name-That-Strip Section

The final portion of Test Section 7 consists of 50 "static" 6-second, Lead II, ECG strips (Figures 7–15 through 7–64). Because a multiple-choice answer format does not easily lend itself to static ECG strip identification and description, the "Name That Strip" section consists of write-in answers.

When interpreting these 50 ECG strips, you do not have a patient present to check pulses, clinically evaluate, or treat. Your task is simply to identify, interpret, and describe each ECG strip. Be as complete—yet concise—as you possibly can, when writing your answer.

Carefully observe each strip, identifying:

1. Primary pacemaker (sinus, atrial, junctional, ventricular, implanted)
2. Rate (atrial *and* ventricular rates, if they are different)
3. Regularity or irregularity of the prevalent rhythm
4. P-wave description (when abnormal)
5. P-R-interval description (when abnormal)
6. QRS-complexes description: WNL (within normal limits) or wider-than-normal (possible bundle branch block)
7. S-T-segment description (when abnormal)
8. T-wave description (when abnormal)
9. Presence of ectopic beats (describe ectopic source and identify unifocal versus multifocal ectopy)
10. Absence of ectopic beats ("pertinent negatives!")

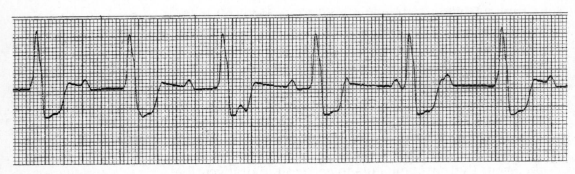

FIGURE 7–14

For Example: ECG Figure 7–14 can most completely and concisely be described as: "A complete (or 'third-degree') AV block, with an accelerated ventricular escape rhythm, at a regular rate of 60/min. An underlying, regular, atrial rate of 110 bpm. No ectopy."

Author's Suggestion: *The first time you perform this exercise, consider describing only the first three to six ECG figures. Then mark your time and check your descriptions against the "answer descriptions" provided in Test Section 7's Answer Key. If your description is not missing any pertinent information, resume your timing and proceed to the remaining ECGs. If you missed some pertinent descriptions, study these three to six ECG answer descriptions before resuming your timed self-test.*

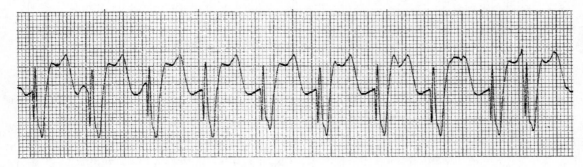

FIGURE 7–15

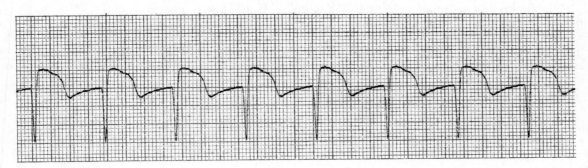

FIGURE 7–16

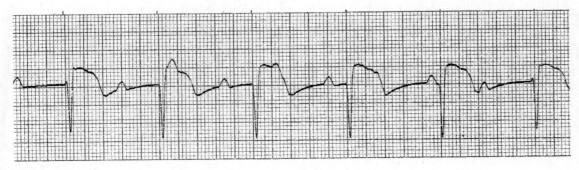

FIGURE 7–17

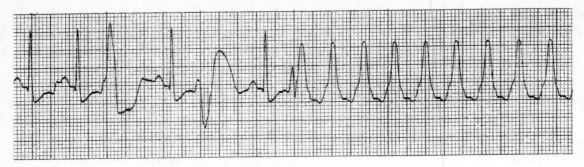

FIGURE 7–18

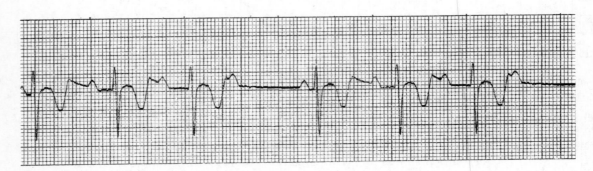

FIGURE 7–19

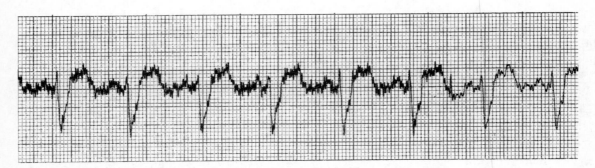

FIGURE 7–20

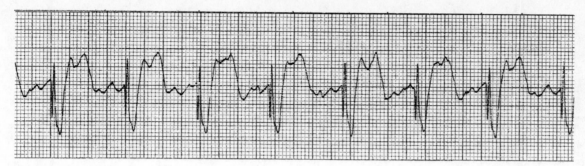

FIGURE 7–21

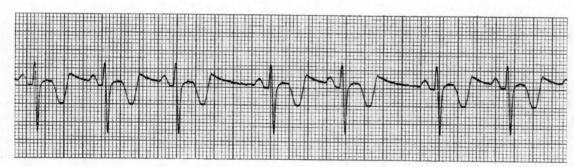

FIGURE 7–22

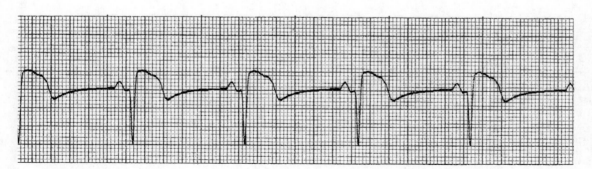

FIGURE 7–23

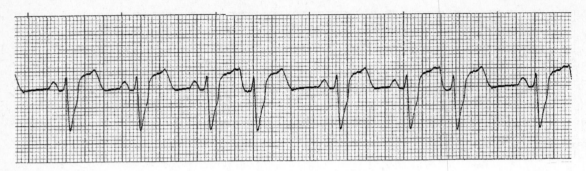

FIGURE 7–24

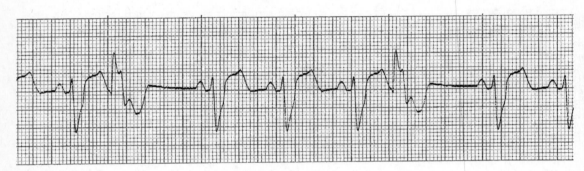

FIGURE 7–25

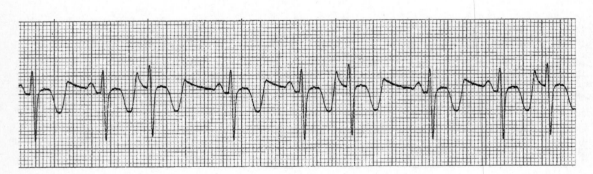

FIGURE 7–26

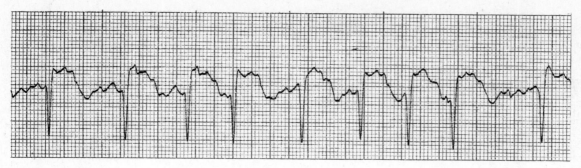

FIGURE 7–27

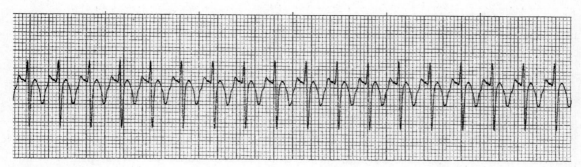

FIGURE 7–28

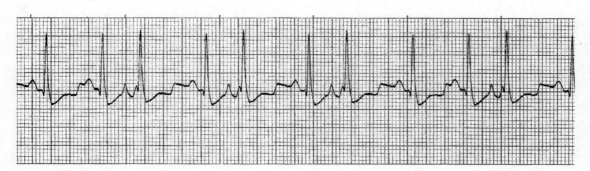

FIGURE 7–29

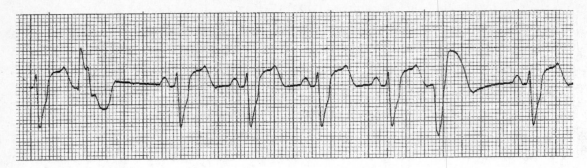

FIGURE 7–30

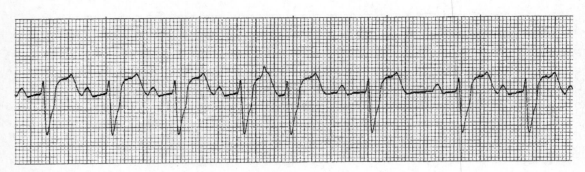

FIGURE 7–31

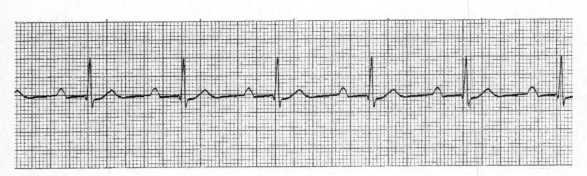

FIGURE 7–32

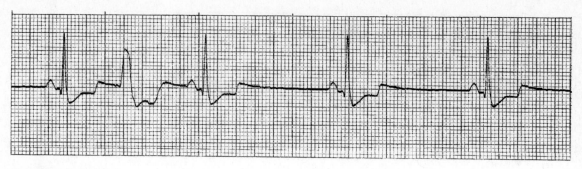

FIGURE 7–33

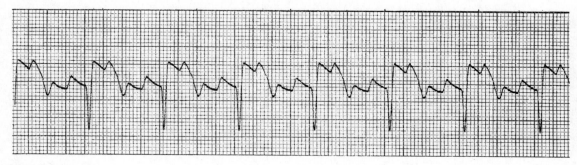

FIGURE 7–34

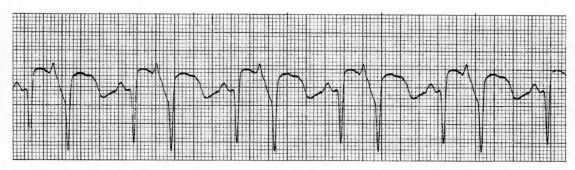

FIGURE 7–35

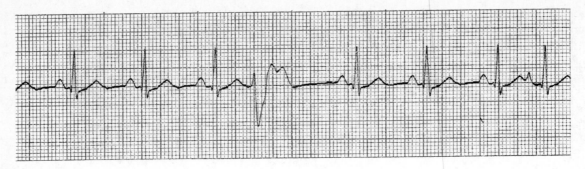

FIGURE 7–36

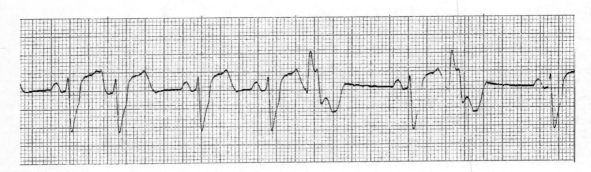

FIGURE 7–37

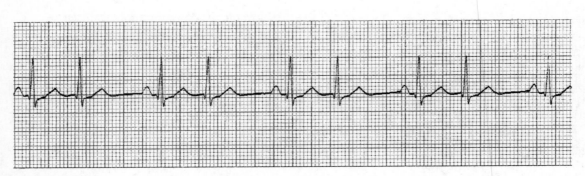

FIGURE 7–38

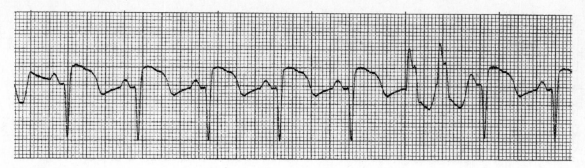

FIGURE 7–39

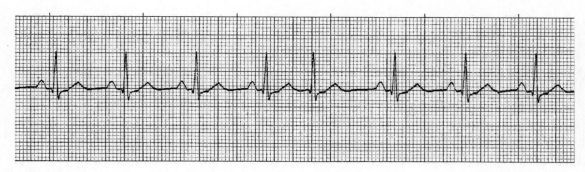

FIGURE 7–40

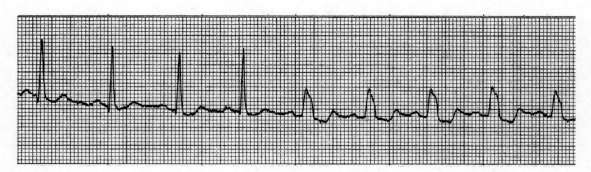

FIGURE 7–41

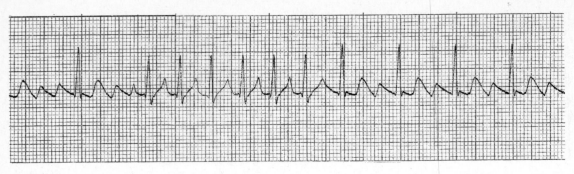

FIGURE 7–42

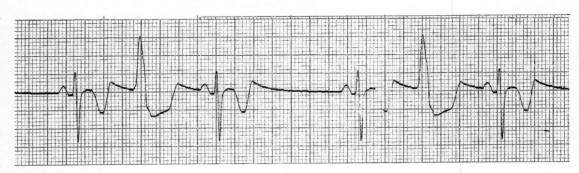

FIGURE 7–43

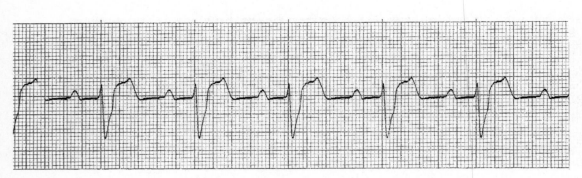

FIGURE 7–44

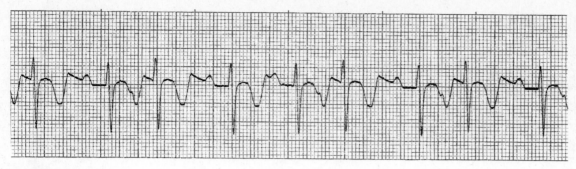

FIGURE 7–45

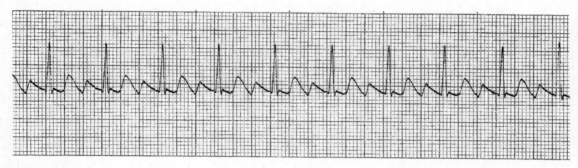

FIGURE 7–46

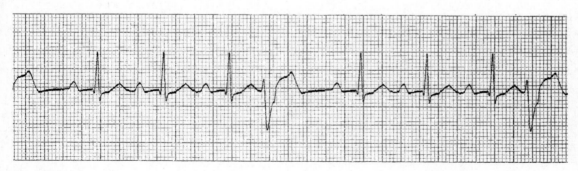

FIGURE 7–47

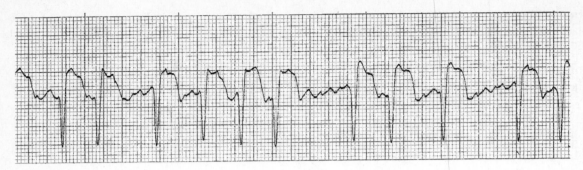

FIGURE 7–48

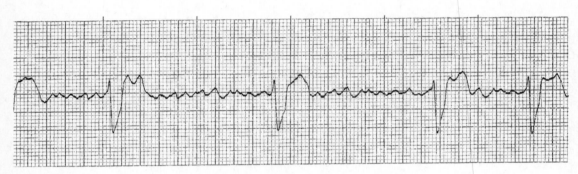

FIGURE 7–49

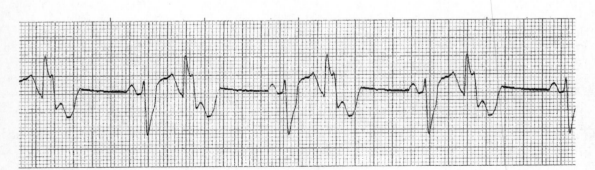

FIGURE 7–50

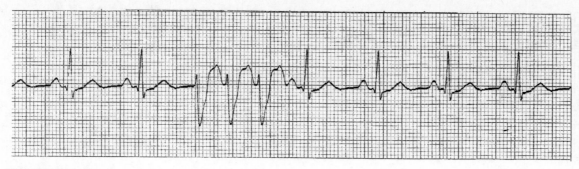

FIGURE 7–51

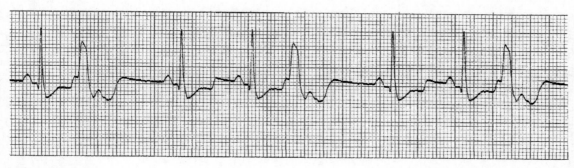

FIGURE 7–52

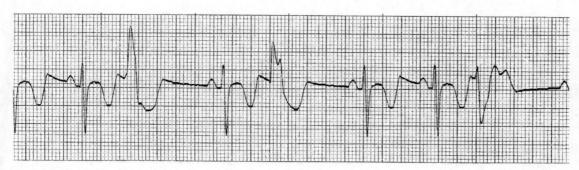

FIGURE 7–53

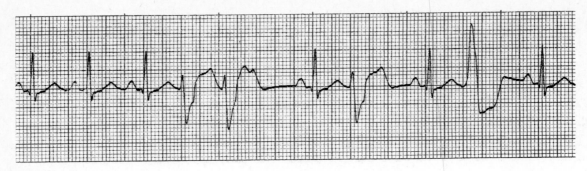

FIGURE 7–54

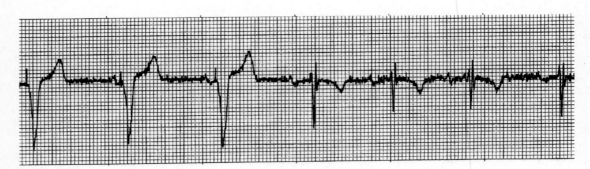

FIGURE 7–55

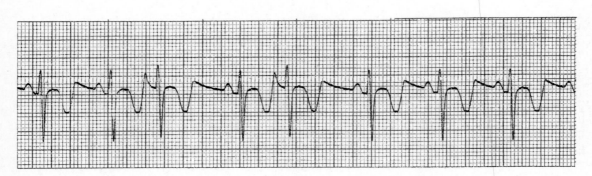

FIGURE 7–56

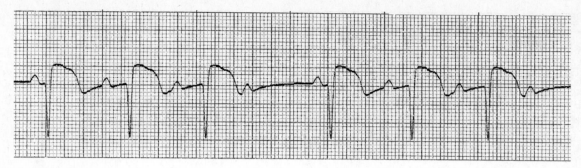

FIGURE 7–57

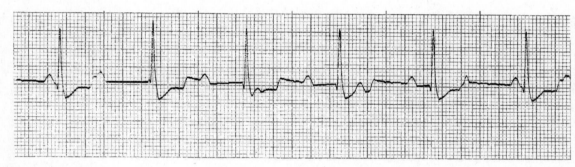

FIGURE 7–58

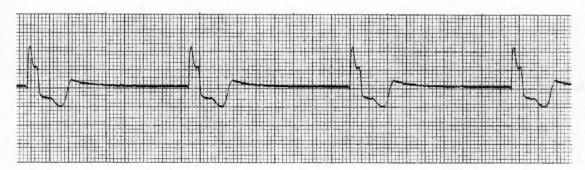

FIGURE 7–59

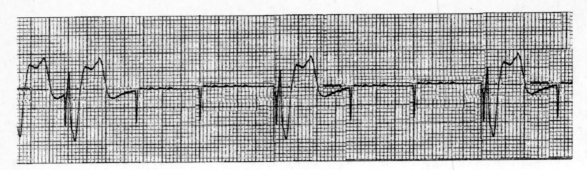

FIGURE 7–60

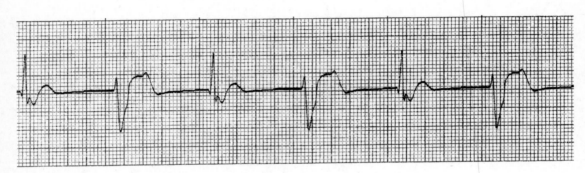

FIGURE 7–61

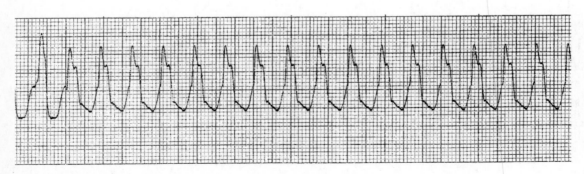

FIGURE 7–62

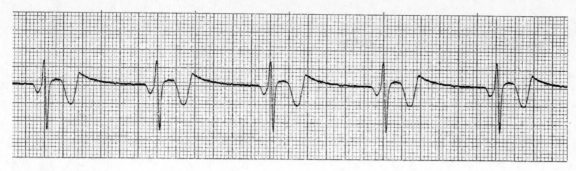

FIGURE 7–63

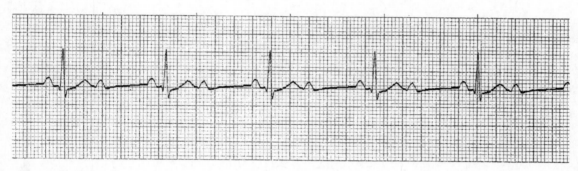

FIGURE 7–64

The Answer Key to Test Section 7 is on page 433.

Test Section Eight

8

Test Section 8's subjects:

- Pulmonary Emergencies
- Neurologic Emergencies
- Allergies and Anaphylaxis

Test Section 8 consists of 120 questions and is allotted 2 hours for completion.

1. Causes of decreased blood oxygen levels include all of the following except
    (a) fluid accumulation in alveolar or interstitial spaces.
    (b) pneumothorax.
    (c) atelectasis.
    (d) hyperventilation.
    (e) pulmonary embolism.

2. Arterial partial pressure of carbon dioxide ($PaCO_2$) is elevated by increased CO_2 production. All of the following are causes of increased CO_2 production, except
    (a) fever.
    (b) respiratory alkalosis.
    (c) muscular exertion.
    (d) shivering.
    (e) metabolic processes resulting in formation of acids.

3. $PaCO_2$ is also elevated by decreased CO_2 elimination. All of the following are causes of decreased CO_2 elimination, except
    (a) chronic obstructive pulmonary diseases.
    (b) ascent to a high altitude.
    (c) drug-induced respiratory depression.
    (d) airway obstruction.
    (e) mechanical dysfunction (muscle injury or impairment).

4. Depending upon the cause, elevated $PaCO_2$ levels can be decreased by any of the following measures, except
    (a) administering high concentrations of supplemental oxygen.
    (b) hyperventilation.
    (c) correcting the mechanical causes of $PaCO_2$ elevation.
    (d) breathing into a paper bag.
    (e) descent from altitude.

5. Chemoreceptors that assist in normal respiratory regulation are located in the
    (a) medulla, aortic arch, and carotid arteries (specifically, the carotid bodies).
    (b) medulla and aortic arch.
    (c) medulla and carotid arteries (specifically, the carotid bodies).
    (d) aortic arch and carotid arteries (specifically, the carotid bodies).
    (e) cerebellum, aortic arch, and carotid arteries (specifically, the carotid bodies).

6. The respiratory-regulating chemoreceptors are normally stimulated by
    (a) decreased PaO_2 levels.
    (b) increased $PaCO_2$ levels.
    (c) decreased pH.
    (d) Answers (a) and (c).
    (e) Answers (a), (b), and (c).

7. Which of the following statements about how arterial concentrations of oxygen or carbon dioxide affect a normal individual's respiratory activity is true?
    (a) A low arterial concentration of CO_2 increases respiratory activity.
    (b) A high arterial concentration of CO_2 decreases respiratory activity.
    (c) Hypoxemia (low PaO_2) is the strongest stimulus for respiratory activity.
    (d) Answers (a) and (c).
    (e) Answers (a), (b), and (c).

8. COPD patients develop a default method of respiratory activity regulation called hypoxic drive. Which of the following statements regarding hypoxic drive is false?
    (a) Aortic arch chemoreceptors become accustomed to high $PaCO_2$ levels.
    (b) Carotid chemoreceptors become accustomed to high $PaCO_2$ levels.
    (c) Decreased levels of PaO_2 stimulate respiratory activity.
    (d) Increased levels of $PaCO_2$ stimulate respiratory activity.
    (e) When PaO_2 levels rise, respiratory activity becomes inhibited.

9. A modified form of respiration, _____ acts to reexpand areas of atelectasis.
    (a) coughing
    (b) sneezing
    (c) sighing
    (d) hiccuping
    (e) grunting

10. A modified form of respiration, _____ has no known physiologic purpose.
    (a) coughing
    (b) sneezing
    (c) sighing
    (d) hiccuping
    (e) grunting

11. A modified form of respiration, _____ serves to protect the airway from obstruction, by expelling foreign material from the lungs.
    (a) coughing
    (b) sneezing
    (c) sighing
    (d) hiccuping
    (e) grunting

12. A modified form of respiration, _____ usually indicates respiratory distress.

  (a) coughing
  (b) sneezing
  (c) sighing
  (d) hiccuping
  (e) grunting

13. A modified form of respiration, _____ is occasionally associated with inferior AMI.

  (a) coughing
  (b) sneezing
  (c) sighing
  (d) hiccuping
  (e) grunting

14. Difficult or labored breathing that occurs whenever the patient becomes supine is called

  (a) orthopnea.
  (b) orthostatic apnea.
  (c) hyperpnea.
  (d) paroxysmal nocturnal dyspnea.
  (e) postural respirations.

15. Difficult or labored breathing that occurs during sleep is called

  (a) orthopnea.
  (b) orthostatic apnea.
  (c) hyperpnea.
  (d) paroxysmal nocturnal dyspnea.
  (e) postural respirations.

16. Palpation of the chest will assist in assessing all of the following, except

  (a) presence/absence of subcutaneous emphysema.
  (b) symmetry/asymmetry of excursion.
  (c) presence/absence of a pneumothorax.
  (d) presence/absence of unilateral or bilateral tactile fremitus.
  (e) presence/absence of structural instability.

17. The tissue layer lining the exterior surface of the lungs is called the

  (a) viscous pleura.
  (b) visceral pleura.
  (c) parietal pleura.
  (d) pneumonic pleura.
  (e) parenchymal pleura.

18. The tissue layer lining the interior surface of the chest wall is called the
    (a) viscous pleura.
    (b) visceral pleura.
    (c) parietal pleura.
    (d) pneumonic pleura.
    (e) parenchymal pleura.

19. Friction rub is an abnormal lung sound associated with pleural disease or pleural inflammation, and sounds like
    (a) pieces of dried leather rubbing together.
    (b) a fine, crackly noise, as if you were rubbing several strands of hair between your thumb and forefinger, next to your ear.
    (c) the dry, rasping noise produced by rubbing your palms together.
    (d) a high-pitched bark.
    (e) a dry whistle.

20. Harsh, rattling noises, heard when auscultating the chest, indicate the presence of thick mucous or other secretions in the throat or bronchi, and are called
    (a) snoring.
    (b) stridor.
    (c) wheezing.
    (d) rhonchi.
    (e) rales.

21. A harsh, high-pitched noise, audible on inspiration and characteristic of upper-airway obstruction, is called
    (a) snoring.
    (b) stridor.
    (c) wheezing.
    (d) rhonchi.
    (e) rales.

22. Fine, moist sounds heard when auscultating the chest indicate the presence of fluid in the smaller airways, and are called
    (a) snoring.
    (b) stridor.
    (c) wheezing.
    (d) rhonchi.
    (e) rales.

Questions 23 through 26 are related to the following scenario, as it develops:

23. You are eating at a restaurant with your family when you hear a woman start to yell, "Frank! Frank!" You observe the yelling woman sitting across from a conscious, medium-sized, middle-aged man. He is sitting in his chair and exhibiting the universal sign of choking. He is cyanotic, making no noise, and has a very distressed expression on his face. Upon reaching him, your first act should be to
   (a) perform 6 to 10 back blows.
   (b) perform 6 to 10 abdominal thrusts.
   (c) ask, "Are you choking? Can you speak? Can you cough?"
   (d) solicit information from bystanders (the yelling woman is hysterical and will be of little help).
   (e) direct a specific person to call 911, and perform finger sweeps of his oral airway.

24. Your treatment of Frank progresses. After your delivery of a tenth abdominal thrust, Frank's airway status remains unchanged. He suddenly loses consciousness and begins to slide from his chair. You gently lower him to the floor. Your next act should be to
   (a) open his airway and attempt your first mouth-to-mouth ventilation.
   (b) perform 6 to 10 chest thrusts while he is supine.
   (c) perform 6 to 10 abdominal thrusts while he is supine.
   (d) direct a specific person to call 911.
   (e) perform "blind" finger sweeps of his oral airway in case the object was dislodged while he was placed on the floor.

25. After your last action, Frank's airway status remains unchanged. Your next step should be to
   (a) open his airway and attempt your first mouth-to-mouth ventilation.
   (b) perform 6 to 10 chest thrusts while he is supine.
   (c) perform 6 to 10 abdominal thrusts while he is supine.
   (d) direct a specific person to call 911.
   (e) perform "blind" finger sweeps of his oral airway in case the object was dislodged while he was placed on the floor.

26. Your last action was unsuccessful. Your next step should be to
   (a) check to see if Frank has a pulse.
   (b) perform 6 to 10 chest thrusts while he is supine.
   (c) perform 6 to 10 abdominal thrusts while he is supine.
   (d) direct a specific person to call 911.
   (e) reposition his airway and again attempt mouth-to-mouth ventilation.

Questions 27 through 29 are related to the following scenario, as it develops:

27. A week later, you are eating at the same restaurant, when a woman runs out of the ladies' room, screaming, "She choked to death! She choked to death!" Upon entering the bathroom you see an obese woman lying on the floor. There is no apparent trauma, and she is very cyanotic. You should immediately

    (a) check for a pulse and direct a specific person to call 911.

    (b) check for breathing and direct a specific person to call 911.

    (c) check for level of consciousness and direct a specific person to call 911.

    (d) perform 6 to 10 abdominal thrusts and direct a specific person to call 911.

    (e) perform 6 to 10 chest thrusts and direct a specific person to call 911.

28. Your first attempt to ventilate this patient is unsuccessful. You should immediately

    (a) perform 6 to 10 abdominal thrusts, perform "blind" finger sweeps of her oral airway, position for an open airway, and attempt to ventilate again.

    (b) perform 6 to 10 chest thrusts, perform "blind" finger sweeps of her oral airway, position for an open airway, and attempt to ventilate again.

    (c) reposition her airway and attempt to ventilate again.

    (d) begin chest compressions (chest compression also acts as a "chest thrust" and may dislodge the airway obstruction).

    (e) perform "blind" finger sweeps of her oral airway, position for an open airway, and attempt to ventilate again.

29. Your second attempt to ventilate this patient is also unsuccessful. Your next act should be to

    (a) perform 6 to 10 abdominal thrusts, perform "blind" finger sweeps of her oral airway, position for an open airway, and attempt to ventilate again.

    (b) perform 6 to 10 chest thrusts, perform "blind" finger sweeps of her oral airway, position for an open airway, and attempt to ventilate again.

    (c) reposition her airway and attempt to ventilate again.

    (d) begin chest compressions (chest compression also acts as a "chest thrust" and may dislodge the airway obstruction).

    (e) perform "blind" finger sweeps of her oral airway, position for an open airway, and attempt to ventilate again.

30. An increased number of mucus-secreting cells in the respiratory epithelium, producing large amounts of sputum, is characteristic of

    (a) emphysema.

    (b) chronic bronchitis.

    (c) asthma.

    (d) Answers (b) and (c).

    (e) Answers (a), (b), and (c).

31. Although the alveoli are not seriously affected by the disease process of _____, alveolar hypoventilation occurs and diminishes gas exchange.

   (a) emphysema
   (b) chronic bronchitis
   (c) asthma
   (d) Answers (b) and (c).
   (e) Answers (a), (b), and (c).

32. Pursed-lip breathing ("puffing") is classically associated with patients suffering from _____, because it helps their expiration.

   (a) emphysema
   (b) chronic bronchitis
   (c) asthma
   (d) Answers (b) and (c).
   (e) Answers (a), (b), and (c).

33. Wheezes may accompany

   (a) emphysema.
   (b) chronic bronchitis.
   (c) asthma.
   (d) Answers (b) and (c).
   (e) Answers (a), (b), and (c).

34. Pneumonia is a respiratory disease caused by a

   (a) bacterial infection.
   (b) virus.
   (c) fungus.
   (d) Answers (a) and (b).
   (e) Answer (a), (b), or (c).

35. Common signs and symptoms of pneumonia include all of the following, except

   (a) fever, chills, and weakness.
   (b) a productive cough.
   (c) chest pain and tachycardia.
   (d) barrel chest and cyanosis.
   (e) wheezes and upper abdominal pain.

36. Management of pneumonia includes all of the following, except

   (a) supplemental high-flow oxygen.
   (b) administration of nebulized bronchodilators.

    (c) cardiac monitoring.

    (d) ice packs applied to the bilateral axillary areas and groin of the febrile patient.

    (e) fluid administration based on hydration status.

37.   Carbon monoxide inhalation causes cellular hypoxia because carbon monoxide

    (a) binds to hemoglobin more strongly than oxygen.

    (b) prevents the inhalation of oxygen.

    (c) alters the density of oxygen molecules, preventing them from crossing cell membranes.

    (d) Answers (a) and (c).

    (e) Answers (a), (b), and (c).

38.   Cherry-red skin coloration has long been associated with carbon monoxide poisoning and occurs

    (a) soon after the onset of euphoria and before loss of consciousness.

    (b) soon after the onset of headache and before loss of consciousness.

    (c) just prior to developing altered mentation.

    (d) very late in the course of poisoning, usually well after loss of consciousness.

    (e) only after death, when lividity is noted.

39.   Which of the following statements regarding management of carbon monoxide (CO) poisoning is false?

    (a) Care-provider safety is the first priority of CO poisoning management.

    (b) Removal of the victim from the exposure site should precede oxygen administration.

    (c) Oxygen should be administered in as high a concentration as possible, even if the CO-poisoned patient has a history of COPD.

    (d) Administration of nebulized bronchodilator will greatly improve oxygen transport into the cells of a CO poisoning victim.

    (e) Severe CO poisoning victims should be transported to a hospital that offers hyperbaric oxygen treatment.

40.   An embolism capable of pulmonary artery obstruction may be composed of any of the following, except

    (a) air.

    (b) fat.

    (c) amniotic fluid.

    (d) a plastic fragment from a venous catheter.

    (e) large glucose molecules.

41. Which of the following conditions or situations increase a patient's likelihood of developing a pulmonary embolus?

(1) aspiration of vomitus   (6) prolonged immobilization
(2) atrial fibrillation   (7) recent surgery
(3) long bone fracture   (8) sickle cell anemia
(4) long-distance travel   (9) thrombophlebitis
(5) pregnancy   (10) use of oral contraceptives

    (a) 1, 2, 3, 4, 5, 6, 7, 8, 9, and 10.
    (b) 2, 3, 4, 5, 6, 7, 8, 9, and 10.
    (c) 1, 2, 3, 6, 7, 8, 9, and 10.
    (d) 1, 2, 3, 6, 7, 9, and 10.
    (e) 2, 3, 6, 7 and 9.

42. Although signs and symptoms will vary, depending upon the size and location of a pulmonary artery obstruction, a patient with a pulmonary embolism may

    (a) report the sudden onset of severe dyspnea, with or without chest pain.
    (b) exhibit tachycardia or tachypnea, with or without hemoptysis.
    (c) have distended jugular veins or hypotension.
    (d) Answers (a) and (b).
    (e) Answers (a), (b), and (c).

43. Appropriate management considerations for patients suffering from pulmonary embolism include all of the following, except

    (a) IV access with D_5W, TKO.
    (b) IV access with LR or NS; flow-rate titrated to the patient's blood pressure.
    (c) Be prepared to intubate and provide positive pressure ventilation.
    (d) Be prepared to perform CPR.
    (e) Rapid transport to a critical-care hospital.

44. Hyperventilation may be caused by many medical emergencies. However, when hyperventilation is purely anxiety induced and is allowed to continue, it will result in

    (a) a decreased $PaCO_2$ level, producing respiratory alkalosis.
    (b) a decreased PaO_2 level, producing respiratory alkalosis.
    (c) an increased PaO_2 level, producing respiratory acidosis.
    (d) a decreased $PaCO_2$ level, producing respiratory acidosis.
    (e) an increased $PaCO_2$ level, producing respiratory acidosis.

45. Signs and symptoms of anxiety-induced hyperventilation syndrome include all of the following, except

    (a) onset of seizures (in patients with a seizure disorder).
    (b) complaints of chest pain.

(c) complaints of numbness or tingling about the mouth or fingers.

(d) rales, rhonchi, or wheezes.

(e) carpopedal spasms.

46. Which of the following methods of oxygen administration to symptomatic patients strongly suspected to be suffering purely from anxiety-induced hyperventilation syndrome is correct?

(a) If a pulse oximeter is unavailable, administer 12–15 LPM of oxygen, via a nonrebreather mask.

(b) If a pulse oximeter is available, have the patient use a paper bag to rebreathe CO_2, until her/his pulse oximeter reading falls below 90% (at sea level).

(c) If a pulse oximeter is available, apply a nonrebreather mask, without administering oxygen, allowing the patient to rebreathe CO_2, until her/his pulse oximeter reading falls below 90% (at sea level).

(d) If a pulse oximeter is available, administer 4 LPM of O_2, via a nonrebreather mask, until the patient's pulse oximeter reading falls below 90% (at sea level).

(e) If a pulse oximeter is unavailable, administer 2 LPM of O_2, via a nasal cannula, until the patient develops cyanosis or an altered level of consciousness.

47. The fundamental structural and functional unit of the nervous system is the nerve cell, which is also called

(a) a nucleus.

(b) a dendrite.

(c) an axon.

(d) a ganglia.

(e) a neuron.

48. Each individual nerve cell has several branches that receive impulses and transmit them to the cell body. Each of these branches is called

(a) a nucleus.

(b) a dendrite.

(c) an axon.

(d) a ganglia.

(e) a neuron.

49. Each nerve cell has a branch that conducts impulses away from the cell body. This branch is called

(a) a nucleus.

(b) a dendrite.

(c) an axon.

(d) a ganglia.

(e) a neuron.

50. Which of the following statements about impulse transmission between nerve cells is true?

   (a) Direct impulse transmission between nerve cells is accomplished by one nerve cell's impulse-sending branch directly connecting with several neighboring nerve cells' impulse-receiving branches.

   (b) Direct transfer of impulses between nerve cells is accomplished by interconnected synapses (junctions formed by the physical attachment of each nerve cell's impulse-sending branch to several of its neighboring nerve cell's impulse-receiving branches).

   (c) A "gap" exists between all nerve cell branches. Nerve impulses must be translated into a chemical form in order to bridge this gap and communicate impulses between nerve cells.

   (d) All of the above statements can be true, depending upon the type of nerve cell.

   (e) None of the above statements are true.

51. The spinal cord connects and communicates with the brain at the brainstem. The brainstem consists of three segments, called the

   (a) medulla oblongata, cerebrum, and cerebellum.
   (b) medulla oblongata, pons, and cerebellum.
   (c) medulla oblongata, pons, and mesencephalon (midbrain).
   (d) mesencephalon (lower brainstem), diencephalon (mid brainstem), and triencephalon (upper brainstem).
   (e) triencephalon (lower brainstem), diencephalon (mid brainstem), and mesencephalon (upper brainstem).

52. Called "the seat of consciousness" and considered the center of higher mental faculties (such as memory, language, and judgment), the _____ comprises the largest part of the human brain.

   (a) cerebellum
   (b) medulla oblongata
   (c) cerebrum
   (d) pons
   (e) diencephalon

53. The brain portion responsible for balance and fine motor coordination is the

   (a) cerebellum.
   (b) medulla oblongata.
   (c) cerebrum.
   (d) pons.
   (e) diencephalon.

54. The portion of the brain that lies immediately behind the human forehead is the

   (a) cerebellum.
   (b) medulla oblongata.

    (c) cerebrum.

    (d) pons.

    (e) diencephalon.

55.   The portion of the brain that lies immediately below the posterior occipital prominence (within the most posterior and inferior portion of the cranium) is the

    (a) cerebellum.

    (b) medulla oblongata.

    (c) cerebrum.

    (d) pons.

    (e) diencephalon.

56.   Involuntary, life-sustaining functions of the respiratory and cardiovascular systems are primarily controlled by centers located within the

    (a) cerebellum.

    (b) medulla oblongata.

    (c) cerebrum.

    (d) pons.

    (e) diencephalon.

57.   The speech center of the human brain is located within the

    (a) temporal lobes of the cerebrum.

    (b) parietal lobes of the cerebrum.

    (c) frontal lobe of the cerebrum.

    (d) cerebellum.

    (e) brainstem.

58.   An individual's special personality traits are stored within the

    (a) temporal lobes of the cerebrum.

    (b) parietal lobes of the cerebrum.

    (c) frontal lobe of the cerebrum.

    (d) cerebellum.

    (e) brainstem.

59.   In the mature adult, the spinal cord is 17–18 inches long and ends at the approximate level of the

    (a) first lumbar vertebra.

    (b) third lumbar vertebra.

    (c) fifth lumbar vertebra.

    (d) seventh lumbar vertebra.

    (e) ninth lumbar vertebra.

60. There are ___ pairs of cranial nerves that originate in the brain and innervate structures outside the brain.

(a) 9

(b) 12

(c) 5

(d) 7

(e) 14

61. Two vascular systems supply blood to the brain. These systems join at the _____ before actually entering the brain.

(a) brachial plexus

(b) Circle of Willis

(c) Junction of Wallace

(d) cranial plexus

(e) cerebral plexus

62. Structural illnesses, lesions, or injuries can cause an altered level of consciousness. All of the following are examples of structural altered LOC causes, except

(a) parasites.

(b) intracranial hemorrhage.

(c) hypoglycemia.

(d) degenerative brain disease.

(e) a brain neoplasm.

63. Toxic-metabolic causes of an altered level of consciousness include all of the following, except

(a) hypoxia.

(b) intracranial hemorrhage.

(c) hyperglycemia.

(d) renal failure.

(e) thiamine deficiency.

64. Which of the following mnemonics are designed to assist in determining a patient's mental status?

(a) AVPU

(b) DECAP-BTLS

(c) OPQRST-ASPN

(d) Answers (a) and (b).

(e) Answers (a) and (c).

65. Specific methods of assessing cerebral function and mental status include evaluation of all of the following, except

(a) quality of speech.

(b) memory and attention span.

    (c) mood.

    (d) posture or gait.

    (e) pulse oxygenation measurement.

66.   Another mnemonic designed to assist assessment of cerebral function and mental status is AEIOU-TIPS. This mnemonic helps you to

    (a) assess the patient's quality of speech production, by having him pronounce the nine most common sounds in the English language.

    (b) consider the multiple possible causes of altered LOC, and pinpoint which treatment protocols to follow.

    (c) remember to assess each of the 9 cranial nerves.

    (d) Answers (a) and (b); this is a dual-use mnemonic.

    (e) Answers (a) and (c); this is a dual-use mnemonic.

67.   The "U" of the AEIOU-TIPS mnemonic stands for

    (a) the "you" sound made by pronouncing the letter "U"; testing for a particular lip-formation and sound-production ability.

    (b) Uremia (kidney failure).

    (c) pUpil contractility test; assessment of the 3rd cranial nerve.

    (d) Answers (a) and (b); this is a dual-use mnemonic.

    (e) Answers (a) and (c); this is a dual-use mnemonic.

68.   Management considerations for treatment of altered level of consciousness include all of the following, except

    (a) IV access D_5W, TKO.

    (b) $D_{50}W$ administration if the patient is hypoglycemic.

    (c) $D_{50}W$ administration if the paramedic is unable to determine whether or not the patient is hypoglycemic.

    (d) naloxone administration if the patient is suspected of narcotic OD.

    (e) thiamine administration if the patient is suspected of chronic alcoholism or appears malnourished.

69.   A stroke, or cerebrovascular accident (CVA), may be caused by any of the following, except

    (a) air embolus occlusion of cerebral vasculature.

    (b) ruptured thoracic aortic aneurysm.

    (c) atherosclerotic plaque or tumor tissue occlusion of cerebral vasculature.

    (d) subarachnoid hemorrhage.

    (e) thrombus occlusion of cerebral vasculature.

70.   Factors that may increase the incidence of stroke include all of the following, except
      (a) hypertension.
      (b) diabetes.
      (c) mid-shaft tibia fractures.
      (d) use of oral contraceptives.
      (e) sickle cell disease.

71.   A patient should immediately be considered a possible CVA victim if she or he exhibits the sudden onset of any one of the following S/Sx, except
      (a) difficulty communicating or understanding communication.
      (b) an unusually severe headache, with no known cause.
      (c) pain, numbness, or tingling in both arms or both legs.
      (d) vision alteration in both eyes.
      (e) difficulty walking due to unilateral leg weakness, loss of balance, or loss of coordination.

72.   Which of the following statements related to evaluating facial droop and arm drift using the Cincinnati Prehospital Stroke Scale (CPSS) is false?
      (a) A patient whose face doesn't move as well on the right as it does on the left is almost always a CVA victim.
      (b) Left-sided facial droop is non-specific for identifying a CVA victim, because a "right-handed" patient's face commonly does not move as well on the left as it does on the right.
      (c) To accurately assess a patient for presence or absence of arm drift, he must have both his eyes closed while holding his arms out in front of him.
      (d) If a patient has relatively equal difficulty holding both arms out in front of her, she is probably not a CVA victim.
      (e) If one of the patient's arms drift, down during the 10-second arm drift tests he or she is probably a CVA vitim.

73.   To rapidly evaluate a patient's ability to speak using the CPSS, the AHA recommends that the patient be asked to say which of the following sentences?
      (a) "How much wood could a woodchuck chuck, if a woodchuck could chuck wood?"
      (b) "Mares eat oats, and does eat oats, and little lambs eat ivy."
      (c) "My dog has fleas."
      (d) "You are the cat's meow."
      (e) "You can't teach an old dog new tricks."

74.   Which of the following statements regarding use of the Los Angeles Prehospital Stroke Screen (LAPSS) is false?
      (a) Specific altered LOC causes (including hypoglycemia, seizure, or history of epilepsy) must be investigated and ruled out when using the LAPSS.
      (b) Whether or not the patient has been confined to a wheelchair is an LAPSS factor.

(c) Whether or not the patient has been bedridden is an LAPSS factor.

(d) The patient's age is a factor when using the LAPSS to assess for CVA.

(e) The LAPSS is less successful than the CPSS at identifying a CVA victim.

75. Which of the following statements regarding the 2005 AHA treatment recommendations for a CVA victim is false?

(a) Any CVA patient with an oxygen saturation less than 92% is hypoxemic, and oxygen should be administered.

(b) If a pulse oximeter indicates that the possible CVA patient is not hypoxemic, oxygen administration is contraindicated (unnecessary oxygen administration may increase intracranial pressure).

(c) Whether or not a pulse oximeter is available, high-flow oxygen should be administered to every emergency patient.

(d) If the suspected CVA patient seizes, standard seizure protocols should be followed.

(e) IV access with NS, titrated to maintaining a normal blood pressure, is recommended.

76. Which of the following statements regarding the administration of $D_{50}W$ to a CVA victim is true?

(a) If a paramedic does not have the ability to determine a confused patient's blood glucose level, the paramedic must administer 25G of IV $D_{50}W$ in order to rule-out hypoglycemia as the confusion's cause.

(b) Studies have shown that hyperglycemic acute CVA patients enjoyed an improved clinical outcome compared to that of normoglycemic patients.

(c) Acute CVA patients who are caused to be hyperglycemic (by IV administration of $D_{50}W$) do not suffer a worse clinical outcome than normoglycemic patients, because hyperglycemia does not cause any acute S/Sx, and it can quickly be corrected by insulin administration once the patient is in the hospital.

(d) Because it is a hypertonic solution, IV administration of $D_{50}W$ will decrease intracranial pressure via hypertonic vascular osmosis. Thus, every suspected CVA victim should receive 25G of IV $D_{50}W$, whether or not the paramedic has the ability to measure the patient's blood glucose level.

(e) All of the above answers are true.

77. Transient ischemic attacks (TIAs) are episodes of strokelike neurological deficits (cerebral dysfunction) that commonly last

(a) only a few minutes, with complete recovery achieved within two hours.

(b) from one week to ten days, with complete recovery achieved after ten days.

(c) anywhere from two or three minutes to several hours but are almost always entirely resolved within 24 hours.

(d) from three to four days, with complete recovery achieved after five days.

(e) less than one hour.

78. Which of the following statements regarding prehospital differentiation between, and prehospital treatment of, TIA and CVA is false?

    (a) TIA signs and symptoms usually have a slow, gradual, onset.

    (b) CVA signs and symptoms are usually of rapid, acute, onset.

    (c) Unless one or more acute neurological deficits are observed to resolve during prehospital care, it is virtually impossible to differentiate between TIA and CVA.

    (d) All TIA patients should be treated as though they were CVA patients, whether or not their neurological deficits begin to improve or resolve during prehospital care.

    (e) All TIA patients should be considered to be at risk for suddenly developing a CVA.

79. Treatment of the CVA/TIA victim includes all of the following emergency care considerations, except

    (a) provide aggressive airway management as needed, to include suction and intubation, should the airway become threatened by excessive saliva production, inability to swallow, or respiratory depression.

    (b) provide gentle, positive pressure hyperventilation (no faster than a rate of 20/min), if the patient is apneic or breathing inadequately.

    (c) transport in a position determined by patient comfort, unless unconscious or hypotensive.

    (d) explain all care procedures and provide continuous verbal assurance, even if the spontaneously breathing patient is unable to speak or appears to be unconscious.

    (e) avoid all rapid modes of transport unless the patient is unconscious. Slow, quiet, and gentle transport to the patient's hospital of record will prevent precipitation of any other embolic incidents en route.

80. Prehospital medication administration considerations when treating the CVA/TIA victim include

    (a) IV access D_5W, TKO.

    (b) IV administration of $D_{50}W$ if hypoglycemic.

    (c) thiamine administration if alcoholic or malnourished.

    (d) Answers (b) and (c).

    (e) Answers (a), (b), and (c).

81. Seizures are most commonly caused by

    (a) hypoxia, hypoglycemia, infections, or other metabolic disorders.

    (b) CVA, TIA, or other vascular disorders.

    (c) head trauma.

    (d) toxin exposure, including alcohol or drug overdose or withdrawal.

    (e) idiopathic epilepsy.

82. _____ seizures cause loss of consciousness and are commonly accompanied by incontinence of urine or feces.

    (a) Jacksonian

    (b) Psychomotor

      (c) Grand mal

      (d) Hysterical

      (e) Petit mal

83.   _____ seizures frequently go unnoticed because of their brief duration and lack of overt motor activity.

      (a) Jacksonian

      (b) Psychomotor

      (c) Grand mal

      (d) Hysterical

      (e) Petit mal

84.   Unexplained attacks of rage or bizarre behavior are indicative of

      (a) Jacksonian seizures.

      (b) psychomotor seizures.

      (c) grand mal seizures.

      (d) hysterical seizures.

      (e) petit mal seizures.

85.   _____ seizures are characterized by uncontrolled movement of one body part, causing unconsciousness only if they progress to become total body seizures.

      (a) Jacksonian

      (b) Psychomotor

      (c) Grand mal

      (d) Hysterical

      (e) Petit mal

86.   _____ seizures, also known as focal motor or focal sensory seizures, are characterized by uncontrolled motor activity of only one side of the body and may or may not progress to become total body seizures.

      (a) Jacksonian

      (b) Psychomotor

      (c) Grand mal

      (d) Hysterical

      (e) Petit mal

87.   Complex partial seizures are not accompanied by loss of consciousness. They may be characterized by altered personality states, staggering, purposeless movement, or automatic behavior. These seizures are also called temporal lobe seizures, or

      (a) Jacksonian seizures.

      (b) psychomotor seizures.

      (c) grand mal seizures.

      (d) petit mal seizures.

      (e) hysterical seizures.

88. _____ seizures are considered simple partial seizures and are often also called focal motor, or focal sensory, seizures.
     (a) Jacksonian
     (b) Psychomotor
     (c) Grand mal
     (d) Hysterical
     (e) Petit mal

89. The peculiar metallic taste that frequently precedes temporal lobe seizures is an example of
     (a) an auditory aura.
     (b) a visual aura.
     (c) an olfactory aura.
     (d) a gustatory aura.
     (e) a tactile aura.

90. An "odd" feeling or an unusual physical sensation sensed in one or more parts of the body preceding a seizure is an example of
     (a) an auditory aura.
     (b) a visual aura.
     (c) an olfactory aura.
     (d) a gustatory aura.
     (e) a tactile aura.

91. Smelling a specific or unpleasant odor prior to a seizure is an example of
     (a) an auditory aura.
     (b) a visual aura.
     (c) an olfactory aura.
     (d) a gustatory aura.
     (e) a tactile aura.

92. The _____ motor activity phase of a seizure is characterized by rapid, rhythmic flexion and extension of the victim's muscles.
     (a) postictal
     (b) clonic
     (c) tonic
     (d) Answer (a) or (b).
     (e) Answer (b) or (c).

93. The _____ motor activity phase of a seizure is characterized by continuous muscle tension as the muscles increasingly contract.
     (a) postictal
     (b) clonic

    (c) tonic

    (d) Answer (a) or (b).

    (e) Answer (b) or (c).

94. Although the duration time for any particular phase of a full body seizure varies from patient to patient, the order of phase progression remains basically the same. Place the following phases in the order that they commonly occur.

    (1) Loss of consciousness   (4) Tonic phase
    (2) Clonic phase   (5) Postictal phase
    (3) Aura   (6) Jaw-clenching

    (a) 1, 2, 3, 4, 5, 6.
    (b) 1, 3, 2, 4, 6, 5.
    (c) 6, 1, 3, 2, 4, 5.
    (d) 3, 1, 2, 6, 4, 5.
    (e) 3, 1, 4, 6, 2, 5.

95. All of the following medications are commonly prescribed to patients with a seizure disorder, except

    (a) phenobarbital.
    (b) carbamazepine.
    (c) cyclobenzaprine.
    (d) valproic acid.
    (e) phenytoin.

96. The trade name for the anticonvulsant _____ is Dilantin®.

    (a) phenobarbital
    (b) carbamazepine
    (c) cyclobenzaprine
    (d) valproic acid
    (e) phenytoin

97. The generic name for the tricyclic anticonvulsant Tegretol® is

    (a) phenobarbital.
    (b) carbamazepine.
    (c) cyclobenzaprine.
    (d) valproic acid.
    (e) phenytoin.

98. Treatment of a patient experiencing a single seizure includes which of the following performances?

    (a) If seizure activity is still occurring upon your arrival, insert a padded tongue blade or "bite block," to prevent oral trauma, thus preventing subsequent aspiration of broken teeth or blood.

    (b) IV access, D_5W, TKO, and determination of blood glucose level.

    (c) Soft restraints as needed, to prevent injury to self and others.

    (d) All of the above.

    (e) None of the above.

99. Status epilepticus is defined as

    (a) two or more seizures occurring within 30 minutes.

    (b) two or more seizures occurring within one hour.

    (c) two or more seizures without an intervening period of consciousness.

    (d) Answers (a) and (c).

    (e) Answers (b) and (c).

100. Status epilepticus is a life-threatening emergency, because it may result in

    (a) respiratory arrest.

    (b) respiratory and metabolic acidosis.

    (c) severe hypertension or increased intracranial pressure.

    (d) Answers (a) and (b).

    (e) Answers (a), (b), and (c).

101. The most common cause of status epilepticus in adults is

    (a) head trauma.

    (b) failure to take prescribed anticonvulsant medications.

    (c) cerebral infection.

    (d) hypoglycemia.

    (e) cardiovascular disease.

102. Management of status epilepticus includes

    (a) IV access, D_5W, TKO, and determination of blood glucose level.

    (b) 5–10 mg diazepam, IV.

    (c) 25 grams of $D_{50}W$, IV, if hypoglycemic.

    (d) Answers (b) and (c).

    (e) Answers (a), (b), and (c).

103. Which of the following statements about syncope is false?

    (a) Syncope may be caused by dysrhythmias or abnormal cardiac function.

    (b) Hypoglycemia can cause syncope.

(c) CVA or TIA can cause syncope.

(d) Anxiety-induced syncope (psychiatric syncope) may cause a loss of consciousness lasting 1–10 minutes.

(e) If, after extensive neurological assessment and testing, the cause of a patient's syncope cannot be determined, the syncope is considered to be idiopathic.

104. Which of the following statements regarding allergic-reaction pathophysiology is false?

(a) Delayed hypersensitivity reactions do not occur until hours or days following exposure to certain drugs or chemicals.

(b) Immediate hypersensitivity reactions are usually more severe than delayed hypersensitivity reactions.

(c) Delayed hypersensitivity reactions occur as a result of cellular immunity and do not involve antigen-antibody interaction.

(d) Immediate hypersensitivity reactions are those that occur the very first time an antigen is introduced into an individual's body.

(e) Immediate hypersensitivity reactions occur when an individual has been previously exposed to an antigen and has developed antibodies that respond in an exaggerated manner to antigen re-exposure.

105. Allergic-reaction fatalities related to medication exposure are most commonly caused by

(a) use of unregulated "street" drugs.

(b) inhaled medications.

(c) acetaminophen ingestion.

(d) use of unregulated vitamins and "health" supplements.

(e) antibiotics.

106. The second most common cause of allergic reaction fatalities are

(a) poison ingestions.

(b) inhaled allergens.

(c) food ingestions (especially nuts).

(d) insect stings.

(e) latex product exposure.

107. Which of the following insects are included in the highly specialized order of insects known as *Hymenoptera*?

(a) Wasps, yellow jackets, and hornets (*Vespidae*).

(b) Honey bees (*Apoidea*).

(c) Fire ants (*Formicoidea*).

(d) Answers (a) and (b).

(e) Answers (a), (b), and (c).

108. The principal chemical mediator of an allergic reaction is histamine. Histamine release causes all of the following, except
   (a) increased gastric-acid production.
   (b) increased intestinal motility.
   (c) bronchodilation.
   (d) vasodilation.
   (e) increased capillary permeability.

109. Severe allergic reactions or anaphylaxis may produce generalized edema of the skin; usually involving the head, neck, face, and upper airway. This condition is known as
   (a) angioneurotic edema.
   (b) urticaria.
   (c) ascites.
   (d) agonalneurotic edema.
   (e) vasobasilar edema.

110. Allergic reactions often produce isolated areas of vasodilation and extravasation that appear as red blotches (with or without white centers) and may or may not be raised (appearing as welts, or "wheals"). These isolated areas of vasodilation and extravasation are called
   (a) angioedemas.
   (b) urticaria.
   (c) hyphemas.
   (d) polyps.
   (e) angioneurotic lesions.

111. The medical term for the "hives" classically produced by allergic reactions is
   (a) angioedema.
   (b) urticaria.
   (c) hyphemas.
   (d) polyps.
   (e) angioneurotic lesions.

112. Anaphylaxis is the most severe form of allergic reaction, and may acutely produce any of the following signs and symptoms, except
   (a) dyspnea, sneezing, coughing, or stridor.
   (b) wheezing, rales, or total respiratory obstruction.
   (c) peripheral vasoconstriction (pallor and diaphoresis).
   (d) tachycardia and hypotension.
   (e) abdominal cramping, nausea, vomiting, or diarrhea.

113. Possible acute signs and symptoms of anaphylaxis also include all of the following, except
   (a) headache.
   (b) seizures.

(c) cyanosis.

(d) facial swelling.

(e) pitting pedal edema.

114. Which of the following statements regarding the airway management of an anaphylaxis patient is false?

(a) The acute onset of stridor indicates impending total airway occlusion and is a valid cue for immediate performance of endotracheal intubation.

(b) Anaphylaxis frequently causes laryngeal irritability, and any airway manipulation may result in laryngospasm.

(c) Laryngospasm may be caused by the use of oropharyngeal airways or of over-sized nasopharyngeal airways when ventilating an anaphylaxis patient.

(d) Because laryngospasm airway occlusion is such a frequent side effect of anaphylaxis, and because epinephrine is so successful at reversing histamine reactions, endotracheal intubation should be delayed until after IV medication administration has failed to resolve anaphylactic airway obstruction or respiratory insufficiency.

(e) If endotracheal intubation is unsuccessful and laryngospasm or edema continues to obstruct the patient's airway, consider immediately performing needle cricothyrotomy or another form of surgical airway access.

115. Which of the following IV solutions should be used for the treatment of anaphylaxis?

(a) Lactated Ringers solution.

(b) Normal Saline (0.9% sodium chloride).

(c) D_5W.

(d) Answer (a) or (b).

(e) Answer (a), (b), or (c).

116. Place the following medications in order, from the most-important to least-important order of administration, when treating a life-threatening anaphylactic reaction. (Select only one form of epinephrine.)

(1) epinephrine 1:1000

(2) epinephrine 1:10,000

(3) dexamethasone, methylprednisolone, or hydrocortisone

(4) diphenhydramine

(5) inhaled β-adrenergic agents (such as albuterol or ipratropium)

    (a) 1, 4, 3, 5.

    (b) 2, 4, 5, 3.

    (c) 1, 2, 3, 4, 5.

    (d) 1, 3, 5.

    (e) 2, 3, 4, 5.

117. The administration dosage for epinephrine 1:1000 in response to allergic reactions is
   (a) 0.3–0.5 mg IM.
   (b) 0.3–0.5 mg IV.
   (c) 3–5 ml SQ.
   (d) Answer (a) or (b).
   (e) Answer (b) or (c).

118. The administration dosage for epinephrine 1:10,000 in response to allergic reactions is
   (a) 0.3–0.5 mg IM.
   (b) 0.3–0.5 mg IV.
   (c) 0.1 mg IV, slowly (over 5 minutes).
   (d) Answer (a) or (b).
   (e) Answer (b) or (c).

119. Which of the following statements regarding IV epinephrine administration to patients suffering severe allergic reactions, anaphylaxis, or cardiac arrest caused by anaphylaxis, is false?
   (a) Incidents of fatal epi overdose have been reported when IV epi was administered for anaphylaxis.
   (b) Patients on β-blockers are more at risk for suffering severe anaphylaxis, and may not respond to lower doses of epi.
   (c) Patients on β-blockers may suffer a paradoxical reaction to standard epi dose administration.
   (d) If hypotension is not corrected by epi administration, aggressive fluid resuscitation may induce pulmonary edema and is concretely contraindicated.
   (e) The "high dose" regimen of epi administration (1 to 3 mg IV, followed after 3 min. by 3 to 5 mg IV, followed after 3 min. by a 4 to 10 μg/min infusion) is now contraindicated in cardiac arrest situations, except for those caused by anaphylaxis, β-blocker overdose, or calcium channel blocker overdose.

120. To mix an epinephrine drip for the treatment of hypotensive anaphylaxis,
   (a) add 1mg of epi 1:10 to 250 ml D$_5$W.
   (b) add 2.5 mg of epi 1:1 to 250 ml D$_5$W.
   (c) add 1 mg of epi 1:1 to 1000 ml NS or D$_5$W.
   (d) add 1 mg of epi 1:1 to 500 ml NS or D$_5$W.
   (e) add 1 mg of epi 1:1 to 250 ml NS or D$_5$W.

The Answer Key to Test Section 8 is on page 441.

Test
Section
Nine

9

Test Section 9's subjects:

- Endocrine Emergencies
- Gastroenteric Emergencies
- Renal/Urologic Emergencies
- Toxicological Emergencies
- Hematological Emergencies

Test Section 9 consists of 120 questions and is allotted 2 hours for completion.

1.   Exocrine glands secrete hormones or chemical substances
        (a) via ducts, and tend to have a systemic (total body) effect.
        (b) via ducts, and tend to have only a local effect.
        (c) into the gastrointestinal system only.
        (d) directly into surrounding tissue capillaries, and tend to have only a local
             effect.
        (e) directly into surrounding tissue capillaries, and tend to have a systemic (total
             body) effect.

2.   Endocrine glands secrete hormones or chemical substances
        (a) via ducts, and tend to have a systemic (total body) effect.
        (b) via ducts, and tend to have only a local effect.
        (c) into the gastrointestinal system only.
        (d) directly into surrounding tissue capillaries, and tend to have only a local
             effect.
        (e) directly into surrounding tissue capillaries, and tend to have a systemic (total
             body) effect.

3.   Which of the following are examples of exocrine glands?
        (1) Adrenal   (5) Sebaceous
        (2) Ovaries   (6) Sudoriferous
        (3) Pituitary   (7) Testes
        (4) Salivary   (8) Thyroid

           (a) 1, 2, 3, 7, and 8.
           (b) 4, 5, and 6.
           (c) 2, 4, 5, 6, and 7.
           (d) 1, 3, and 8.
           (e) 2 and 7.

4.   The _____ is/are composed of tissue with both exocrine and endocrine
      functions.
        (a) ovaries
        (b) testes
        (c) adrenal glands
        (d) pancreas
        (e) thyroid gland

5.   Once called the "master gland" of the body, the _____ was recently
      discovered to be dependent upon a chemical relationship with the hypothalamus,
      which enables it to influence the functions of so many other glands.
        (a) cerebrum
        (b) adrenal gland

    (c) pituitary gland

    (d) thyroid gland

    (e) parathyroid gland

6.   Epinephrine and norepinephrine are secreted by the

    (a) ovaries or testes.

    (b) adrenal glands.

    (c) pituitary gland.

    (d) thyroid gland.

    (e) pancreas.

7.   Increased blood calcium levels are stimulated by a hormone secreted by the pea-sized

    (a) ovaries or testes.

    (b) posterior pituitary gland.

    (c) adult thalamus.

    (d) thyroid gland.

    (e) parathyroid glands.

8.   The body's basic metabolic rate is regulated by hormones secreted by the

    (a) ovaries or testes.

    (b) posterior pituitary gland.

    (c) adrenal glands.

    (d) thyroid gland.

    (e) parathyroid glands.

9.   Estrogen is a hormone secreted by the

    (a) adrenal glands.

    (b) ovaries.

    (c) testes.

    (d) Answers (a) and (b).

    (e) Answers (b) and (c).

10.   Progesterone is a hormone secreted by the

    (a) adrenal glands.

    (b) ovaries.

    (c) testes.

    (d) Answers (a) and (b).

    (e) Answers (b) and (c).

11.   Testosterone is a hormone secreted by the

    (a) adrenal glands.

    (b) ovaries.

    (c) testes.

    (d) Answers (a) and (c).

    (e) Answers (b) and (c).

12. Cushing's syndrome is an endocrine disorder caused by the hypersecretion of glucocorticoids by the

(a) ovaries or testes.

(b) adrenal glands.

(c) pituitary gland.

(d) thyroid gland.

(e) pancreas.

13. The antidiuretic hormone (ADH) stimulates increased reabsorption of water into the body's blood volume and is secreted by the

(a) kidneys.

(b) adrenal glands.

(c) pituitary gland.

(d) thyroid gland.

(e) parathyroid glands.

14. Follicle-stimulating hormone (FSH) affects the function of the

(a) ovaries or testes.

(b) pituitary gland.

(c) adrenal glands.

(d) thyroid gland.

(e) parathyroid glands.

15. FSH is secreted by the

(a) ovaries or testes.

(b) pituitary gland.

(c) adrenal glands.

(d) thyroid gland.

(e) parathyroid glands.

16. Oxytocin is a hormone that stimulates the uterus to contract and the mammary glands to release milk. Oxytocin is secreted by the

(a) ovaries.

(b) testes.

(c) pituitary gland.

(d) Answers (a) and (b).

(e) Answers (a) and (c).

17. Luteinizing hormone (LH) stimulates release of estrogen, progesterone, and testosterone, affecting the activity of the

(a) ovaries or testes.

(b) adrenal glands.

(c) pituitary gland.

(d) thyroid gland.

(e) parathyroid glands.

18.   The adrenal glands are located
       (a) within the cerebrum.
       (b) within the triencephalon of the brainstem.
       (c) anterior and lateral to the trachea.
       (d) on top of each kidney.
       (e) below each kidney.

19.   The pituitary gland is located
       (a) within the cerebrum.
       (b) within the triencephalon of the brainstem.
       (c) anterior and lateral to the trachea.
       (d) on top of each kidney.
       (e) below each kidney.

20.   The thyroid gland is located
       (a) within the cerebrum.
       (b) within the triencephalon of the brainstem.
       (c) anterior and lateral to the trachea.
       (d) on top of each kidney.
       (e) below each kidney.

21.   Alpha cells within the islets of Langerhans secrete
       (a) glucagon.
       (b) glycogen.
       (c) glucose.
       (d) insulin.
       (e) follicle-stimulating hormone.

22.   Beta cells within the islets of Langerhans secrete
       (a) glucagon.
       (b) glycogen.
       (c) glucose.
       (d) insulin.
       (e) follicle-stimulating hormone.

23.   The complex carbohydrate form produced by, and stored within, the liver
       is called
       (a) glucagon.
       (b) glycogen.
       (c) glucose.
       (d) glutose.
       (e) glytose.

24. The carbohydrate form normally released into the blood to fuel the body is called
   (a) glucagon.
   (b) glycogen.
   (c) glucose.
   (d) glutose.
   (e) glytose.

25. The pancreatic hormone that stimulates an increase in blood sugar levels is called
   (a) glucagon.
   (b) glycogen.
   (c) glucose.
   (d) insulin.
   (e) somatostatin.

26. The pancreatic hormone responsible for increasing cellular uptake sugar is called
   (a) glucagon.
   (b) glycogen.
   (c) glucose.
   (d) insulin.
   (e) somatostatin.

27. The process by which a stored carbohydrate form is broken down, so as to create the carbohydrate form normally released into the blood, is called
   (a) glycogenolysis.
   (b) glycogenasis.
   (c) glycogenosis.
   (d) gluconeogenesis.
   (e) glutinasis.

28. The process by which new carbohydrate molecules are synthesized from noncarbohydrate sources is called
   (a) glycogenolysis.
   (b) glycogenasis.
   (c) glycogenosis.
   (d) gluconeogenesis.
   (e) glutinasis.

29. Diabetes mellitus is a disorder caused by
   (a) the decreased responsiveness of body cells to insulin.
   (b) inadequate insulin production.
   (c) excessive insulin production.
   (d) Answers (a) and (b).
   (e) Answers (a), (b), and (c).

30.   If no insulin were present in the body,

(a) the body would immediately begin to metabolize fat, which would block all sugar from being able to enter the cells.

(b) no sugar would be able to enter the body's cells.

(c) the amount of sugar that could enter the cells would be far less than that required to meet the body's average energy demands.

(d) the body would immediately begin to use osmosis to bring glycogen into the cells.

(e) the body would immediately begin to use diffusion to bring glycogen into the cells.

31.   Ketone bodies are compounds produced during the

(a) normal metabolic synthesis of glucose into glycogen, for energy production.

(b) abnormal metabolic synthesis of glucose into glycogen, for energy production.

(c) normal synthesis of adipose cells into a form of glycogen, for energy production.

(d) anabolism (construction) of glycogen from fatty acids, for use as an emergency energy source.

(e) catabolism (breakdown) of fatty acids, for use as an emergency energy source.

32.   Which of the following statements regarding Type I Diabetes Mellitus is false?

(a) Type I diabetes is characterized by very low production of insulin.

(b) Type I diabetes is often characterized by the complete absence of insulin production.

(c) Type I diabetes is also called IDDM (insulin-dependent diabetes mellitus).

(d) Type I diabetes is far less common than Type II diabetes.

(e) Type I diabetes is also called adult-onset diabetes due to the average age of patients when they are diagnosed with the disease.

33.   Which of the following statements regarding Type II Diabetes Mellitus is false?

(a) Type II diabetes is characterized by moderately low production of insulin.

(b) Type II diabetes is characterized by a decreased responsiveness of body cells to the insulin present within the body.

(c) Type II diabetes is also called NIDDM (non-insulin-dependent diabetes mellitus).

(d) Diabetic ketoacidosis emergencies are far less common in Type II diabetics than in Type I.

(e) Hyperglycemic hyperosmolar nonketotic diabetic emergencies are far less common in Type II diabetics than in Type I.

34. Which of the following statements regarding a hyperglycemic hyperosmolar nonketotic (HHNK) diabetic emergency is false?

    (a) Insulin activity typically remains present during a HHNK diabetic emergency.

    (b) Glucagon activity typically remains present during a HHNK diabetic emergency.

    (c) Osmotic diuresis causes serious dehydration in patients suffering from a HHNK diabetic emergency.

    (d) Orthostatic hypotension is significantly indicative of a patient suffering from a HHNK diabetic emergency.

    (e) The mortality rate for diabetic ketoacidosis is much higher than that for HHNK diabetic emergencies.

35. The highest blood sugar levels (up to 1000 ml/dL) are most often measured in patients suffering from

    (1) diabetic ketoacidosis.

    (2) hypoglycemia.

    (3) a hyperglycemic hyperosmolar nonketotic diabetic emergency.

        (a) 1 and 2.

        (b) 2 and 3.

        (c) 1.

        (d) 2.

        (e) 3.

36. Any unconscious patient, or any patient with an altered level of consciousness, may be suffering from

    (1) diabetic ketoacidosis.

    (2) hypoglycemia.

    (3) a hyperglycemic hyperosmolar nonketotic diabetic emergency.

        (a) 1, 2, or 3.

        (b) 1 or 3.

        (c) 1.

        (d) 2.

        (e) 3.

37. Diabetic emergency signs and symptoms developing over a period of 12–24 hours most likely indicate that the patient is suffering from

    (1) diabetic ketoacidosis.

    (2) hypoglycemia.

    (3) a hyperglycemic hyperosmolar nonketotic diabetic emergency.

        (a) 1, 2, or 3.

        (b) 1 or 3.

        (c) 1.

        (d) 2.

        (e) 3.

38. Very slow onset of diabetic emergency signs and symptoms, requiring several days to manifest, is most commonly associated with

(1) diabetic ketoacidosis.
(2) hypoglycemia.
(3) a hyperglycemic hyperosmolar nonketotic diabetic emergency.

    (a) 1, 2, and 3.
    (b) 1 and 3.
    (c) 1.
    (d) 2.
    (e) 3.

39. A rapid onset of diabetic emergency signs and symptoms is most commonly associated with

(1) diabetic ketoacidosis.
(2) hypoglycemia.
(3) a hyperglycemic hyperosmolar nonketotic diabetic emergency.

    (a) 1, 2, and 3.
    (b) 1 and 3.
    (c) 1.
    (d) 2.
    (e) 3.

40. Excessive fluid intake, excessive urine output, and abnormally large food intake (excessive hunger) are symptoms commonly associated with patients suffering from

(1) diabetic ketoacidosis.
(2) hypoglycemia.
(3) a hyperglycemic hyperosmolar nonketotic diabetic emergency.

    (a) 1, 2, or 3.
    (b) 1 or 3.
    (c) 2 or 3.
    (d) 2.
    (e) 3.

41. The medical terms for excessive fluid intake, excessive urine output, and abnormally large food intake are

    (a) megaimbibo, multiliquidus, and multiamplus.
    (b) hyperimbibo, hyperliquidus, and hyperamplus.
    (c) hyperdipsia, multiuria, and polyamplus.
    (d) polyimbibo, polyuria, and polydipsia.
    (e) polydipsia, polyuria, and polyphagia.

42.   Complaints of abdominal pain are classically associated with
    (1) diabetic ketoacidosis.
    (2) hypoglycemia.
    (3) a hyperglycemic hyperosmolar nonketotic diabetic emergency.

        (a) 1, 2, and 3.
        (b) 2 and 3.
        (c) 1.
        (d) 2.
        (e) 3.

43.   A diabetic patient complaining of headache is most likely suffering from
    (1) diabetic ketoacidosis.
    (2) hypoglycemia.
    (3) a hyperglycemic hyperosmolar nonketotic diabetic emergency.

        (a) 1, 2, or 3.
        (b) 1 or 3.
        (c) 1.
        (d) 2.
        (e) 3.

44.   Tachycardia is associated with patients suffering from
    (1) diabetic ketoacidosis.
    (2) hypoglycemia.
    (3) a hyperglycemic hyperosmolar nonketotic diabetic emergency.

        (a) 1, 2, or 3.
        (b) 2 or 3.
        (c) 1.
        (d) 2.
        (e) 3.

45.   Cool, diaphoretic skin is associated with patients suffering from
    (1) diabetic ketoacidosis.
    (2) hypoglycemia.
    (3) a hyperglycemic hyperosmolar nonketotic diabetic emergency.

        (a) 1, 2, or 3.
        (b) 1 or 3.
        (c) 1.
        (d) 2.
        (e) 3.

46. Warm, dry skin is associated with patients suffering from
  (1) diabetic ketoacidosis.
  (2) hypoglycemia.
  (3) a hyperglycemic hyperosmolar nonketotic diabetic emergency.

    (a) 1, 2, or 3.
    (b) 1 or 3.
    (c) 1.
    (d) 2.
    (e) 3.

47. Kussmaul's respirations are associated with patients suffering from
  (1) diabetic ketoacidosis.
  (2) hypoglycemia.
  (3) a hyperglycemic hyperosmolar nonketotic diabetic emergency.

    (a) 1, 2, and 3.
    (b) 1 and 3.
    (c) 1.
    (d) 2.
    (e) 3.

48. A fruity, acetone-like breath odor is associated with patients suffering from
  (1) diabetic ketoacidosis.
  (2) hypoglycemia.
  (3) a hyperglycemic hyperosmolar nonketotic diabetic emergency.

    (a) 1, 2, and 3.
    (b) 1 and 3.
    (c) 1.
    (d) 2.
    (e) 3.

49. A diabetic-emergency seizure is most likely to be caused by
  (1) diabetic ketoacidosis.
  (2) hypoglycemia.
  (3) a hyperglycemic hyperosmolar nonketotic diabetic emergency.

    (a) 1, 2, or 3.
    (b) 1 or 3.
    (c) 1.
    (d) 2.
    (e) 3.

50. Vomiting is rarely associated with patients suffering from

    (1) diabetic ketoacidosis.

    (2) hypoglycemia.

    (3) a hyperglycemic hyperosmolar nonketotic diabetic emergency.

        (a) 1 or 2.

        (b) 1 or 3.

        (c) 1.

        (d) 2.

        (e) 3.

51. Management of all types of diabetic emergencies include all of the following treatment measures, except

    (a) IV access D_5W, TKO (obtain blood samples as able).

    (b) If blood sugar level cannot be measured, administer IV $D_{50}W$ to all unconscious patients.

    (c) Administer IV $D_{50}W$ to all patients with a blood sugar level less than 60 mg/dL.

    (d) Administer 100 mg thiamine IV to all patients who receive $D_{50}W$ and appear malnourished or are suspected of alcoholism.

    (e) If IV access is unobtainable, and the patient is hypoglycemic (or blood sugar level cannot be measured), administer glucagon, IM.

52. Mallory-Weiss syndrome is a

    (a) disease of the pancreas.

    (b) disease of the liver.

    (c) tear in the lining of the stomach, caused by explosive coughing.

    (d) tear in the esophagus, usually caused by vomiting.

    (e) rupture of the small intestine, caused by severe overeating.

53. Esophageal varices are most frequently caused by

    (a) chronic alcohol consumption, eroding away the lining of the esophagus.

    (b) liver damage or disease, creating a high-pressure backup of blood flow, which results in engorged and dilated esophageal veins.

    (c) overly aggressive intubation attempts.

    (d) Answers (a) and (b).

    (e) Answers (a) and (c).

54. Acutely ruptured esophageal varices may produce any of the following signs or symptoms, except

    (a) melena and "rebound" abdominal pain.

    (b) bright red, bloody emesis (sometimes explosive, and in copious amounts).

    (c) complaints of dysphagia.

    (d) complaints of a "burning" or "tearing" sensation in the chest or throat.

    (e) signs and symptoms of hypovolemic shock.

55. Which of the following statements regarding treatment of a patient suspected to be suffering from ruptured esophageal varices is false?

    (a) If hypotensive, place the atraumatic patient in the left laterally recumbent position.

    (b) Place the atraumatic patient in a high semi-Fowler's position, if he remains normotensive in that position.

    (c) Establish two IVs, using large-bore (14- to 16-gauge) angiocaths.

    (d) If "blood tubing" (an IV administration tubing that contains a filter) is available, use this for at least one of the two IVs.

    (e) Prehospital placement of a nasogastric tube is of high priority, and may precede advanced airway placement, so as to improve the subsequent success of (or entirely avoid the need for) endotracheal intubation.

56. Bleeding within the upper GI tract should be suspected when the patient reports or exhibits

    (a) bright red emesis.

    (b) coffee-ground emesis.

    (c) black and tarry-colored stool.

    (d) Answers (a) and (b).

    (e) Answers (a), (b), and (c).

57. The medical term for vomiting blood is

    (a) hematemesis.

    (b) hemoptysis.

    (c) hemovomitus.

    (d) hemolysis.

    (e) hemoemesis.

58. Acute gastroenteritis is defined as inflammation of the

    (a) stomach.

    (b) small intestine.

    (c) large intestine.

    (d) Answers (a) and (b).

    (e) Answers (a), (b), and (c).

59. Which of the following signs and symptoms may be directly related to acute gastroenteritis?

    (1) melena   (5) diarrhea

    (2) hemoptysis   (6) tachycardia and hypotension

    (3) hematemesis   (7) fever

    (4) hematochezia (8) hypovolemia

        (a) 1, 2, 3, 4, 5, 6, 7, and 8.

        (b) 1, 3, 4, 5, 6, 7, and 8.

        (c) 2, 5, 6, 7, and 8.

        (d) 2, 5, 6, and 8.

        (e) 5, 6, and 8.

60. Which of the following is classified as an idiopathic inflammatory bowel disorder?

(a) Diverticulitis.
(b) Crohn's disease.
(c) Ulcerative colitis.
(d) Answers (a) and (b).
(e) Answers (b) and (c).

61. Which of the following is caused by the inflammation of small outpouchings of the intestinal mucosal lining?

(a) Diverticulitis.
(b) Crohn's disease.
(c) Ulcerative colitis.
(d) Answers (a) and (b).
(e) Answers (b) and (c).

62. When the blood supply to an area of the bowel becomes obstructed, resulting in tissue necrosis, it is called

(a) a hernia.
(b) a volvulus.
(c) a bowel infarction.
(d) an intussusception.
(e) an adhesion.

63. When a section of intestine protrudes through its protective sheath, producing a partial or complete bowel obstruction, it is called an intestinal

(a) hernia.
(b) volvulus.
(c) infarction.
(d) intussusception.
(e) adhesion.

64. Which of the following signs are classically suggestive of bowel obstruction?

(a) Emesis containing bile.
(b) Emesis that looks and smells like feces.
(c) Absent bowel sounds.
(d) Answers (a) and (c).
(e) Answers (a), (b), and (c).

65. The medical term for a kidney infection is

(a) diverticulitis.
(b) cholecystitis.
(c) appendicitis.
(d) pyelonephritis.
(e) renal calculi.

66. The medical term for an inflammation of the gallbladder is
    (a) diverticulitis.
    (b) cholecystitis.
    (c) appendicitis.
    (d) pyelonephritis.
    (e) renal calculi.

67. Periumbilical pain that radiates or migrates to the right lower quadrant of the abdomen is most frequently associated with
    (a) diverticulitis.
    (b) cholecystitis.
    (c) appendicitis.
    (d) pyelonephritis.
    (e) renal calculi.

68. Unilateral flank pain that radiates or migrates forward into the lower abdominal quadrant of that side or downward into the groin is most frequently associated with
    (a) diverticulitis.
    (b) cholecystitis.
    (c) appendicitis.
    (d) pyelonephritis.
    (e) renal calculi.

69. Fever accompanied by complaint of lower back pain, with urinary burning and frequency, is often associated with
    (a) diverticulitis.
    (b) cholecystitis.
    (c) appendicitis.
    (d) pyelonephritis.
    (e) renal calculi.

70. Upper right quadrant abdominal pain that occurs after a meal high in fat content and that may or may not radiate to the right shoulder is often associated with
    (a) diverticulitis.
    (b) cholecystitis.
    (c) appendicitis.
    (d) pyelonephritis.
    (e) renal calculi.

71. Your patient is lying very still, with knees flexed and drawn close to the chest. This posture is classically suggestive of

  (a) meningitis.
  (b) pyrexia.
  (c) peritonitis.
  (d) pyelonephritis.
  (e) renal calculi.

72. Your patient is grimacing and will not sit down, preferring to pace. This behavior is classically suggestive of

  (a) meningitis.
  (b) pyrexia.
  (c) peritonitis.
  (d) pyelonephritis.
  (e) renal calculi.

73. After opening the door to let you in, your patient slowly walks back to sit on the couch. You notice that she walks with her back hunched up, leaning slightly to one side. She appears febrile and is complaining of back pain. This behavior is classically suggestive of

  (a) meningitis.
  (b) pyrexia.
  (c) peritonitis.
  (d) pyelonephritis.
  (e) renal calculi.

74. The presence of rebound tenderness classically and most frequently indicates

  (a) peritoneal irritation.
  (b) abdominal trauma.
  (c) a bowel obstruction.
  (d) an aortic aneurysm.
  (e) a sexually transmitted disease.

75. Functions of the kidneys include

  (a) maintenance of the body's fluid volume and pH.
  (b) elimination of metabolic waste products.
  (c) control of arterial blood pressure.
  (d) Answers (a) and (b).
  (e) Answers (a), (b), and (c).

76. Renal failure is often caused by

  (a) Type I Diabetes Mellitus.
  (b) Type II Diabetes Mellitus.

(c) uncontrolled, or inadequately controlled, hypertension.

(d) Answers (a) and (c).

(e) Answers (a), (b), and (c).

77. Renal failure results in an increased blood level of urea, a chemical produced by the metabolism of protein by the liver and usually excreted by the kidneys. This increased level of urea in the blood is called

(a) uremia.

(b) hematourea.

(c) oligurea.

(d) polyurea.

(e) megalourea.

78. The medical term for the decreased production or elimination of urine is

(a) anuria.

(b) hematuria.

(c) oliguria.

(d) polyuria.

(e) hypouria.

79. The medical term for the presence of blood in urine is

(a) hematouria.

(b) hematuria.

(c) oliguria.

(d) polyuria.

(e) hemouria.

80. The medical term for the entire lack or absence of urine production or elimination is

(a) anuria.

(b) nocturia.

(c) oliguria.

(d) hypouria.

(e) nonuria.

81. The medical term for excessive urine production or elimination frequency is

(a) polyphagia.

(b) polydipsia.

(c) oliguria.

(d) polyuria.

(e) megalouria.

82. The accumulation of serous fluid (edema) within the abdomen is called
  (a) peritonitis.
  (b) peristalsis.
  (c) ascites.
  (d) gastrotic edema.
  (e) interstitial nephritis.

83. Signs and symptoms of renal failure may include all of the following, except
  (a) tented skin and frequent or excessive urination.
  (b) hypertension, tachycardia, and ECG signs of hyperkalemia.
  (c) hypotension, tachycardia, and ECG signs of hyperkalemia.
  (d) edema of the face, hands, and feet.
  (e) abdominal edema.

84. Common signs and symptoms of a patient suffering from chronic renal failure include all of the following except
  (a) multiple areas of variously aged ecchymosis.
  (b) c/o severe itching, and multiple areas of variously aged scratches.
  (c) absent bowel sounds.
  (d) a dried, "frosty-white" substance on the skin.
  (e) jaundiced skin color.

85. When managing a patient suffering from renal failure, it is important to remember that the renal-failure patient
  (a) requires higher initial doses of medication than a patient with normally functioning kidneys.
  (b) requires more frequent repeat bolus doses (or higher maintenance drip rates) to sustain a therapeutic blood level of medication than does the patient with normally functioning kidneys.
  (c) is more susceptible to toxic responses to EMS-administered drugs than the patient with normally functioning kidneys, even at normal administration doses.
  (d) should never receive a volume-expanding fluid; only D_5W should be used for IV access.
  (e) Answers (a), (b), and (d).

86. Methods of renal dialysis include
  (a) hemodialysis via an external arteriovenous fistula.
  (b) hemodialysis via an indwelling catheter.
  (c) peritoneal dialysis.
  (d) Answers (a) and (b).
  (e) Answers (a), (b), and (c).

87.   Important assessment and management considerations for a dialysis patient include
      (a) measuring blood pressure only on an arm without a dialysis shunt.
      (b) obtaining IV access on an arm without a dialysis shunt.
      (c) if IV access is unsuccessful elsewhere, in life-threatening emergencies (if allowed by your service protocols) the dialysis shunt may be used for IV access but only if a "bruit" can be auscultated over the shunt area.
      (d) Answers (a) and (b).
      (e) Answers (a), (b), and (c).

88.   Complications related to renal dialysis include all of the following, except
      (a) antigen-antibody reactions to chronic dialysate introduction.
      (b) local infection at the site of an arteriovenous fistula.
      (c) thrombosis production and thrombosis-related emergencies.
      (d) peritonitis.
      (e) seizures.

89.   Which of the following statements regarding urinary tract infection (UTI) is false?
      (a) Fluid infusion will stimulate renal function and increase the discomfort of all UTI patients. Thus, only D_5W should be used for UTI patient IV access.
      (b) Complaint of pelvic pain is common to both men and women with UTI.
      (c) The patient may describe a strong or foul odor of the urine.
      (d) Complaints of painful urination, urinary urgency, burning, or difficulty on attempts to urinate commonly accompany UTI.
      (e) In general, UTI is more common in women than in men.

90.   Your 4-year-old patient is alert (has an intact gag reflex) and has ingested approximately 6 ounces of unleaded gasoline. You should consider administration of
      (a) ipecac to induce vomiting.
      (b) activated charcoal.
      (c) an acidic substance, to neutralize the gasoline.
      (d) an alkaline substance, to neutralize the gasoline.
      (e) copious amounts of milk, to dilute the gasoline.

91.   The adult dose of activated charcoal is _____ mixed with water.
      (a) 50–75 grams
      (b) 500–750 grams
      (c) 1 g/kg
      (d) Answer (a) or (c).
      (e) Answer (b) or (c).

92. The pediatric dose of activated charcoal is _____ mixed with water.
   (a) 50–75 grams
   (b) 500–750 grams
   (c) 1 g/kg
   (d) Answer (a) or (c).
   (e) Answer (b) or (c).

93. Activated charcoal products often contain sorbitol, a cathartic, which will
   (a) increase gastric motility (intestinal emptying) and minimize the amount of time a toxin remains in the GI tract.
   (b) enhance the emetic qualities of activated charcoal, minimizing the amount of time a toxin remains in the stomach.
   (c) enhance the absorption capabilities of activated charcoal, so that less charcoal needs to be ingested.
   (d) Answers (a) and (b).
   (e) Answers (b) and (c).

94. When managing the victim of a toxic inhalation, your highest priority should be to
   (a) intubate and hyperventilate the unconscious patient.
   (b) remove the patient from the toxic environment.
   (c) perform the usual primary and secondary exam.
   (d) remove contaminated clothing.
   (e) ensure personal safety prior to accessing the unconscious patient.

95. After accomplishing the toxic-inhalation management task indicated in the previous question, your next highest priority should be to
   (a) intubate and hyperventilate the unconscious patient.
   (b) remove the patient from the toxic environment.
   (c) perform the usual primary and secondary exam.
   (d) remove contaminated clothing.
   (e) ensure personal safety prior to accessing the unconscious patient.

96. Organophosphate poisoning stimulates the parasympathetic nervous system, causing signs and symptoms that include all of the following, except
   (a) excessive salivation, nausea, and vomiting.
   (b) diarrhea and diaphoresis.
   (c) tachycardia.
   (d) dilated pupils.
   (e) hypotension.

97. Organophosphate poisoning management includes all of the following, except
   (a) wearing protective clothing and respiratory protection.
   (b) removal of patient clothing.

(c) patient decontamination with soap and water.

(d) 1.0 mg atropine IV administration, every 3 to 5 minutes, to a maximum dose of 3 mg.

(e) 2-4 mg atropine IV administration, repeated as ordered, even up to 40 mg or more.

98. N-Acetylcysteine is the antidote for an overdose of

(a) acetylsalicylic acid.

(b) acetaminophen.

(c) ibuprofen.

(d) Answers (a) and (b).

(e) Answers (b) and (c).

99. Signs and symptoms of acute acetylsalicylic acid overdose may include any of the following, except

(a) multiple areas of various-aged ecchymosis.

(b) hyperventilation.

(c) hyperthermia.

(d) dysrhythmias and cardiac failure.

(e) lethargy, altered level of consciousness, and unconsciousness.

100. Management considerations for a tricyclic drug overdose include all of the following, except

(a) cardiac monitoring.

(b) 0.2 mg flumazenil, slow IVP.

(c) 1–2 mEq/kg sodium bicarbonate, IV.

(d) aggressive airway management.

(e) ventilatory support.

101. Which of the following statements regarding treatment of *Hymenoptera* stings is true?

(a) Remove the stinger by using forceps or tweezers (not your fingers).

(b) Apply ice directly to the injection/bite site.

(c) Be alert for allergic reactions and anaphylactic shock.

(d) Answers (a) and (c).

(e) Answers (a), (b), and (c).

102. Observation of a violin-shaped marking is associated with the identification of

(a) a black widow spider.

(b) a brown recluse spider.

(c) a nonpoisonous spider.

(d) a poisonous scorpion.

(e) a nonpoisonous scorpion.

103. Observation of a yellow- or red-colored hourglass-shaped marking is associated with the identification of
  (a) a black widow spider.
  (b) a brown recluse spider.
  (c) a nonpoisonous spider.
  (d) a poisonous scorpion.
  (e) a nonpoisonous scorpion.

104. There is no specific antivenin or antiserum for
  (a) a black widow spider bite.
  (b) a brown recluse spider bite.
  (c) a scorpion sting.
  (d) Answers (a) and (b).
  (e) Answers (b) and (c).

105. Analgesic administration may increase the venom toxicity of
  (a) a black widow spider bite.
  (b) a brown recluse spider bite.
  (c) a scorpion sting.
  (d) Answers (a) and (b).
  (e) Answers (a) and (c).

106. Consider administration of diazepam or calcium gluconate for treatment of severe muscle spasms secondary to
  (a) a black widow spider bite.
  (b) a brown recluse spider bite.
  (c) a scorpion sting.
  (d) Answers (a) and (b).
  (e) Answers (a) and (c).

107. The venom of the _____ contains a neurotoxin that can produce respiratory and skeletal muscle paralysis.
  (a) rattlesnake
  (b) copperhead snake
  (c) cottonmouth snake
  (d) coral snake
  (e) water moccasin

108. "Red touch yellow, kill a fellow; red touch black, venom lack," accurately refers to the distinctive appearance (and dangerousness) of the two different kinds of
  (a) rattlesnake.
  (b) copperhead snake.
  (c) cottonmouth snake.
  (d) coral snake.
  (e) water moccasin.

109. Which of the following statements regarding treatment of a snake bite is true?

    (a) Apply an arterial tourniquet proximal to the bite, and a venous tourniquet distal to the bite.

    (b) Apply ice, a cold pack, or commercial snake bite freon spray to the wound.

    (c) Immobilize the bitten limb with a splint.

    (d) Elevate the bitten limb to diminish edema and minimize tissue damage.

    (e) As soon as possible, make an X incision over each bite mark and apply suction, using a commercial snake bite suction cup (not your mouth).

110. Blood type B contains

    (a) antibody A, antibody B, and no antigen.

    (b) antigen A, antigen B, and no antibody.

    (c) antigen B and antibody A.

    (d) antigen A and antibody B.

    (e) no antigen and no antibody.

111. Blood type O contains

    (a) antibody A, antibody B, and no antigen.

    (b) antigen A, antigen B, and no antibody.

    (c) antigen B and antibody A.

    (d) antigen A and antibody B.

    (e) no antigen and no antibody.

112. Blood type AB contains

    (a) antibody A, antibody B, and no antigen.

    (b) antigen A, antigen B, and no antibody.

    (c) antigen B and antibody A.

    (d) antigen A and antibody B.

    (e) no antigen and no antibody.

113. A person who has blood type ___ is called a "universal donor."

    (a) D

    (b) A

    (c) B

    (d) AB

    (e) O

114. A person who has blood type ___ is called a "universal recipient."

    (a) R

    (b) A

    (c) B

    (d) AB

    (e) O

115. When a person with Rh negative blood is first exposed to Rh positive blood
   (a) no adverse reaction occurs.
   (b) the person becomes sensitized and is thereafter able to produce Rh antibodies.
   (c) a severe, often fatal, allergic reaction occurs.
   (d) Answers (a) and (b).
   (e) Answers (b) and (c).

116. Transfusion reaction signs and symptoms include all of the following, except
   (a) facial flushing.
   (b) sinus bradycardia or A-V blocks.
   (c) hyperventilation.
   (d) urticaria.
   (e) fever and chills.

117. Management considerations for treatment of a transfusion-reaction patient include all of the following, except
   (a) stop the transfusion, change all the IV tubing, resume IV therapy with D_5W TKO.
   (b) furosemide administration to promote diuresis.
   (c) diphenhydramine administration to counteract histamine-related side effects.
   (d) low-dose dopamine administration to maintain renal perfusion.
   (e) IV epinephrine administration if anaphylactic reaction ensues.

118. Signs and symptoms of circulatory overload after a blood transfusion include all of the following, except
   (a) pulmonary edema.
   (b) chest pain.
   (c) dyspnea.
   (d) tachycardia.
   (e) hypotension.

119. A condition known as _____ can occur secondary to dehydration and increases the patient's risk of thrombosis formation.
   (a) anemia
   (b) leukocytosis
   (c) polycythemia
   (d) leukemia
   (e) sickle cell disease

120. _____ can cause episodes of *vasoocclusive crisis*, which are often extremely painful. Morphine or another parenteral analgesic should be administered in the field as soon as IV access is obtained.

    (a) Anemia

    (b) Leukocytosis

    (c) Polycythemia

    (d) Leukemia

    (e) Sickle cell disease

The Answer Key to Test Section 9 is on page 446.

10

Test
Section
Ten

Test Section 10's subjects:

- Alcoholism and Substance Abuse
- Environmental Conditions
- Infectious and Communicable Diseases
- Behavioral and Psychiatric Disorders
- Abuse and Assault
- Patients with Special Challenges

Test Section 10 consists of 103 questions and is allotted 1 hour and 45 minutes for completion. (One question is unusually long, and should be allotted at least 3 minutes.)

1. All of the following drugs are narcotics, except
    (a) Darvon®.
    (b) codeine.
    (c) Demerol®.
    (d) heroin.
    (e) cocaine.

2. Larger than average doses of naloxone will be required to manage an overdose of synthetic narcotics, such as
    (a) Darvon®.
    (b) codeine.
    (c) Demerol®.
    (d) heroin.
    (e) cocaine.

3. β-blocker medication administration for treatment of tachycardias is absolutely contraindicated if the patient is suspected of having overdosed on
    (a) amphetamines.
    (b) hallucinogens.
    (c) cocaine.
    (d) Answers (a) and (c).
    (e) Answers (a), (b), and (c).

4. Diazepam may be administered to individuals suffering seizure activity secondary to an overdose of
    (a) amphetamines.
    (b) hallucinogens.
    (c) cocaine.
    (d) Answers (a) and (c).
    (e) Answers (a), (b), and (c).

5. Chronic alcohol abuse can directly result in all of the following diseases or conditions, except
    (a) emphysema, chronic bronchitis, and asthma.
    (b) malnutrition and hypoglycemia.
    (c) pancreatitis.
    (d) cirrhosis of the liver.
    (e) a subdural hematoma.

6.   All of the following diseases or conditions may "mimic" the S/Sx of alcohol intoxication, except

(a) an acute abdomen.

(b) hypoglycemia.

(c) sepsis.

(d) diabetic ketoacidosis.

(e) a subdural hematoma.

7.   Which of the following statements regarding alcohol withdrawal syndrome is false?

(a) An individual who is not an "alcoholic" may suffer from alcohol withdrawal syndrome, if her/his blood alcohol level acutely falls after an episode of excessive use.

(b) Although it is seriously psychologically and physically debilitating, alcohol withdrawal syndrome is rarely a "fatal" physiological condition.

(c) The "DTs" (delirium tremens), usually develop on the second or third day after an alcoholic abruptly discontinues alcohol ingestion.

(d) "Rum fits" (seizures) may occur, usually within the first 24–36 hours after an alcoholic discontinues alcohol ingestion.

(e) Hallucinations, visual or tactile, accompany the DTs.

8.   Which of the following statements regarding alcohol withdrawal syndrome S/Sx and management considerations is false?

(a) S/Sx of withdrawal may occur within hours after an alcoholic abruptly discontinues alcohol ingestion.

(b) Alcohol withdrawal syndrome S/Sx commonly last only 2–5 days.

(c) Seizures or DTs are signs of a serious medical emergency. Aggressive prehospital management is warranted.

(d) Any patient exhibiting S/Sx of DTs should be considered to be "fatally" ill.

(e) IV diazepam is the recommended prehospital treatment for severe DTs or seizure activity related to alcohol withdrawal.

9.   S/Sx of alcohol withdrawal syndrome include all of the following, except

(a) bradycardia (sinus, or A-V blocks).

(b) nausea, vomiting, diaphoresis.

(c) generalized weakness and orthostatic hypotension.

(d) irritability, anxiety, or depression.

(e) hypertension.

10. Which of the following S/Sx and condition descriptions represent those of Korsakoff's psychosis?

(1) hallucinations	(5) related to chronic alcohol intake
(2) memory loss or disorientation	(6) related to a thiamine deficiency
(3) bilateral "foot drop"	(7) commonly a reversible condition
(4) delusions	(8) commonly an irreversible condition

   (a) 1, 2, 3, 4, 5, 6, and 7.
   (b) 1, 2, 3, 4, 5, 6, and 8.
   (c) 3, 5, 6, and 8.
   (d) 2, 5, 6, and 7.
   (e) 5, 6, and 8.

11. Which of the following S/Sx and condition descriptions represent those of Wernicke's syndrome?

(1) hallucinations	(5) related to chronic alcohol intake
(2) memory loss or disorientation	(6) related to a thiamine deficiency
(3) bilateral "foot drop"	(7) commonly a reversible condition
(4) delusions	(8) commonly an irreversible condition

   (a) 1, 2, 3, 4, 5, 6, and 7.
   (b) 1, 2, 3, 4, 5, 6, and 8.
   (c) 3, 5, 6, and 8.
   (d) 2, 5, 6, and 7.
   (e) 5, 6, and 8.

12. Which of the following statements regarding administration of thiamine is false?
   (a) Thiamine is a B vitamin extremely important to cellular carbohydrate metabolism.
   (b) Without an adequate amount of thiamine present, the body cannot obtain the full benefit of the glucose available in the blood stream.
   (c) The brain is extremely sensitive to thiamine deficiency.
   (d) Administration of thiamine when the patient is not malnourished or alcoholic may cause a severe (often fatal) B-vitamin overdose.
   (e) Administration of $D_{50}W$ without the concomitant administration of thiamine may precipitate Wernicke's or Korsakoff's syndromes in the alcoholic patient.

13. The body's primary thermoregulatory control mechanism is located in the
   (a) temporal cerebrum.
   (b) frontal cerebrum.
   (c) cerebellum.
   (d) triencephalon.
   (e) hypothalamus.

14. Heat loss from the body that occurs because of sweating is attributed to the mechanism of
    (a) radiation.
    (b) conduction.
    (c) convection.
    (d) evaporation.
    (e) respiration.

15. Cold packs placed upon areas of the body surface effect body heat loss via the mechanism of
    (a) radiation.
    (b) conduction.
    (c) convection.
    (d) evaporation.
    (e) respiration.

16. Heat loss from the body when air currents pass over its surface is attributed to the mechanism of
    (a) radiation.
    (b) conduction.
    (c) convection.
    (d) evaporation.
    (e) respiration.

17. Moist, warm skin commonly accompanies
    (a) heat stroke.
    (b) heat cramps.
    (c) heat exhaustion.
    (d) Answers (a) and (b).
    (e) Answers (a) and (c).

18. Confusion or an altered level of consciousness commonly accompanies
    (a) heat stroke.
    (b) heat cramps.
    (c) heat exhaustion.
    (d) Answers (a) and (c).
    (e) Answers (b) and (c).

19. Cool, clammy skin accompanies
    (a) heat stroke.
    (b) heat cramps.
    (c) heat exhaustion.
    (d) Answers (a) and (c).
    (e) None of the above.

20. Hyperkalemia may accompany exertional
(a) heat stroke.
(b) heat cramps.
(c) heat exhaustion.
(d) Answers (a) and (c).
(e) Answers (b) and (c).

21. If the patient is alert and able to swallow, oral administration of one or more salt tablets is appropriate to the treatment of
(a) heat stroke.
(b) heat cramps.
(c) heat exhaustion.
(d) Answers (a), (b), and (c).
(e) None of the above.

22. If the patient is alert and able to swallow, oral administration of a "sports" drink or other beverage containing sodium is appropriate to the treatment of
(a) heat stroke.
(b) heat cramps.
(c) heat exhaustion.
(d) Answers (a), (b), and (c).
(e) None of the above.

23. Emergency care of a patient suffering from a serious heat-related illness includes removal of clothing and
(1) covering with sheets soaked in tepid water.
(2) covering with sheets soaked in ice water.
(3) placing ice packs in the axillary areas, groin, and bilateral to the neck.
(4) establishing 1 or 2 large-bore IV's, with LR or NS initially infused wide open.
(5) if hypotension is unresponsive to a 1000 ml LR or NS infusion, administering a vasopressor, titrated to a systolic blood pressure of 100 mmHg.

(a) 1 or 2, 3, 4, and 5.
(b) 2 or 3, 4 and 5.
(c) 2, 3, 4 and 5.
(d) 1 and 4.
(e) 4 and 5.

24. When measured orally, the average normal body temperature is
(a) 37 degrees centigrade (C).
(b) 98.6 degrees Fahrenheit (F).
(c) 44 degrees centigrade (C).
(d) Answers (a) and (b).
(e) Answers (b) and (c).

25. To measure the patient's core temperature, _____ thermometers may be used.

    (a) tympanic or rectal

    (b) skin-strip or tympanic

    (c) only rectal

    (d) only tympanic

    (e) skin-strip, tympanic, or rectal

26. The medical term for an elevated body temperature caused by an infection is

    (a) hyperthermia.

    (b) hypothermia.

    (c) pyrexia.

    (d) pathogenia.

    (e) hyprexia.

27. Emergency care of a patient suffering from a serious fever includes removal of external clothing and

    (1) exposing them to ambient air.

    (2) sponging them with cold water.

    (3) immersing them in cool water.

    (4) covering them with sheets soaked in ice water.

    (5) placing ice packs in the axillary areas, groin, and bilateral to the neck.

        (a) 1.

        (b) 1 and 2.

        (c) 3, followed by 4.

        (d) 4.

        (e) 4 and 5.

28. Mild hypothermia is best defined as a patient symptomatic of hypothermia, with a core temperature of

    (a) less than 44° Celsius (98.6° Fahrenheit).

    (b) less than 37° Celsius (98.6° Fahrenheit).

    (c) 34 to 36° Celsius (93.2° to 96.8° Fahrenheit).

    (d) 30 to 34° Celsius (86° to 93.2° Fahrenheit).

    (e) less than 30° Celsius (90° Fahrenheit).

29. Moderate hypothermia is best defined as a patient symptomatic of hypothermia, with a core temperature of

    (a) less than 37° Celsius (95.6° Fahrenheit).

    (b) 34 to 36° Celsius (90° to 92° Fahrenheit).

    (c) 30 to 34° Celsius (86° to 93.2° Fahrenheit).

    (d) less than 30° Celsius (90° Fahrenheit).

    (e) less than 20° Celsius (80° Fahrenheit).

30.   Severe hypothermia is best defined as a patient symptomatic of hypothermia, with
a core temperature of
   (a) 34° to 36° Celsius (90° to 92° Fahrenheit).
   (b) 30° to 34° Celsius (86° to 93.2° Fahrenheit).
   (c) less than 30° Celsius (86° Fahrenheit).
   (d) less than 27° Celsius (80.6° Fahrenheit).
   (e) less than 10° Celsius (70° Fahrenheit).

31.   Research indicates that shivering ceases and pupils dilate when a patient's core
temperature falls to
   (a) 37° Celsius (98.6° Fahrenheit).
   (b) 36° Celsius (96.8° Fahrenheit).
   (c) 34° Celsius (93.2° Fahrenheit).
   (d) 32° Celsius (89.6° Fahrenheit).
   (e) 30° Celsius (86° Fahrenheit).

32.   The ECG of a hypothermic patient may exhibit
   (a) J waves, also called "Jackson" waves.
   (b) J waves, also called "Osborn" waves.
   (c) O waves, also called "Osborn" waves.
   (d) prominent U waves, indicative of cold-induced hyperkalemia.
   (e) prominent U waves, indicative of cold-induced hypokalemia.

33.   Although not diagnostically useful, question 32's waves may be observed in lead II
   (a) between the P wave and the QRS.
   (b) on the upslope of the R wave.
   (c) on the downslope of the R wave or at the junction of the QRS and T wave.
   (d) after the T wave, prior to the next P wave.
   (e) in the middle of the P wave.

34.   The most common dysrhythmia associated with moderate hypothermia is
   (a) sinus bradycardia.
   (b) atrial fibrillation.
   (c) ventricular tachycardia.
   (d) ventricular fibrillation.
   (e) Answer (a) or (c).

35.   When a patient is moderately to severely hypothermic, ventricular fibrillation may
be induced by
   (a) rough handling.
   (b) "rewarming shock."
   (c) orotracheal or nasotracheal intubation stimulation.
   (d) Answers (a) and (b).
   (e) Answers (a), (b), and (c).

36. Treatment of mild hypothermia includes removal of all wet clothing and

    (a) protection from further heat loss.

    (b) placement of insulated heat packs in the axillary areas, groin, and bilateral to the neck.

    (c) oral administration of warmed, sweetened fluids, if the patient is alert and able to swallow.

    (d) Answers (a) and (c).

    (e) Answers (a), (b), and (c).

37. Treatment of moderate and severe hypothermia includes administration of warmed and humidified oxygen, and

    (a) oral administration of warmed, sweetened fluids, if the patient is alert and able to swallow.

    (b) placement of insulated heat packs in the axillary areas, groin, and bilateral to the neck.

    (c) IV access (preferably a "central" location, such as EJ) and warmed NS infusion.

    (d) Answers (b) and (c).

    (e) Answers (a), (b), and (c).

Author's note: The following question is unusually long, and will require more than the standard 1 minute allowed per question. Allow at least 3 minutes to answer this question.

38. Your patient is found lying, pulseless and apneic, on a snow bank at the edge of a parking lot next to a lake. You are able to park your vehicle within 5 feet of him. First, from the following list of cardiac arrest treatment procedures, select only those procedures you would perform, *no matter what the hypothermic patient's core temperature was.* Second, indicate the order in which you would perform the treatment procedures you selected.

    (1) Initiate CPR and continue as appropriate.

    (2) Intubate, confirm endotracheal placement, and secure the ET tube.

    (3) Check ECG rhythm; V-fib (ventricular fibrillation) discovered.

    (4) Remove all wet garments and wrap in warm blankets.

    (5) Rapidly but gently place the patient in the ambulance.

    (6) Defibrillate once.

    (7) Defibrillate a second time if first defib is unsuccessful and V-fib persists.

    (8) Defibrillate a third time if second defib is unsuccessful and V-fib persists.

    (9) Establish IV access.

    (10) Administer first-round IV medications, per standard 2005 AHA ACLS Tx protocols.

    (11) If V-fib persists after first-round IV medication administration, defibrillate up to three more times before a second round of medications.

        (a) 1, 4, 5, 3, 6, and 9.

        (b) 5, 4, 1, 3, 6, 7, 8, 2, and 9.

        (c) 4, 5, 3, 6, 1, 9, 10, and 11.

        (d) 5, 4, 1, 3, 6, 2, and 9.

        (e) 1, 3, 6, 7, 8, 5, 4, 9, and 10.

39. Your patient is apneic, pulseless, and in ventricular fibrillation. She has a core temperature of 31° Celsius (87.8° Fahrenheit). According to core-temperature-based 2005 AHA ACLS hypothermic cardiac arrest treatment protocols, which of the following procedures should you perform when caring for her?

(1) Ventilate gently with BVM (intubation may precipitate or sustain V-fib).

(2) Intubate and ventilate with warmed, humidified oxygen.

(3) IV access and warm fluid administration.

(4) IV access and cool (room temperature) fluid administration.

(5) ACLS medication administration, per standard AHA treatment protocols.

(6) ACLS medication administration, but with increased time intervals between doses.

(7) 1 defibrillation attempt.

(8) A 2nd defibrillation, immediately after the 1st defibrillation, if V-fib persists.

(9) A 3rd defibrillation, immediately after the 2nd defibrillation, if V-fib persists.

(10) A 4th defibrillation, if V-fib persists after "first round" ACLS medication administration.

(11) A 5th defibrillation immediately after the 4th defibrillation, if V-fib persists.

(12) A 6th defibrillation immediately after the 5th defibrillation, if V-fib persists.

    (a) 2, 3, 6, and 7.

    (b) 1, 4, 5, 7, 8, and 9.

    (c) 2, 3, 5, and 7 through 12.

    (d) 2, 3, and 7.

    (e) 1, 4, 5, and 7.

40. Your patient is apneic, pulseless, and in ventricular fibrillation. He has a core temperature of 29° Celsius (84.2° Fahrenheit). According to core-temperature-based 2005 AHA ACLS hypothermic cardiac arrest treatment protocols, which of the care procedure options provided for question 39 should you perform when caring for him?

    (a) 2, 3, 6, and 7.

    (b) 1, 4, 6, and 7 through 12.

    (c) 1, 4, 7, 8, and 9.

    (d) 2, 3, and 7.

    (e) 2, 3, 7, 8, and 9.

41. Which of the following statements regarding the pathophysiology of frostbite is false?

    (a) Initial exposure to subfreezing temperatures results in a reflex external vasodilation, shunting increased circulation to peripheral body parts (such as toes and fingers, ears, and the nose), accelerating surface tissue heat loss, and ultimately damaging surface cellular tissue.

    (b) Subfreezing temperature exposure results in ice-crystal formation within intracellular fluid, causing cellular damage.

(c) The expansion of forming ice crystals results in further structural damage of the frozen cells.

(d) When cellular tissue is frozen, biochemical electrolyte alterations occur in the surrounding intracellular spaces, resulting in additional structural damage.

(e) Frozen tissues suffer from anoxia and ischemia, which may result in necrosis.

42. Which of the following statements regarding treatment of frostbite is true?

(a) Rubbing or massaging the frozen body part is extremely painful but is necessary to accomplish rapid thawing, and to prevent further crystallization damage.

(b) Keep the thawed part dependent (below heart level) to enhance circulation to the thawed tissues.

(c) Allow the frozen part to remain frozen if there is any possibility that, should it be thawed, it might subsequently become exposed to subfreezing temperatures again.

(d) Do not administer analgesia to a patient suffering from mere frostbite.

(e) Assist the patient in moving and "working" the frozen part, in order to increase circulation to the frozen area, especially after the area has thawed.

43. Drowning can be caused by any kind of liquid submersion/immersion event, and is

(a) usually an entirely preventable cause of accidental (unintentional) injury or death.

(b) legally defined as a death that occurs within 24 hours of a submersion/immersion event.

(c) usually survived with a good outcome if the victim spontaneously regains circulation and breathing before reaching the ED.

(d) Answers (a) and (c).

(e) Answers (a), (b), and (c).

44. The phrase "near drowning"

(a) is no longer considered a valid diagnosis or victim description. Anyone should be considered a "drowning" incident victim whether or not she/he survives or dies following a submersion/immersion incident.

(b) indicates that death did not occur following a submersion/immersion event.

(c) describes a death that occurred more than 24 hours after the submersion/immersion event.

(d) describes a death that occurred more than 48 hours after the submersion/immersion event.

(e) Answers (b) and (c) or (d), depending upon the local medical examiner's opinion.

45. Compared to the osmotic properties of human blood, fresh water is

(a) hypertonic.

(b) isotonic.

(c) hypotonic.

(d) Answer (a) or (b), depending upon the individual's blood type.

(e) Answer (b) or (c), depending upon the individual's blood type.

46. Compared to the osmotic pressure of human blood, salt water is

    (a) hypertonic.

    (b) isotonic.

    (c) hypotonic.

    (d) Answer (a) or (b), depending upon the individual's blood type.

    (e) Answer (b) or (c), depending upon the individual's blood type.

47. Fresh-water aspiration produces

    (a) hemodilution, a relative reduction of RBC (red blood cell) concentration, and expansion of blood plasma volume.

    (b) surfactant destruction and atelectasis.

    (c) hypovolemia, a relative increase of RBC concentration, and pulmonary edema.

    (d) Answers (a) and (b).

    (e) Answers (b) and (c).

48. Which of the following statements regarding factors that influence resuscitation success and near drowning survival is false?

    (a) The colder the water, the greater the probability of successful resuscitation and survival.

    (b) Beyond 60 minutes of submersion, resuscitation is unlikely.

    (c) Children survive longer submersion times with a greater probability of successful resuscitation and survival.

    (d) Cleanliness of the water affects the probability of survival.

    (e) The "mammalian diving reflex" accelerates the pulse rate, improving the probability of successful resuscitation and survival by increasing cardiac output to all vital organs.

49. Which of the following statements regarding treatment of unconscious drowning victims is false?

    (a) Only a trained rescue swimmer, secured by a safety line, should be allowed to enter the water.

    (b) Treat the patient for spinal injury if there is any indication of trauma, if the cause of submersion is unknown, or if alcohol is involved.

    (c) Ventilation of the patient (mouth-to-mouth resuscitation) should be immediately initiated, even while still in the water.

    (d) Upon removal from the water, the Heimlich maneuver should be performed 4–6 times prior to beginning mechanical ventilation, to facilitate lung drainage and enable more effective oxygen exchange.

    (e) All victims of a submersion/immersion emergency require hospital admission for observation, because complications may not occur during the first 24 hours.

50. Barotrauma is best defined as injury that occurs

    (a) because of changes in surrounding atmospheric pressure.

    (b) because the chest has become hyperexpanded (like a barrel).

(c) because the chest has become hypercompressed (losing its barrel shape).

(d) Answers (a) and (c).

(e) Answer (b) or (c).

51. Which of the following statements regarding injuries associated with fresh- or salt-water diving is false?

(a) Injuries that may occur during ascent include pneumomediastinum and pneumothorax.

(b) Barotrauma to the middle ear may occur during descent to depth, especially if the diver has an upper respiratory tract infection.

(c) At the bottom of the dive, if nitrogen narcosis develops, the diver may incur trauma resulting from euphoria or impaired judgment causing increased risk taking.

(d) Barotrauma to the middle ear or lung parenchyma may occur during ascent.

(e) Injuries that occur while divers are still on the water surface (direct trauma, or entanglement in lines or aquatic matter) are not classified as "diving" injuries.

52. All of the following conditions may occur secondary to a diver holding her/his breath during ascent, except

(a) lung tissue trauma caused by expansion of trapped air.

(b) cardiovascular circulatory obstruction resulting from an air embolism.

(c) a stroke (CVA) resulting from an air embolism.

(d) a pneumothorax.

(e) nitrogen narcosis.

53. Which of the following statements about decompression illness ("the bends") is false?

(a) The bends occur from nitrogen bubbles forming in the blood and body tissues.

(b) The bends may occur as a result of an ascent from even only 3–6 feet below the water surface.

(c) Principle S/Sx of the bends include joint or abdominal pain.

(d) Principle S/Sx of the bends include fatigue, paresthesias, and CNS disturbances.

(e) Patients suffering from the bends may not seek treatment for up to 12–24 hours after the dive.

54. Treatment of a patient suffering from a decompression illness includes all of the following, except

(a) treatment of seizures with IV diazepam.

(b) protection from excessive heat, cold, wetness, or noxious fumes.

(c) oral administration of fruit juices or "sports" drinks, if the patient is alert and able to swallow.

(d) high semi-Fowlers positioning to avoid further CNS disturbances.

(e) transportation to the nearest 911-participating emergency department, whether or not it is a facility capable of providing recompression via a hyperbaric oxygen chamber.

55. Which of the following statements regarding pulmonary overpressure diving accidents is false?
  (a) They are the most serious form of ascent-related barotrauma.
  (b) They occur only during ascent from 33 feet (or more) below the water surface.
  (c) Principle S/Sx of this kind of diving accident include hemoptysis and chest pain.
  (d) The most common S/Sx of this kind of diving accident are dyspnea and diminished breath sounds in an area of the chest.
  (e) Treatment of pulmonary overpressure diving accidents is the same as that for pneumothorax; hyperbaric oxygen recompression is usually not necessary.

56. Which of the following statements regarding arterial gas embolism (AGE) is false?
  (a) S/Sx of AGE are abrupt and severe, usually occurring rapidly after ascent, commonly within 10 minutes.
  (b) Principle S/Sx of AGE include c/o "sharp" or "tearing" pain in one or more parts of the body.
  (c) AGE may rapidly produce stroke-like S/Sx.
  (d) Confusion, vertigo, visual disturbances, or loss of consciousness may accompany AGE.
  (e) Treatment of AGE is the same as that for any other embolism emergency; hyperbaric oxygen recompression is usually not necessary.

57. Treatment of patients suffering from arterial gas embolism includes all of the following, except
  (a) administration of 100% oxygen via nonrebreather mask, unless the patient's level of consciousness suggests endotracheal intubation.
  (b) consideration of corticosteroid administration.
  (c) IV fluid administration titrated to blood pressure.
  (d) high semi-Fowlers positioning to avoid further CNS disturbances.
  (e) rapid transport to the nearest emergency department of a hospital capable of providing emergency recompression via a hyperbaric oxygen chamber.

58. Which of the following statements regarding nitrogen narcosis (the "raptures of the deep") is false?
  (a) Nitrogen narcosis occurs only while at a deep dive depth (greater than 33 feet below the water surface).
  (b) To avoid nitrogen narcosis, deep divers' oxygen supply is mixed with helium.
  (c) The cure for nitrogen narcosis is ascent to a more shallow depth or the water surface.
  (d) Any altered level of consciousness S/Sx may be caused by nitrogen narcosis.
  (e) If nitrogen narcosis is suspected, rapidly transport the patient to the nearest emergency department of a hospital capable of providing emergency recompression via a hyperbaric oxygen chamber.

59.   Which of the following statements regarding ascent to high altitude is false?

    (a) Individuals with chronic diseases are extremely susceptible to significant side effects or illnesses from rapid ascent to a higher altitude, such as flying in a pressurized cabin to a skiing-vacation location.

    (b) If an individual is relatively healthy, rapid ascent to a higher altitude is entirely unlikely to produce significant side effects or high-altitude-related illnesses.

    (c) During ascent from sea level, the atmospheric barometric pressure decreases.

    (d) During ascent from sea level, the partial pressure of atmospheric oxygen decreases.

    (e) Gradual ascent to altitude (over several days, preferably weeks) allows individuals with chronic illnesses an opportunity to gradually acclimatize to high altitudes, thus avoiding development of high-altitude-illness-related S/Sx.

60.   S/Sx of the mild form of acute mountain sickness (AMS) include all of the following, except

    (a) lightheadedness or breathlessness.

    (b) weakness.

    (c) headache.

    (d) an altered level of consciousness.

    (e) nausea and vomiting.

61.   S/Sx of the severe form of AMS include all of the following, except

    (a) shortness of breath.

    (b) profound weakness.

    (c) increased urine output.

    (d) an altered level of consciousness.

    (e) nausea and excessive vomiting.

62.   Which of the following statements regarding high-altitude pulmonary edema (HAPE) is false?

    (a) In the early stages, descent to a lower altitude will reverse all of the S/Sx caused by HAPE.

    (b) Changes in blood flow at high altitudes are what cause the increased pulmonary pressure and hypertension associated with HAPE.

    (c) Children are more susceptible to HAPE than are adults.

    (d) HAPE emergencies are frequently frightening and sometimes painful but rarely result in death.

    (e) S/Sx of HAPE range from mild SOB with dry cough and exertional weakness, to severe weakness, cyanosis, frothy sputum production, and unconsciousness.

63. Which of the following statements regarding high-altitude cerebral edema (HACE) is false?

    (a) Descent to a lower altitude will not reverse any of the S/Sx produced by HACE.

    (b) HACE usually occurs in untreated individuals suffering from AMS or HAPE.

    (c) Headache, nausea, and vomiting are infrequent S/Sx of HACE.

    (d) Altered level of consciousness is a primary S/Sx of HACE.

    (e) The exact cause of HACE remains unknown.

64. Place the following forms of ionizing radiation in order, from least penetrating and least dangerous to most penetrating and most dangerous.

    (1) Neutron exposure   (3) Gamma rays

    (2) Beta particles   (4) Alpha particles

        (a) 1, 2, 3, 4.

        (b) 1, 3, 2, 4.

        (c) 4, 2, 1, 3.

        (d) 1, 4, 2, 3.

        (e) 4, 2, 3, 1.

65. Which of the following statements regarding radiation exposure is false?

    (a) The longer the exposure duration, the greater the radiation received.

    (b) The greater the distance from the source, the smaller the amount of radiation received.

    (c) The more material located between the source and the victim (shielding), the less radiation received.

    (d) The stronger the radiation source, the more distance or shielding required for safety.

    (e) All adverse radiation exposure S/Sx require weeks, months, even years to manifest. Thus any acute onset of S/Sx at the site of a radiation accident indicates injuries or illnesses unrelated to radiation.

66. Which of the following statements regarding radiation injury management risks is false?

    (a) Patients internally contaminated (having ingested or inhaled radioactive particles) will remain "radioactive" for up to three days.

    (b) A "clean" accident occurs when the victim is exposed to radiation but does not come in contact with radioactive dust, liquids, gases, or smoke.

    (c) A "dirty" accident occurs when the victim is exposed to radioactive dust, liquids, gases, or smoke.

    (d) Once properly decontaminated, patients who have been exposed to ionizing radiation pose no hazard to rescue personnel.

    (e) After treating an externally contaminated radiation victim, radiation decontamination of rescue personnel, vehicles, and all equipment is required.

67. The immune system is the body's defense against disease. All of the following are major components of the immune system, except
       (a) leukocytes.
       (b) lymphocytes.
       (c) antigens.
       (d) antibodies.
       (e) macrophages.

68. The lymphatic system is comprised of all of the following, except
       (a) the spleen.
       (b) the thymus.
       (c) the adrenal glands.
       (d) lymph nodes.
       (e) lymphatic ducts.

69. The most essential component of the lymphatic system is the
       (a) spleen.
       (b) thymus.
       (c) adrenal glands.
       (d) lymph nodes.
       (e) lymphatic ducts.

70. All of the following statements represent mandates established by the Ryan White Act, a federal law, except
       (a) when exposure to an infectious disease is suspected, agencies and health care workers have specific rights and responsibilities.
       (b) an exposed care provider has a right to information about the infection status of any patient suspected to have exposed the provider to an infectious disease.
       (c) employers are required to notify their employees as to what to do if an exposure occurs.
       (d) individuals suspected to have exposed a care provider to an infectious disease are required to submit to testing, to determine whether or not they have the suspected infectious disease.
       (e) each health care agency is required to designate an infectious disease control officer to serve as a contact person for exposed personnel and to maintain the agency's infection control program.

71. Which of the following statements regarding infectious diseases is false?
    (a) Because viruses are smaller than bacteria, viral infectious diseases respond more readily to antibiotic treatment than do bacterial infectious diseases.
    (b) Pinworms are an infectious disease generally treated with a single antibiotic dose.
    (c) Hookworms are an infectious disease generally treated with a single antibiotic dose.
    (d) Trichinosis is an infectious disease most often contracted by eating raw or inadequately cooked pork products and is treated with an antibiotic.
    (e) Patients taking broad-spectrum antibiotics may develop fungal infections.

72. Of the following infectious diseases, which poses the lowest transmission threat to unvaccinated or previously unexposed health care providers who utilize standard BSI precautions?
    (a) Tuberculosis.
    (b) Pneumonia.
    (c) Mumps.
    (d) Chickenpox.
    (e) Meningitis.

73. Which of the following statements regarding pneumonia is true?
    (a) True pneumonia is caused only by a bacterial infection.
    (b) The absence of a fever rules out a diagnosis of pneumonia.
    (c) Even though otherwise healthy, after treating a patient with pneumonia, a health-care provider should immediately seek acute immunization.
    (d) Pneumonia bacteria have failed to develop drug-resistant strains.
    (e) Pneumonia may be spread by direct contact with linens containing respiratory excretions.

74. Meningitis is an infectious nervous system disease that
    (a) produces inflammation of the membranes surrounding the brain.
    (b) produces inflammation of the membranes surrounding the spinal cord.
    (c) can be caused by either bacterial or viral infections.
    (d) Answers (a) and (b).
    (e) Answers (a), (b), and (c).

75. Which of the following statements regarding meningitis transmission risks is false?
    (a) Because transmission of meningitis is primarily via respiratory droplets, observing universal precautions and wearing a mask will adequately prevent the health care provider from contracting the disease.
    (b) The peak months of meningitis occurrence are those of low temperature and humidity (the winter months in the continental U.S.).

    (c) In otherwise healthy individuals, viral meningitis is a self-limited disease that has no specific treatment and generally resolves in 7–10 days.

    (d) The postexposure medication available to prevent care providers from developing meningitis is frequently accompanied by uncomfortable, even severe, side effects.

    (e) To minimize the risk of meningitis-exposed health care providers spreading the disease to their other patients, postexposure medication should be initiated within 24 hours of exposure.

76. Which of the following statements regarding meningitis S/Sx is false?

    (a) An upper respiratory disease or otitis media may precede meningitis S/Sx.

    (b) Headache and vomiting are potential S/Sx of meningitis.

    (c) An altered level of consciousness is a potential S/Sx of meningitis.

    (d) Pediatric fever and seizures are potential S/Sx of meningitis.

    (e) Fever in newborns is most likely caused by teething or "milk fever," and rarely is a cause for concern or suspicion of meningitis.

77. Which of the following statements regarding TB (tuberculosis) is false?

    (a) Although commonly called a "TB Test," besides identifying individuals already carrying the TB complex, subcutaneous injection of the tuberculin factor acts as a vaccine and prevents TB-complex-free individuals from becoming susceptible to TB.

    (b) TB is the most common preventable adult infectious disease in the world.

    (c) Although most commonly transmitted via airborne respiratory droplets, contact with mucous membranes or broken skin may also transmit TB.

    (d) TB may be transmitted to humans from cattle, swine, primates, or badgers.

    (e) A single occupational exposure to a patient with TB is highly unlikely to result in TB transmission to the care provider who uses universal precautions.

78. Which of the following statements regarding methods of protection against TB transmission is false?

    (a) Opening one or more windows of the ambulance will provide increased ventilation of the patient compartment, and significantly minimize the transmission risk of TB.

    (b) NIOSH/OSHA standards for health care providers currently call for the use of N95 respirators (special masks) when treating an individual suspected of having TB.

    (c) NIOSH/OSHA standards currently no longer require health care providers to use the HEPA (high efficiency particulate respirator) masks when treating an individual suspected of having TB.

    (d) Simple paper surgical masks will not prevent TB-contaminated air from reaching the EMT.

    (e) Respirator mask failure to protect health care providers is most frequently caused by health care providers failing to use the correct mask.

79. When a patient has "active" TB, which of the following S/Sx may be observed?

   (a) Hemoptysis and night sweats.
   (b) Chills, fever, fatigue, or weight loss.
   (c) Nonproductive cough.
   (d) Answers (a) and (b).
   (e) Answers (a), (b), and (c).

80. Hepatitis is an inflammation of the liver that can directly be caused by

   (1) excessive alcohol consumption.   (4) fungi.
   (2) medications.   (5) parasites.
   (3) viruses or bacteria.   (6) excessive sexual activity.

   (a) 1, 2, 3, 4, 5, and 6.
   (b) 1, 2, 3, 4, and 5.
   (c) 3, 4, and 5.
   (d) 1, 3, and 6.
   (e) 1 and 3.

81. Hepatitis may be transmitted via

   (1) shaking hands.   (4) blood.
   (2) sexual contact.   (5) needle sticks.
   (3) saliva.   (6) feces.

   (a) 1, 2, 3, 4, 5, and 6.
   (b) 2, 3, 4, 5, and 6.
   (c) 3, 4, 5, and 6.
   (d) 2, 5, and 6.
   (e) 2 and 5.

82. Which of the following statements regarding hepatitis is false?

   (a) Viral hepatitis can exist on unwashed hands for as long as four hours.
   (b) Once blood has dried on a surface, the hepatitis B virus is no longer viable
        and transmittable.
   (c) Many hepatitis A and B infections are asymptomatic, and recognizable only
        via liver function studies.
   (d) Hepatitis B is much more contagious than HIV.
   (e) No effective vaccination against hepatitis C exists.

83. A patient who complains of fever, nausea or vomiting, headache, and a stiff neck is
   most likely suffering from

   (a) hepatitis.
   (b) AIDS.
   (c) meningitis.
   (d) Answers (a) and (b).
   (e) Answers (b) and (c).

84.   A patient who complains of fever, night sweats, and recent weight loss may be suffering from
      (a) tuberculosis.
      (b) AIDS.
      (c) meningitis.
      (d) Answers (a) and (b).
      (e) Answers (b) and (c).

85.   HIV/AIDS has been documented to have been transmitted via
      (a) contact with blood, semen, or vaginal secretions.
      (b) breast milk.
      (c) contact with skin surfaces (e.g., by shaking hands).
      (d) Answers (a) and (b).
      (e) Answers (a), (b), and (c).

86.   HIV virus transmission via _____ has occasionally been documented.
      (1) contact with lacrimal fluid (tears)   (4) contact with saliva
      (2) contact with amniotic fluid   (5) contact with bronchial secretions
      (3) urine contact   (6) mosquito bites

      (a) 5 and 6.
      (b) 3, 4, and 5.
      (c) 3, 4, 5, and 6.
      (d) 1, 2, 3, 4, 5, and 6.
      (e) None of the above have ever been documented as a means of HIV transmission.

87.   Prior to the HIV/AIDS epidemic, Kaposi's sarcoma was quite rare. Kaposi's sarcoma is a cancerous skin lesion that consists of
      (a) a raised and reddened welt or lump on the skin.
      (b) a purple splotch on the skin.
      (c) an elongated, irregular, raised strip of blue coloration, similar to that of a varicose vein.
      (d) a depressed area of skin that is red in color.
      (e) one or more circular areas of depressed skin, that is/are yellow in color.

88.   When used in conjunction with discussion of HIV or AIDS patients, the medical term/abbreviation PCP, refers to
      (a) phencyclidine.
      (b) polymorphic cystic paralysis.
      (c) polymorphic cutaneous polyps.
      (d) *Pneumocystis carinii* pneumonia.
      (e) polycystic polyps.

89. Which of the following statements regarding "chickenpox," a.k.a. varicella zoster virus (VZV), is false?
    (a) VZV is a member of the "herpes" family of viruses.
    (b) "Shingles" is a form of VZV.
    (c) VZV is significantly more lethal to pediatric patients than to adult patients.
    (d) If a health care provider did not have chickenpox during childhood and has been vaccinated against it, adult VZV vaccination is still available and is highly recommended.
    (e) If a health care provider has never been vaccinated and has never had the disease, she/he may still successfully be protected from VZV infection if vaccinated within 3 days after exposure to an individual with chickenpox.

90. Which of the following statements regarding sexually transmitted diseases (STDs) is false?
    (a) Syphilis may be transmitted by kissing.
    (b) STDs are communicable diseases that are transmitted only by sexual contact with the carrier.
    (c) Men are usually asymptomatic carriers of *Trichomonas vaginalis*, the protozoan parasite that is a frequent cause of vaginitis.
    (d) STD immunization is not available.
    (e) Chlamydia may be transmitted via common use (or unprotected handling) of infected linen.

91. Which of the following statements regarding risk factors for suicide is false?
    (a) The more specific and detailed the plan, the greater the risk of a successful suicide.
    (b) Typically, women are more successful at committing suicide than men.
    (c) A bullet wound is the most frequent method of successful suicide.
    (d) A history of prior attempts increases the potential for a successful suicide.
    (e) The presence of depression increases the potential for a successful suicide.

92. Which of the following statements regarding partner abuse is true?
    (a) In heterosexual couples, a battered man is more likely to report physical abuse than a battered woman, because men typically feel more "empowered" than women.
    (b) Because of their more sensitive natures, lesbian couples do not engage in physical abuse of their partner.
    (c) Abuse victims often believe that they are "responsible for" (even "deserving of") the physical or emotional abuse to which they are subjected.
    (d) Verbal abuse is much less dangerous than physical abuse, and never generates any sort of acutely dangerous medical emergency.
    (e) Because of their more sensitive natures, homosexual men do not engage in physical abuse of their partner.

93.   Which of the following statements regarding elder abuse is false?
   (a) Most abused elders are over the age of 80.
   (b) Elder abuse situations are primarily limited to low-income socioeconomic homes.
   (c) Elder abuse frequently results when the elder is no longer able to be independent, and the family has financial or emotional difficulty providing care.
   (d) Unexplained trauma is the primary S/Sx when elders are abused.
   (e) Inconsistencies between the patient-provided and caretaker-provided histories may be important clues to report to authorities.

94.   Which of the following statements regarding common risk characteristics for child abuse is false?
   (a) Girls are more often abused than boys.
   (b) Physically handicapped children are at high risk for abuse.
   (c) Developmentally disabled children are at high risk of abuse.
   (d) Twins are at high risk of abuse.
   (e) Premature infants are at high risk of abuse.

95.   Which of the following statements regarding common characteristics of individuals who abuse children is false?
   (a) The abuser is usually a parent or an individual in the role of a parent.
   (b) Overall, men and women are equally likely to abuse children.
   (c) When the mother is the parent who spends the most time with the child, she is frequently identified as the abuser.
   (d) The abuser often was an abused child.
   (e) The vast majority of abusers come from a low-income socioeconomic group.

96.   Conditions unrelated to child abuse that may produce signs easily mistaken for those of child abuse include
   (a) chickenpox (mistaken for cigarette burns).
   (b) car seat abrasions (linear burns).
   (c) hematological disorders (bruises of various ages in several different body areas).
   (d) Answers (b) and (c).
   (e) Answers (a), (b), and (c).

97.   S/Sx that should prompt strong suspicion of child abuse include all of the following, except
   (a) a child less than 2 years old with an obvious, or suspected, upper extremity fracture.
   (b) multiple types of injuries of various ages.
   (c) any apparently healthy and atraumatic infant reported to have "died in his sleep."
   (d) distinctive burn or bruise patterns, suggestive of intentional infliction.
   (e) any injury that does not seem to "fit" with the caretaker's description of its cause.

98. Which of the following observations should not prompt strong suspicion of child abuse?

    (a) The caretaker's account does not match the nature or severity of the injury.

    (b) The child volunteers a description of the cause of injury, and the caretaker agrees with the description.

    (c) The caretaker suggests that the child intentionally injured herself/himself.

    (d) An inappropriate delay in seeking help occurred.

    (e) The caretaker appears unconcerned about the nature or severity of injury.

99. All of the following are potential signs of neglect (a form of child abuse), except

    (a) an excessively unclean or unhealthy living environment.

    (b) the child is inappropriately dressed for the situation or weather.

    (c) the child's clothing is excessively dirty or torn.

    (d) the child has areas of minor diaper rash, and is "fussy."

    (e) the child appears underweight or malnourished.

100. Which of the following statements regarding treatment of the sexual assault victim is true?

    (a) Psychological support concerns are more important than performing a medical evaluation (individuals do not die of rape).

    (b) The same assessment and treatment methods as those used for other patients should be employed but with a heightened sense of protecting the patient's privacy.

    (c) All clothing must be removed from the patient and placed in a plastic bag ("hazmat" bags are usually handy to use).

    (d) Give any items of evidence that you collect during the course of your care to the triage nurse for storage in a separate location, to avoid compounding the victim's distress at seeing people examine them.

    (e) After ascertaining the absence of serious injuries, washing the patient's face and hands has been found to be psychologically soothing and is something that can easily be accomplished enroute to the emergency department.

101. Which of the following statements regarding treatment and interaction with a blind patient is true?

    (a) Seeing Eye dogs are often dangerously protective of their masters. Unless a life-threatening injury exists, never touch the blind patient until after you have touched and reassured the Seeing Eye dog.

    (b) Counsel the patient gently but firmly about the fact that animals are never allowed in ambulances or hospitals, and then ask her/him where to find the cane they use.

(c) If the patient is ambulatory, do not take her/his arm; instead, have the patient hold onto your arm.

(d) Extensively explaining everything you are doing to a blind patient often causes her/him to feel "crippled" and should be avoided. Speak to them just as you would speak to any other patient.

(e) All of the above statements are true.

102. Urinary tract medical devices that may be encountered when responding to emergencies involving patients who receive "home care" include all of the following, except

(a) a Texas catheter: a condomlike sheath that slides onto the penis, with a tube connected to the distal end, through which urine is directed into a collection bag.

(b) a Kansas catheter: a tube transversely inserted through the abdominal wall (just below the umbilicus), directly into the bladder; the catheter directs urine into a drainage collection bag.

(c) a Foley catheter: a tube inserted into the urethra via the urethra's external opening, extending up into the bladder, where it is anchored by means of an inflatable balloon; the tube directs urine into a drainage collection bag.

(d) a suprapubic catheter: a tube surgically inserted directly into the bladder, entering the patient's body just above the pubic bone; the catheter directs urine into a drainage collection bag.

(e) a urostomy device: a surgical opening, created when the patient's bladder is unable to effectively collect urine, that diverts urine from the urinary tract to a stoma, where a urine collection system can be attached.

103. _____ is a highly contagious disease that leads to permanent paralysis of infected muscle groups.

(a) Multiple sclerosis (MS)
(b) Muscular dystrophy (MD)
(c) Poliomyelitis
(d) Spina bifida
(e) Myasthenia gravis

The Answer Key to Test Section 10 is on page 451.

Test
Section
Eleven

Test Section 11's subjects:

- Gynecology
- Obstetrics
- Neonatology
- Pediatrics
- Geriatrics

Test Section 11 consists of 120 questions and is allotted 2 hours for completion.

1.   Pelvic inflammatory disease (PID)
      (a) is an acute infection of the uterus, ovaries, or fallopian tubes.
      (b) may be a chronic infection, and may progress to sepsis.
      (c) may involve the peritoneum and intestines.
      (d) Answers (a) and (c).
      (e) Answers (a), (b), and (c).

2.   S/Sx of PID include all of the following except
      (a) a rigid abdomen and rebound tenderness.
      (b) hypertension (may progress to hypertensive crisis).
      (c) fever, chills, and nausea or vomiting.
      (d) a shuffling gait.
      (e) a malodorous vaginal discharge.

3.   A term often used when referring to the entire area of external female genitalia, the *perineum*, is best defined as the skin-covered muscular tissue that is found
      (a) just below the symphysis pubis, anterior to the vaginal opening.
      (b) on either side of the vaginal opening.
      (c) between the anus and the vaginal opening.
      (d) between the urethra and the vaginal opening.
      (e) posterior to the anus.

4.   The _____ is the interior lining of the uterus.
      (a) myometrium
      (b) perimetrium
      (c) endometrium
      (d) parietalmetrium
      (e) interometrium

5.   The _____ is the thick, muscular, middle layer of the uterine wall.
      (a) myometrium
      (b) perimetrium
      (c) endometrium
      (d) parietalmetrium
      (e) interometrium

6.   The obstetric medical term *gravida* is used when referring to the number of times a woman has
      (a) been pregnant.
      (b) given birth (delivered a viable infant).
      (c) had a miscarriage or abortion.
      (d) had sex.
      (e) had a cesarean section.

7. The obstetric medical term *para* (parity) is used when referring to the number of times a woman has

    (a) been pregnant.
    (b) given birth (delivered a viable infant).
    (c) had a miscarriage or abortion.
    (d) had sex.
    (e) had a cesarean section.

8. The obstetric medical term *abortion* refers to

    (a) only those pregnancies that were purposefully terminated.
    (b) only those pregnancies that were accidentally terminated.
    (c) only those pregnancies that were illegally terminated.
    (d) any pregnancy that ended (for any reason) prior to 20 weeks of gestation.
    (e) any pregnancy that resulted in the delivery of a dead infant.

9. It is your patient's due date, and she is in labor. She has two three-year-old sons (twins) and has had one miscarriage. You should describe her as being

    (a) P3, G1, M1 (para 3, gravida 1, miscarriage 1).
    (b) P2, G2, M1 (para 2, gravida 2, miscarriage 1).
    (c) G3, P2, M1 (gravida 3, para 2, miscarriage 1).
    (d) G3, P1, Ab1 (gravida 3, para 1, abortion 1).
    (e) G2, P1, Ab1 (gravida 2, para 1, abortion 1).

10. The normal duration of a human pregnancy is

    (a) 36 weeks from the last day of the patient's LMP (9 calendar months).
    (b) 36 weeks from the first day of the patient's LMP (9 calendar months).
    (c) 40 weeks from the first day of the patient's LMP (10 lunar months).
    (d) 45 weeks from the first day of the patient's LMP (9 lunar months).
    (e) 45 weeks from the last day of the patient's LMP (9 lunar months).

11. Which of the following statements regarding pregnancy and medical disorders is false?

    (a) Previously diagnosed diabetes may become unstable during pregnancy.
    (b) Pregnancy may cause the onset of diabetes in a previously nondiabetic patient.
    (c) Pregnancy may induce hypertension in the previously normotensive patient.
    (d) Patients with heart disease may develop CHF when pregnant.
    (e) Even if well controlled by medication, mothers who have seizure disorders prior to pregnancy are likely to experience a seizure during delivery.

12. An ectopic pregnancy is best defined as when a fertilized ovum is implanted
    (a) anywhere outside of the uterus.
    (b) anywhere within a fallopian tube.
    (c) anywhere within the abdominal cavity.
    (d) at the mouth of the cervix.
    (e) anywhere within the endometrium.

13. The most common site of an ectopic implantation is
    (a) on an ovary.
    (b) within a fallopian tube.
    (c) within the abdominal cavity.
    (d) at the mouth of the cervix.
    (e) within the endometrium.

14. The placenta performs all of the following functions except
    (a) providing an extra cushion of protection (diminishing the likelihood of fetal trauma).
    (b) transfer of gases (fetal respiration).
    (c) transport of nutrients and excretion of wastes.
    (d) transfer of heat.
    (e) hormone secretion.

15. Premature separation of the placenta from the uterine wall is called
    (a) spontaneous abortion.
    (b) placentectomy.
    (c) placenta previa.
    (d) abruptio placentae.
    (e) ectopic placenta.

16. Occasionally a placenta develops too low in the uterus and partially or completely covers the opening of the cervix. This condition is called
    (a) spontaneous abortion.
    (b) placentectomy.
    (c) placenta previa.
    (d) abruptio placentae.
    (e) ectopic placenta.

17. Any painless, bright red vaginal bleeding that occurs late in pregnancy is considered to be _____, until proven otherwise.
    (a) a spontaneous abortion
    (b) an ectopic pregnancy
    (c) placenta previa
    (d) an abruptio placenta
    (e) a uterine rupture

18.   Which of the following statements about the S/Sx of abruptio placentae is false?

  (a) Complete abruptio placentae is accompanied by pain, and may not produce external vaginal bleeding.

  (b) Partial abruptio placentae is not always accompanied by pain, and may not produce external vaginal bleeding.

  (c) Complete abruptio placentae can produce massive hemorrhage and profound maternal hypotension.

  (d) Abruptio placentae is always accompanied by pain and vaginal bleeding.

  (e) Bleeding without pain is possible in some cases of abruptio placentae.

19.   S/Sx of pre-eclampsia include any of the following except

  (a) pregestational history of a hypertensive disorder and a currently prescribed antihypertensive medication.

  (b) edema of the hands and face.

  (c) visual disturbances.

  (d) major motor seizure activity.

  (e) headache.

20.   S/Sx of eclampsia include any of the following except

  (a) pregnancy-induced hypertension.

  (b) edema of the hands and face.

  (c) visual disturbances.

  (d) major motor seizure activity.

  (e) pregestational history of a seizure disorder and a currently prescribed anticonvulsant medication.

21.   Management considerations for a patient with pre-eclampsia include all of the following except

  (a) oxygen administration.

  (b) IV access with NS.

  (c) IV administration of magnesium sulfate.

  (d) IV administration of diazepam.

  (e) left laterally recumbent positioning.

22.   Management considerations for a patient with eclampsia include all of the following except

  (a) oxygen administration.

  (b) IV access with NS.

  (c) IV administration of magnesium sulfate.

  (d) IV administration of diazepam.

  (e) Trendelenburg positioning.

23.   Pre-eclampsia commonly occurs during
    (a)   the last 10 weeks of gestation.
    (b)   labor.
    (c)   the first 48 hours postpartum.
    (d)   Answer (a) or (b).
    (e)   Answer (a), (b), or (c).

24.   Supine hypotensive syndrome usually occurs in the third trimester, when the large abdominal mass of the gravid uterus
    (a)   demands more of the maternal blood supply than can be compensated for by the already increased maternal cardiac output, when the patient is placed supine.
    (b)   compresses the inferior vena cava when the patient is placed supine, thereby reducing venous return.
    (c)   compresses the abdominal aorta when the patient is placed supine, thereby obstructing cardiac output.
    (d)   Answer (a) or (b).
    (e)   Answer (a) or (c).

25.   In the absence of trauma, when transporting a patient in her third trimester of pregnancy, the patient should be placed in _____, in order to prevent supine hypotensive syndrome.
    (a)   a left laterally recumbent position, or placed supine with her right hip elevated
    (b)   the Trendelenburg position
    (c)   the semi-Fowler's position
    (d)   a right laterally recumbent position, or placed supine with her left hip elevated
    (e)   the reverse-Trendelenburg position

26.   In the presence of trauma, when transporting a patient in her third trimester of pregnancy, the patient should be placed supine upon a long backboard,
    (a)   without any deviation from standard trauma treatment protocols; the threat of maternal spine injury supercedes that of the rare possibility of causing supine hypotensive syndrome.
    (b)   with the head of the board elevated.
    (c)   with the left side of the board elevated.
    (d)   with the right side of the board elevated.
    (e)   with the foot of the board elevated.

27.   Which of the following statements about Braxton Hicks contractions is false?
    (a)   Intermittent uterine contraction may occur as early as the 13th week of gestation.
    (b)   Braxton Hicks contractions are not "true" labor contractions. If the patient is preterm and has a history of Braxton Hicks episodes, ambulance transportation and care is not required.

(c) Intermittent uterine contractions are believed to enhance placental circulation.

(d) Braxton Hicks contractions are usually painless, and irregular in occurrence.

(e) A patient suspected of having Braxton Hicks contractions should be treated as though she is experiencing actual labor and may soon deliver.

28. Full dilation of the cervix signifies
  (a) the beginning of the fourth stage of labor.
  (b) the end of the second stage of labor.
  (c) the end of the first stage of labor.
  (d) the beginning of the first stage of labor.
  (e) the end of the third stage of labor.

29. The birth of the baby signifies
  (a) the beginning of the fourth stage of labor.
  (b) the end of the second stage of labor.
  (c) the end of the first stage of labor.
  (d) the beginning of the first stage of labor.
  (e) the end of the third stage of labor.

30. The delivery of the placenta signifies
  (a) the end of the fourth stage of labor.
  (b) the end of the second stage of labor.
  (c) the end of the first stage of labor.
  (d) the beginning of the first stage of labor.
  (e) the end of the third stage of labor.

31. The classic sign or symptom that delivery is imminent and that the paramedic should immediately prepare for a field delivery is
  (a) an uncontrollable maternal urge to push, or the visible crowning of the fetus.
  (b) uterine contractions that are 5 minutes apart.
  (c) rupture of the amniotic sac.
  (d) uterine contractions of 15–30 seconds in duration.
  (e) the maternal report that uterine contractions have been 5 minutes apart for the past 20 minutes.

32. Factors that should prompt immediate patient transport, in spite of indications that delivery is imminent, include all of the following situations except
  (a) a patient whose amniotic sac ruptured over 24 hours earlier.
  (b) any patient who has previously delivered a baby by cesarean section.
  (c) a prolapsed umbilical cord.
  (d) a single-limb presentation.
  (e) when the patient's contractions have been 2–3 minutes apart for more than 20 minutes of scene time, but delivery has not occurred.

33. The classic sign of fetal distress is _____, and should prompt immediate patient transport, in spite of indications that delivery is imminent.
  (a) when the patient's abdomen becomes "soft" between contractions
  (b) a fetal heart rate of less than 110/minute
  (c) a fetal heart rate of less than 90/minute
  (d) a fetal heart rate of greater than 100/minute
  (e) a fetal heart rate of greater than 110/minute

34. As delivery begins, most infants present with their head first and their face down. This is called
  (a) the vertex position.
  (b) a breech presentation.
  (c) an occipital presentation.
  (d) the crown presentation.
  (e) a posterior presentation.

35. As delivery begins, if the baby's buttocks are the first structure to be seen, this is called
  (a) the vertex position.
  (b) a breech presentation.
  (c) an occipital presentation.
  (d) the crown presentation.
  (e) a posterior presentation.

36. As the baby's head emerges, if the amniotic sac has not ruptured you should
  (a) use your fingers to pinch and tear open the sac, then push the sac away from the baby's nose and mouth before performing airway suction.
  (b) continue with the delivery; the sac will break on its own once the baby is completely delivered, and suction can be performed at that time.
  (c) use a sterile scalpel to quickly cut open the sac, and provide suction before the baby takes its first breath.
  (d) leave the sac alone until delivery, unless the baby begins to breathe.
  (e) delay further progress of the delivery, and transport immediately for an emergency cesarean section.

37. After the baby's head emerges, you discover that the umbilical cord is wrapped around its neck. You gently attempt to slip the cord over the baby's head and shoulder but find that it is too tightly wrapped to do this. You should
  (a) continue with the delivery; the cord will naturally unwind as the baby's body emerges.
  (b) forcefully pull the cord over the baby's head before the baby is strangled.

   (c) clamp or tie the cord in two places, approximately 2 inches apart and cut the cord between the clamps or ties before continuing with the delivery.

   (d) not delay freeing the baby to clamp or tie-off the cord. Immediately cut the cord and remove it before the baby is strangled.

   (e) delay further progress of the delivery and transport immediately for an emergency cesarean section.

38. Which of the following statements regarding steps in assisting a normal delivery is false?

   (a) Once the infant's head has emerged, instruct the mother to stop pushing and suction the infant's airway. Instruct the mother to resume pushing after suction is completed.

   (b) Once the infant's head has emerged, gently guide the infant's head downward, to allow delivery of the upper shoulder.

   (c) After upper shoulder delivery, gently guide the infant's body upward, to allow delivery of the lower shoulder.

   (d) Once the head and shoulders have delivered, the rest of the infant's body will rapidly emerge. Do not drop the infant.

   (e) Immediately after delivery, quickly elevate the infant above the level of the vagina (usually by placing the infant up on the mother's abdomen) and prepare to cut the umbilical cord.

39. Which of the following statements regarding delivery of the placenta and postpartum hemorrhage is false?

   (a) There is no need to wait for placental delivery before transporting your patients.

   (b) Gentle umbilical cord traction will speed placental delivery and the delivery of your patients to the emergency department and advanced neonatal care.

   (c) Gentle massage of the mother's abdomen will assist in uterine contraction and diminish maternal blood loss after delivery.

   (d) Never "pack" the vagina by inserting a sterile sanitary napkin within it.

   (e) Allowing the mother to begin breastfeeding will diminish maternal blood loss after delivery.

40. The placenta must be transported with the patient to the emergency department, because

   (a) some religious sects require a special burial ceremony for the placenta, especially if the child dies prior to reaching the age of 10 days old.

   (b) some alternative-lifestyle sects require maternal ingestion of the placenta, after the birth of a healthy child.

   (c) the placenta must be examined for missing parts or pieces. Should placental parts or pieces be allowed to remain within the uterus, maternal infection or continued hemorrhage may occur.

   (d) Answers (a) and (b).

   (e) Answers (b) and (c).

41. Which of the following statements regarding cephalopelvic disproportion (something that requires delivery via cesarean section) is false? Cephalopelvic disproportion

- (a) is when the baby's head is too large to pass through the maternal pelvis.
- (b) is when the maternal pelvis is too small to allow the baby's head to deliver.
- (c) is often associated with diabetic mothers, women who have delivered several children, or gestation periods that have exceeded the due date for delivery.
- (d) is often associated with women of petite stature, or women with contracted pelvises.
- (e) rarely ever occurs in women delivering their first child.

42. Which of the following abnormal-presentation, or delivery-complication, situations absolutely cannot be delivered in the field?

(1) vertex presentation	(5) twins or other multiple births
(2) breech presentation	(6) prolapsed cord
(3) single-arm presentation	(7) cephalopelvic disproportion
(4) single-leg presentation	(8) shoulder dystocia

- (a) 1, 2, 3, 4, 5, 6, 7, and 8.
- (b) 2, 3, 4, 5, 6, 7, and 8.
- (c) 3, 4, 5, 6, and 7.
- (d) 3, 4, 6, and 7.
- (e) 4, 6, 7, and 8.

43. Which of the following statements regarding fetal feces expulsion in utero is false?

- (a) Fetal feces may be observed within amniotic fluid as yellow-green, light green, greenish-black, or brown matter.
- (b) Fecal matter expulsion into amniotic fluid is a natural function that usually begins during the 20th week of every infant's gestation and is an entirely normal event.
- (c) Expulsion of fecal matter into amniotic fluid is considered a sign that the fetus has experienced a hypoxic event in utero.
- (d) Aspiration of amniotic fluid containing fetal fecal matter is accompanied by a high risk of fetal morbidity.
- (e) Amniotic fluid that appears to have the color and consistency of dark "pea soup," is a dire sign and indicates a high risk for fetal morbidity.

44. If the amniotic fluid appears to have the color and consistency of "pea soup,"

- (a) reassure observing family members that this is entirely normal for most births.
- (b) bag-valve mask ventilate the infant with 100% oxygen prior to suctioning.
- (c) suction the infant's nose first, then its mouth, and ventilate the infant to stimulate natural breathing.
- (d) suction the infant's nose first, then its mouth, prior to providing normal stimulus to breathe.
- (e) intubate the infant and thoroughly suction the lungs, prior to stimulating breathing or performing bag-valve mask ventilations.

45.   Which of the following statements regarding multiple-birth situations is true?

    (a)   If the mother reports that she is carrying more than one fetus, prehospital delivery is absolutely contraindicated. Delay further progress of the delivery and immediately transport for an emergency cesarean section.

    (b)   Provide suction and warming, assure breathing, clamp and cut the umbilical cord of each delivered infant, prior to delivering the next infant.

    (c)   If any form of uterine contraction continues after the vaginal delivery of one infant, this indicates a multiple-birth situation. Delay further progress of the delivery and immediately transport for an emergency cesarean section.

    (d)   Answers (a) and (c).

    (e)   None of the above.

46.   Which of the following statements regarding uterine rupture is false?

    (a)   Uterine rupture usually occurs at the onset of labor.

    (b)   Because of today's advanced medical procedures, uterine rupture rarely results in maternal or fetal death.

    (c)   Blunt abdominal trauma may cause uterine rupture prior to labor.

    (d)   Uterine rupture may occur during labor, especially if uterine contractions occur for an exceptionally long period of time or if the mother has previously delivered a baby via a vertical incision cesarean section.

    (e)   Cephalopelvic disproportion may result in uterine rupture.

47.   Signs and symptoms of uterine rupture include all of the following except

    (a)   complaints of excruciating abdominal pain.

    (b)   signs of systemic shock.

    (c)   cessation of abdominal pain, after the rupture has occurred.

    (d)   sudden intensification of uterine contractions (contractions suddenly becoming harder and faster than they were before) or sudden onset of hard and strong uterine contractions (when none existed before).

    (e)   a tender, rigid abdomen, possibly accompanied by rebound tenderness.

48.   Uterine inversion occurs when the uterus literally turns inside-out upon delivery of an infant or placenta, and extrudes from the cervix. Uterine inversion

    (a)   is accompanied by the tearing and rupture of ligaments and blood vessels associated with the uterus, producing profound, life-threatening maternal shock.

    (b)   can be caused by pulling on an umbilical cord.

    (c)   can be caused by aggressive attempts to express the placenta when the uterus is relaxed.

    (d)   Answers (a) and (c).

    (e)   Answers (a), (b), and (c).

49. Management of uterine inversion includes all of the following except

    (a) one attempt to manually replace the uterus, while the mother is still in the "birth position," by using the palm of your hand to push the fundus of the extruded uterus toward the vagina.

    (b) one attempt to facilitate "natural" detachment of the placenta from the uterus, by exerting gentle traction on the umbilical cord portion attached to the extruded placenta.

    (c) high-flow oxygen administration and rapid transportation.

    (d) initiation of two or more large-bore NS or LR IVs.

    (e) gently encompassing all extruded tissue within saline-soaked towels.

50. Pulmonary embolism (PE) is one of the most common causes of maternal death. Pregnancy-related PE

    (a) occurs more frequently following a cesarean section than following a vaginal delivery.

    (b) is usually accompanied by the complaint of a sudden onset of severe dyspnea, with or without complaint of "sharp" chest pain.

    (c) may occur at any time during any pregnancy.

    (d) Answers (a) and (b).

    (e) Answers (a), (b), and (c).

51. Place the following newborn management tasks in the order in which they should be performed (the order of their primary performance importance):

    (1) umbilical cord cutting   (4) drying and warming

    (2) suction of the nares   (5) APGAR scoring

    (3) suction of the oral airway

        (a) 1, 2, 3, 4, 5.

        (b) 2, 3, 4, 5, 1.

        (c) 2, 3, 5, 4, 1.

        (d) 3, 2, 4, 1, 5.

        (e) 5, 1, 2, 3, 4.

52. The proximal clamp (or tie) is placed on the umbilical cord at approximately

    (a) 4 inches (10 centimeters) from the baby.

    (b) 6 inches (15 centimeters) from the baby.

    (c) 8 inches (20 centimeters) from the baby.

    (d) 10 inches (26 centimeters) from the baby.

    (e) 12 inches (31 centimeters) from the baby.

53. The distal clamp (or tie) is placed on the umbilical cord at approximately

    (a) 6 inches (15 centimeters) from the baby.

    (b) 8 inches (20 centimeters) from the baby.

    (c) 10 inches (26 centimeters) from the baby.

    (d) 12 inches (31 centimeters) from the baby.

    (e) 14 inches (35 centimeters) from the baby.

54. Factors that stimulate the newborn to take its first breaths include all of the following except

   (a) hypoxia or mild acidosis.
   (b) drying.
   (c) severing the umbilical cord.
   (d) airway suctioning.
   (e) hypothermia.

55. Suctioning of the newborn's mouth and nose should be performed

   (a) immediately upon emergence of the head.
   (b) immediately after birth.
   (c) only after the umbilical cord is cut.
   (d) Answers (a) and (b).
   (e) Answer (a), (b), or (c), depending upon local protocols.

56. APGAR assessment and scoring system of the newborn should be performed

   (a) only in the ambulance, on the way to the hospital.
   (b) at 1 and 5 minutes after the infant's birth.
   (c) at 1 and 5 minutes after the placenta delivers.
   (d) at 5 and 10 minutes after the placenta delivers.
   (e) before any resuscitation is performed; resuscitation changes the initial APGAR.

57. The two "A"s of the APGAR mnemonic stand for

   (a) Appearance (skin color) and Apnea (absent respirations).
   (b) Activity (extremity movement) and Altered level of consciousness.
   (c) Appearance (skin color) and Activity (extremity movement, muscle tone).
   (d) Appearance (skin color) and Activity (heart rate).
   (e) Activity (extremity movement, muscle tone) and Abnormality (presence of obvious congenital anomalies).

58. The "P" of the APGAR mnemonic stands for

   (a) Pulse (heart rate).
   (b) Pupils (equal or unequal).
   (c) Pink (skin color).
   (d) Pallor (skin color).
   (e) Purple (cyanosis).

59. The G of the APGAR mnemonic stands for

   (a) Gasping (poor respiratory effort).
   (b) Gurgling (indication of aspiration).
   (c) Gross deformity (presence of obvious congenital anomalies).
   (d) Grimace (irritability or crying response to stimulus).
   (e) Groaning (poor crying ability).

60. The R of the APGAR mnemonic stands for
(a) Rigor (stillborn infant).
(b) Rapid (heart rate or respiratory rate).
(c) Rate (heart rate or pulse).
(d) Respiratory effort (rate and effort of respirations).
(e) Robust (healthy, crying infant).

61. The maximum (best) APGAR score possible is
(a) 5.
(b) 10.
(c) 15.
(d) 25.
(e) 50.

62. The minimum (worst) APGAR score possible is
(a) 0.
(b) 3.
(c) 5.
(d) 6.
(e) 10.

63. *Meconium staining* is a phrase that describes
(a) the observable presence of fetal feces in the amniotic fluid.
(b) the "port-wine"-colored birth marks that are strikingly apparent in newborns (especially Caucasian infants) but fade as the infant ages.
(c) the "port-wine"-colored birth marks that are slightly observable in newborns but become even darker as the infant ages.
(d) the erythemic (reddened) areas of an infant's body, which are often strikingly apparent when labor was lengthy or exceptionally strenuous but which fade as the infant ages.
(e) the greenish-tinge of an infant's skin when labor was lengthy or exceptionally strenuous, which fades as the infant ages.

64. Which of the following vital signs is the most important indicator of whether or not a neonate is distressed?
(a) Skin color.
(b) Respiratory rate.
(c) Heart rate.
(d) Skin temperature.
(e) Umbilical cord pulsations.

65.   You just delivered a baby boy who has a spontaneous respiratory rate of 30 and a heart rate of 90. You should

(a) do nothing more than maintain the infant's warmth; these are entirely "normal" newborn vital signs, and oxygen administration is optional.

(b) maintain the infant's warmth and administer supplemental oxygen.

(c) begin positive pressure ventilations.

(d) begin chest compressions.

(e) Answers (c) and (d).

66.   You just delivered a baby girl who has a spontaneous respiratory rate of 30 and a heart rate of 50. You should

(a) do nothing more than maintain the infant's warmth; these are entirely "normal" newborn vital signs, and oxygen administration is optional.

(b) maintain the infant's warmth and administer supplemental oxygen.

(c) begin positive pressure ventilations.

(d) begin chest compressions.

(e) Answers (c) and (d).

67.   You just delivered a baby boy who has a spontaneous respiratory rate of 40, a heart rate of 120, and cyanosis of the trunk and limbs. You should

(a) do nothing more than maintain the infant's warmth; these are entirely "normal" newborn vital signs, and oxygen administration is optional.

(b) maintain the infant's warmth and administer supplemental oxygen.

(c) begin positive pressure ventilations.

(d) begin chest compressions.

(e) Answers (b) and (c).

68.   You just delivered a baby girl who had a heart rate of 70 but was apneic in spite of appropriate stimulation. You have performed positive pressure ventilations, providing 100% oxygen, for 30 seconds. Her heart rate remains at 70. You should

(a) discontinue positive pressure ventilations and check for a spontaneous respiratory rate.

(b) continue positive pressure ventilations for at least another 30 seconds before stopping to check for a spontaneous respiratory rate.

(c) continue positive pressure ventilations until her heart rate increases to 100, without pausing to check for a spontaneous respiratory rate.

(d) continue positive pressure ventilations until you deliver her to the emergency department, without pausing to check for a spontaneous respiratory rate.

(e) begin chest compressions.

69. When a newborn requires mechanical ventilation assistance, the ventilatory rate should be _____ ventilations per minute.
- (a) 60–100
- (b) 14–24
- (c) 20–30
- (d) 12–20
- (e) 40–60

70. The approximate rate of newborn chest compression performance is _____ times per minute.
- (a) 50
- (b) 60
- (c) 80
- (d) 100
- (e) 120

71. When performing CPR on a newborn, the compression/ventilation ratio is
- (a) 5 chest compressions to each ventilation (5:1).
- (b) 15 chest compressions to 2 ventilations (15:2).
- (c) 5 chest compressions to 2 ventilations (5:2).
- (d) 3 chest compressions to each ventilation (3:1).
- (e) 3 chest compressions to 2 ventilations (3:2).

72. The "inverted pyramid" protocol for newborn resuscitation specifies which of the following as being the most vital care provision (the first steps for providing newborn resuscitation)?
- (a) Provision of supplemental oxygen.
- (b) Chest compressions.
- (c) Intubation.
- (d) Drying, warming, positioning, suctioning, and tactile stimulation.
- (e) Bag-valve mask ventilations.

73. A premature infant is best defined as any infant born before
- (a) 9 calendar months of gestation have transpired.
- (b) the 50th week of gestation.
- (c) the 45th week of gestation.
- (d) the 37th week of gestation.
- (e) 10 lunar months of gestation have transpired.

74. Which of the following statements regarding sudden infant death syndrome (SIDS) is true?
- (a) The incidence of SIDS is greatest during hot summer months.
- (b) SIDS is more common in female infants than in male infants.

    (c) Because of today's medical treatment advances, low-birth-weight infants are at no greater risk for SIDS than are normal-birth-weight infants.

    (d) Studies strongly suggest that placing an infant in a prone sleep position significantly contributes to SIDS deaths.

    (e) All of the above are true.

75.   SIDS victims commonly may exhibit all of the following except

    (a) a normal state of nutrition and hydration.

    (b) frothy fluids in and around the mouth and nostrils, which may be blood-tinged.

    (c) emesis within the airway or about the infant.

    (d) multiple bruises upon the infant's body, appearing to be of different ages.

    (e) an unusual (sometimes disturbing) body position at time of death.

76.   Upon your arrival, you find an apneic, pulseless infant, with very cold skin in spite of being in a warm environment. The infant's neck is stiff and hard to move, and there is obvious dependent lividity. You should

    (a) assure the parents and other family members that you will provide all possible resuscitation efforts and that there is a very good chance the infant will recover.

    (b) quickly determine the infant's name and use that name every time you refer to the infant.

    (c) provide BLS only (so as to appear "busy") while extricating the infant from the house to the ambulance for a "slow code" transport to the emergency department.

    (d) Answers (a) and (b).

    (e) Answers (a) and (c).

77.   The pediatric dose for IV/IO administration of epinephrine is

    (a) 0.1 mL/kg of epi 1:1000.

    (b) 0.01 mg/kg.

    (c) 0.1 mL/kg of epi 1:10,000.

    (d) Answer (a) or (b).

    (e) Answer (b) or (c).

78.   The pediatric dose for IV/IO administration of atropine is

    (a) 0.01 mg/kg.

    (b) 0.02 mg/kg.

    (c) minimum dose of 0.1 mg.

    (d) Answers (a) and (c).

    (e) Answers (b) and (c).

79. The pediatric dose for endotracheal tube administration of epinephrine is
    (a) 0.1 mL/kg of epi 1:1000.
    (b) 0.01 mg/kg.
    (c) 0.1 mL/kg of epi 1:10,000.
    (d) Answer (a) or (b).
    (e) Answer (b) or (c).

80. The pediatric dose for IV/IO administration of lidocaine is
    (a) 0.1 mg/kg.
    (b) 0.5 mg/kg.
    (c) 1.0 mg/kg.
    (d) 1.5 mg/kg.
    (e) 2.0 mg/kg.

81. The pediatric dose for IV/IO administration of amiodarone is
    (a) only administered after IV/IO lidocaine has failed to convert V-fib to a non-shockable rhythm.
    (b) 5 mg/kg, repeated up to 15 mg/kg.
    (c) 10 mg/kg, repeated to a maximum delivery of 300 mg.
    (d) Answers (a) and (b).
    (e) Answer (b) or (c), depending upon local protocol.

82. In a normal, healthy state, an infant's fontanelle
    (a) feels tight and may bulge above the level of the skull surface.
    (b) is level with the skull surface or appears slightly sunken and may pulsate.
    (c) is lower than the level of the skull surface and appears sunken.
    (d) sinks with each pulse and then returns to the level of the skull surface.
    (e) bulges with each pulse and then returns to the level of the skull surface.

83. In the presence of increased intracranial pressure, an infant's fontanelle
    (a) feels tight and may bulge above the level of the skull surface.
    (b) is level with the skull surface or appears slightly sunken and may pulsate.
    (c) is lower than the level of the skull surface and appears sunken.
    (d) sinks with each pulse and then returns to the level of the skull surface.
    (e) bulges with each pulse and then returns to the level of the skull surface.

84. If the newborn or infant is dehydrated (hypovolemic), the fontanelle
    (a) feels tight and may bulge above the level of the skull surface.
    (b) is level with the skull surface or appears slightly sunken and may pulsate.
    (c) is lower than the level of the skull surface and appears sunken.
    (d) sinks with each pulse and then returns to the level of the skull surface.
    (e) bulges with each pulse and then returns to the level of the skull surface.

85. Which of the following statements regarding pediatric febrile seizures is false?
     (a) Seizures are a normal occurrence for any febrile child and do not require transportation to the hospital if the fever "broke" (abated) after the seizure.
     (b) Febrile seizures most commonly occur between the ages of 6 months and 6 years.
     (c) All pediatric patients who experience a seizure must be transported to an emergency department, so that they may be evaluated for a potentially serious underlying illness or injury.
     (d) Febrile seizures seem more related to a rapid rate of temperature increase rather than to any specifically "high" degree of fever.
     (e) Suspect fever as the cause of seizure if the child's temperature is above 103°F (39.2°C).

86. The pediatric dose of diazepam for infants and children up to the age of 5 years old is
     (a) 0.2–0.5 mg every 2 to 5 minutes, to a maximum of 2.5 mg.
     (b) 1 mg every 2 to 5 minutes, to a maximum of 5 mg.
     (c) 2–5 mg every 5 minutes, to a maximum of 10 mg.
     (d) 0.01–0.02 mg/kg every 5 minutes, to a maximum of 1 mg.
     (e) 0.2–0.5 mg/kg every 5 minutes, to a maximum of 10 mg.

87. The pediatric dose of diazepam for children 5 years of age or older is
     (a) 0.2–0.5 mg every 2 to 5 minutes, to a maximum of 2.5 mg.
     (b) 1 mg every 2 to 5 minutes, to a maximum of 5 mg.
     (c) 2–5 mg every 5 minutes, to a maximum of 10 mg.
     (d) 0.01–0.02 mg/kg every 5 minutes, to a maximum of 1 mg.
     (e) 0.2–0.5 mg/kg every 5 minutes, to a maximum of 10 mg.

88. Which of the following statements regarding pediatric dehydration is true?
     (a) Because of their youth and the resilience of their body systems, children rarely suffer from dehydration.
     (b) Any pediatric patient displaying signs and symptoms of dehydration should be suspected to be a victim of neglect or abuse, and the proper authorities should be notified immediately upon arrival at the emergency department.
     (c) Because of the danger of fluid overload in the pediatric patient, dehydration should be treated with D_5W on a microdrip infusion set only.
     (d) The initial fluid bolus for treatment of a pediatric patient suspected of dehydration is 20 ml/kg, IV or IO.
     (e) All of the above are true.

89. A bacterial infection of the bloodstream is called
   (a) meningitis.
   (b) sepsis.
   (c) Reye's syndrome.
   (d) anaphylaxis.
   (e) epiglottitis.

90. Aspirin administration to pediatric patients is contraindicated because of its association with the incidence of
   (a) meningitis.
   (b) sepsis.
   (c) Reye's syndrome.
   (d) anaphylaxis.
   (e) epiglottitis.

91. S/Sx such as fever (with or without seizure), irritability, and a history of recent illness (with or without N/V/D), are often associated with pediatric
   (a) meningitis.
   (b) sepsis.
   (c) Reye's syndrome.
   (d) Answers (a) and (b).
   (e) Answers (a), (b), and (c).

92. A full or bulging fontanelle may be indicative of
   (a) meningitis.
   (b) chicken pox.
   (c) measles.
   (d) Answers (a) and (b).
   (e) None of the above.

93. Classic indications of _____ include complaints of severe headache and evidence of a stiff neck.
   (a) meningitis
   (b) sepsis
   (c) Reye's syndrome
   (d) Answers (a) and (b).
   (e) Answers (a), (b), and (c).

94. Classic indications of _____ include a child who seemed to be recovering from an unremarkable viral illness but who suddenly develops severe N/V and soon begins to exhibit unusually hostile or altered behaviors.
   (a) meningitis
   (b) sepsis

    (c) Reye's syndrome
    (d) Answers (a) and (b).
    (e) Answers (a), (b), and (c).

95.   A recent history of upper respiratory tract or ear infection may be associated with
    (a) meningitis.
    (b) sepsis.
    (c) Reye's syndrome.
    (d) Answers (a) and (b).
    (e) Answers (a), (b), and (c).

96.   Laryngotracheobronchitis is more commonly referred to as
    (a) bronchiolitis.
    (b) epiglottitis.
    (c) croup.
    (d) Any of the above.
    (e) None of the above.

97.   Prominent expiratory wheezing is classically characteristic of
    (a) bronchiolitis.
    (b) epiglottitis.
    (c) croup.
    (d) Any of the above.
    (e) None of the above.

98.   _____ is often accompanied by a fever.
    (a) Bronchiolitis
    (b) Epiglottitis
    (c) Croup
    (d) Any of the above.
    (e) None of the above.

99.   A harsh, "bark"-like cough is classically characteristic of
    (a) bronchiolitis.
    (b) epiglottitis.
    (c) croup.
    (d) Answer (a) or (b).
    (e) Answer (b) or (c).

100.   Difficulty in swallowing, evidenced by excessive drooling, is classically characteristic of
    (a) bronchiolitis.
    (b) epiglottitis.
    (c) croup.
    (d) Answers (a) and (b).
    (e) Answers (b) and (c).

101. Stridor may accompany
   (a) bronchiolitis.
   (b) epiglottitis.
   (c) croup.
   (d) Answers (a) and (b).
   (e) Answers (b) and (c).

102. Attempted prehospital visualization of the posterior oropharynx is absolutely contraindicated in suspected cases of
   (a) bronchiolitis.
   (b) epiglottitis.
   (c) croup.
   (d) Answers (a) and (b).
   (e) Answers (b) and (c).

103. Humidified oxygen should be administered in all cases of
   (a) bronchiolitis.
   (b) epiglottitis.
   (c) croup.
   (d) Answers (a) and (b).
   (e) Answers (a), (b), and (c).

104. Which of the following statements regarding the pediatric airway and pediatric endotracheal intubation is false?
   (a) The pediatric tongue is larger in relationship to the oropharynx than that of the adult.
   (b) The pediatric glottal opening is found higher and more anterior than that of an adult.
   (c) There is much less risk of pediatric intubation stimulus triggering a vagal response because infants and children are more resilient than adults.
   (d) Pediatric laryngoscope blades will fit any adult-sized laryngoscope handle.
   (e) The narrowest portion of the pediatric airway is the cricoid cartilage.

105. Which of the following statements regarding infant hypoglycemia is false?
   (a) Respiratory illnesses may cause hypoglycemia in any infant.
   (b) Hypothermia may cause hypoglycemia in any infant.
   (c) Hypoxia may cause hypoglycemia in any infant.
   (d) Hypoglycemia is more common in large-for-gestational-age infants because of greater body mass requiring greater amounts of glycogen stores.
   (e) CNS hemorrhage may cause hypoglycemia in any infant.

106. Which of the following statements regarding pediatric hypoglycemia is false?

    (a) Nondiabetic children rarely ever become hypoglycemic.

    (b) Strenuous activity may cause hypoglycemia in diabetic children.

    (c) Strenuous activity may cause hypoglycemia in nondiabetic children.

    (d) Illness may cause hypoglycemia in all children.

    (e) All ill infants and children should be tested for hypoglycemia.

107. Which of the following statements regarding treatment of pediatric hypoglycemia is false?

    (a) Oral glucose or sugar-filled fluids may be administered to conscious and alert pediatric patients.

    (b) Glucagon administration is ineffective for treatment of hypoglycemia when insufficient glycogen stores are present.

    (c) $D_{25}W$ can be prepared by mixing $D_{50}W$ 1:1 with normal saline.

    (d) $D_{25}W$ can be prepared by mixing 25 ml of $D_{50}W$ with 50 ml of D_5W.

    (e) $D_{25}W$ may only be administered via IV or IO routes.

108. Which of the following statements regarding geriatric patients and incontinence is false?

    (a) Bladder capacity, urinary flow rate, and the ability to delay elimination appear to naturally decline with age.

    (b) Incontinence is primarily caused by the unavoidable age-related loss of sphincter control and rarely leads to any kind of medical or traumatic emergency.

    (c) Cerebral or spinal lesions may cause loss of sphincter control.

    (d) Diabetes may cause geriatric loss of sphincter control.

    (e) Bowel incontinence occurs less often than urinary incontinence but can seriously impair geriatric activity levels.

109. Which of the following statements regarding geriatric patients and elimination difficulties is true?

    (a) Urinary or bowel elimination difficulties may be a sign of serious underlying illnesses or conditions.

    (b) Many medications (narcotics, diuretics, iron supplements, and others) can cause constipation.

    (c) Transient ischemic attacks or syncopal episodes can be caused by a constipated geriatric patient exerting strong effort to have a bowel movement.

    (d) Answers (a) and (b).

    (e) Answers (a), (b), and (c).

110. Which of the following statements regarding assessment and care of the geriatric trauma patient is false?

(a) Geriatric trauma patients may not exhibit tachycardia during the early stages of hypovolemic shock.

(b) Because of chronic dehydration and the higher arterial pressures required to perfuse geriatric vital organs, geriatric trauma patients require larger volumes and more rapid administration rates of IV fluid than younger patients.

(c) Positive pressure ventilation performed too aggressively may rupture geriatric pulmonary parenchyma and cause a pneumothorax.

(d) Because geriatric patients typically have higher blood pressures than younger patients, a systolic blood pressure of 110 may be indicative of hypotension, or even shock, in the geriatric trauma patient.

(e) Physical deformities (arthritis, spinal abnormalities, atrophied limbs) will require modification of standard immobilization techniques.

111. Which of the following statements regarding assessment and care of the geriatric trauma patient is false?

(a) Geriatric fractures may occur as a result of "minor" events not even considered traumatic in younger patients, such as sneezing.

(b) Signs and symptoms of geriatric brain injury may require days, or even weeks, to develop. Frequently, the patient will deny memory of any trauma history.

(c) Geriatric patients have diminished capacities to sense temperature. Thus they are less susceptible to hypo- or hyperthermia than are younger patients.

(d) In the geriatric trauma patient, sudden neck movement, even without the presence of a cervical fracture, may cause injury to the cervical spinal cord.

(e) Following even a "minor" trauma mechanism, any geriatric patient exhibiting confusion should be considered for rapid transportation to a trauma center.

112. Loss of bone strength caused by _____ (the demineralization that occurs in the elderly musculoskeletal structure) makes geriatric patients more susceptible to fractures.

(a) osteoporosis

(b) kyphosis

(c) hypertrophy

(d) lordosis

(e) spondylosis

113. Degeneration of vertebral body size occurs with age. This degeneration is called

(a) osteoporosis.

(b) kyphosis.

(c) hypertrophy.

(d) lordosis.

(e) spondylosis.

114. Many elderly people develop a "hunchback" appearance because of rheumatoid arthritis, vertebral degeneration, or poor posture. This exaggerated curve, or angulation, of the thoracic spine is called

   (a) osteoporosis.
   (b) kyphosis.
   (c) hypertrophy.
   (d) lordosis.
   (e) spondylosis.

115. Which of the following statements regarding geriatric CNS assessment is false?

   (a) Decreased creative and cognitive mental functions occur because of the reduction in brain weight and size that naturally accompanies aging.
   (b) The reduction in brain weight and size that naturally accompanies aging makes geriatric patients more susceptible to subdural and epidural hemorrhage even when the force of trauma is minimal.
   (c) Any geriatric patient who acts confused or seems to have an altered level of consciousness should be considered to have a serious underlying illness or injury until proven otherwise at the emergency department.
   (d) Any geriatric patient who acts confused or seems to have an altered level of consciousness requires evaluation of his/her blood glucose level as soon as possible.
   (e) Hypo- or hyperthermia are among the most common causes of geriatric altered mental status.

116. Which of the following statements regarding geriatric myocardial infarction is true?

   (a) Confusion or the complaint of "dizziness" may be the only S/Sx of geriatric MI.
   (b) Dyspnea or the complaint of "weakness" may be the only S/Sx of geriatric MI.
   (c) Geriatric MI may be entirely unaccompanied by chest pain or discomfort.
   (d) All of the above are true.
   (e) None of the above are true.

117. The leading cause of disability among United States citizens aged 65 and older is

   (a) Parkinson's disease.
   (b) Alzheimer's disease.
   (c) Meniere's disease.
   (d) Osteoarthritis.
   (e) Marfan's syndrome.

118. The medical condition in which the lens of the eye develops a cloudlike opacity, losing its clearness and significantly interfering with vision, is called

   (a) cataracts.
   (b) dysphagia.
   (c) glaucoma.
   (d) intraocular hydrocephalus.
   (e) chronic hyphema.

119. The medical condition that produces increased intraocular pressure, resulting in diminished sight, and that often leads to blindness is called

   (a) cataracts.
   (b) dysphagia.
   (c) glaucoma.
   (d) intraocular hydrocephalus.
   (e) chronic hyphema.

120. _____ is an inner ear disorder characterized by vertigo, nerve deafness, and a roaring or buzzing in the ear.

   (a) Tinnitus
   (b) Meniere's disease
   (c) Ocular senescence
   (d) Ocular hypertrophy
   (e) Auricular senescence

The Answer Key to Test Section 11 is on page 455.

Test
Section
Twelve

Test Section 12's subjects:

- Ambulance Operations
- Medical-Incident Command
- Rescue Awareness and Operations
- Hazardous-Materials Incidents
- Crime-Scene Awareness
- Fibrinolytic Therapy Administration Eligibility
- 2005 AHA BLS CPR Guidelines

Test Section 12 consists of 120 questions and is allotted 2 hours for completion.

1.   Periodic calibration is required for all of the following pieces of ambulance
     equipment, except
        (a) glucometers.
        (b) oxygen storage tanks.
        (c) pulse oximeters.
        (d) cardiac monitors.
        (e) automated external defibrillators.

2.   The vast majority of collisions involving ambulances occur
        (a) after sundown.
        (b) on urban side streets.
        (c) on major highways.
        (d) in parking lots.
        (e) at intersections.

3.   While responding to the scene of a patient in cardiac arrest, you approach a school
     bus offloading children with its red flashing lights on. Which of the following
     statements indicates the most appropriate way of responding to this situation?
        (a) Wait for the children to stop exiting the bus, and then proceed with caution.
        (b) Turn your lights and siren off and carefully proceed past the bus.
        (c) Do not proceed until the bus driver manually signals you to do so.
        (d) Do not proceed until after the bus driver has turned off the red flashing
             lights.
        (e) Pass the bus as you would any other vehicle, carefully observing for
             pedestrians (as you would when passing any other vehicle).

4.   Which of the following statements about the use of lights and sirens is true?
        (a) Use of lights and sirens diminishes ETAs by up to 50%.
        (b) Most motorists automatically slow their driving speed when they hear a siren
             approaching and begin looking for the emergency vehicle.
        (c) Motorists are more inclined to yield when the same siren tone is continuously
             sounded and it becomes louder.
        (d) Patient anxiety decreases when sirens are used, because sirens signify that
             "emergency care" is approaching or being rendered.
        (e) Paramedic anxiety increases when sirens are used going to or from a call.

5.   When responding to an emergency, a police escort should be used
        (a) during rush hours, when traffic is most congested.
        (b) when responding to a call where CPR is in progress.
        (c) when you are unfamiliar with the call's area of location.
        (d) Answers (a) and (c).
        (e) Answers (a), (b), and (c).

6.   When following an emergency police escort through city streets, you should
   (a) turn off the ambulance siren, so that motorists are not confused by two
       different sirens.
   (b) keep the ambulance siren exactly the same as that used by the police escort,
       so that motorists are not confused by two different sirens.
   (c) stay at least one block behind the escort vehicle so that motorists realize there
       are two emergency vehicles approaching.
   (d) follow the escort as closely as possible, so that motorists cannot pull in
       between your vehicle and the police vehicle.
   (e) be aware that following an escort vehicle is an extremely hazardous endeavor,
       and be even more cautious than you normally would be before following the
       escort vehicle through any intersection.

7.   When approaching an intersection while en route to an emergency call, you should
   (a) carefully pass stopped vehicles on the left (driver's) side.
   (b) carefully pass stopped vehicles on the right (passenger's) side.
   (c) avoid making eye contact with other drivers so as not to become distracted
       by their expressions.
   (d) Answer (a) or (b).
   (e) Answers (a) or (b) and (c).

8.   You are the first emergency unit arriving at the scene of a multiple-vehicle collision
       on an interstate highway. You can see smoke coming from beneath one of the
       vehicles in the midst of the wreckage. You should park your ambulance
   (a) just before the first wrecked vehicle you approach.
   (b) just beyond the wrecked vehicle most distant from your approach.
   (c) angled across the highway so as to block all oncoming traffic from danger and
       to protect the scene occupants.
   (d) immediately next to the smoking vehicle, as that vehicle's occupants will be
       the first you'll need to assess or extricate.
   (e) at least 100 feet upwind (and preferably uphill) from the smoking vehicle.

9.   You are dispatched to an auto accident. Immediately upon your arrival, however,
       you learn that what actually occurred was a high-speed auto-*pedestrian* accident.
       The scene is approximately 20 minutes away (by ground) from the trauma center.
       The trauma center has a helicopter available. You should
   (a) evaluate the pedestrian for level of consciousness and degree of injury prior
       to considering the need for helicopter transportation.
   (b) complete your initial assessment and primary treatment of the pedestrian
       prior to considering the need for helicopter transportation.
   (c) put the helicopter on standby while you evaluate the pedestrian for level of
       consciousness and degree of injury; the helicopter may not be needed.
   (d) call for helicopter transportation immediately after learning of the MOI, even
       before you begin patient assessment and treatment.
   (e) transport the patient by ground ambulance; 20 minutes is not a long enough
       ground transportation ETA to warrant the expensive use of a helicopter.

10. A helicopter is en route to your rural emergency trauma scene, and you are just completing intubation of your unconscious multisystem trauma patient. In preparation for helicopter transportation, you should

    (a) use IV fluid or sterile water, rather than air, to inflate the ET balloon.

    (b) deflate all chambers of the previously applied PASG garment, to allow for pressure changes in flight.

    (c) convert all IV bags to pressure infuser bags, if you carry them.

    (d) Answers (a) and (c).

    (e) Answers (a), (b), and (c).

11. When approaching a helicopter with a patient, it is important to

    (a) loosely drape a blanket over the patient's entire (secured) body, so as to protect her/him from debris kicked up by rotor wash.

    (b) wait until a flight crew member indicates that you may approach the helicopter, even if several minutes have transpired since the aircraft landed.

    (c) keep the ambulance stretcher at its highest elevation, so as to avoid tipping the stretcher while crossing uneven terrain.

    (d) Answers (a) and (b).

    (e) Answers (a), (b), and (c).

12. You have summoned a helicopter to the scene of a wilderness accident. The pilot selects the safest available landing site, but it is on a slight slope. You should approach the helicopter

    (a) from the downhill side, at the front of the aircraft.

    (b) from the uphill side, at the front of the aircraft.

    (c) from the downhill side, at the tail of the aircraft.

    (d) from the uphill side, at the tail of the aircraft.

    (e) from the uphill side, at the middle of the aircraft.

13. A Multiple Casualty Incident (MCI) can be defined as any incident that

    (a) produces multiple casualties.

    (b) significantly taxes the resources available to the local emergency system.

    (c) presents one or more serious scene management challenges.

    (d) Answer (a) or (b).

    (e) Answer (a), (b), or (c).

14. The most widely recognized and utilized system of MCI triage is the START system. START is an acronym that means

    (a) Safety Triage Assessment Removal Treatment.

    (b) Simple Treatment Assessment Removal Triage.

    (c) Safety Treatment Assessment Reporting Termination.

    (d) Simple Triage And Rapid Transport.

    (e) Scene Treatment And Rapid Transportation.

The following six questions involve this MCI scenario: A suspended walkway within a shopping mall has entirely collapsed, apparently because of structural instability. No other sections of the walkway remain suspended, so there is no danger of further collapse. At the time of its collapse, an unknown number of people were on or below the walkway.

15.   To initiate the START triage system, you should first use a bullhorn to announce,

   (a)   "Everyone who is injured should stay where they are. We will come to you!"
   (b)   "Everyone who is injured should hold up an arm or call for help, now!"
   (c)   "If you are not injured, please leave the area now!"
   (d)   "If you are not injured, and you have first aid knowledge, come to me now!"
   (e)   "If you can walk, get up and come to me now, please."

16.   You begin your systematic triage and encounter a patient who is unconscious and apneic. Using the START triage system, you should immediately

   (a)   tag the patient black (dead) and move on to find more viable patients.
   (b)   tag the patient red (critical) and move on to find more viable patients.
   (c)   tag the patient green (a "pass-over," nonviable patient) and move on to find more viable patients.
   (d)   check for a pulse before tagging the patient.
   (e)   open the patient's airway and assess for spontaneous breathing.

17.   You assess the next patient and determine that he is unconscious, has an open airway with a respiratory rate of 30, and a rapid, weak, radial pulse. You should immediately

   (a)   remove the patient to the nearest ambulance and initiate definitive treatment.
   (b)   initiate and maintain spinal immobilization, waiting until others can assume spinal immobilization and treatment before resuming triage.
   (c)   tag the patient green (is breathing, has a pulse and can wait for additional treatment) and move on to the next patient.
   (d)   initiate and provide all appropriate treatment for that patient, forgoing further patient triage (avoiding abandonment charges) and continue to provide treatment for that patient until his care is assumed by another paramedic.
   (e)   tag the patient red (critical), and move on to the next patient.

18.   You assess the next patient and determine that she is awake, has a respiratory rate of 28, and a rapid radial pulse. Your next START triage system step is to

   (a)   ask her if her neck hurts.
   (b)   ask her if her chest hurts.
   (c)   tag her with a "black" tag (she is "critical"), and move on to the next patient.
   (d)   ask her to speak her name.
   (e)   ask her to grip both your hands.

19. According to the START triage system, patients designated as being those of the highest priority for rapid treatment and ambulance transport, are identified by a tag with _____ as the bottom color.
     (a) black
     (b) blue
     (c) yellow
     (d) green
     (e) red

20. According to the START triage system, patients designated as being those who require treatment but do not require ambulance transportation are identified by a tag with _____ as the bottom color.
     (a) black
     (b) blue
     (c) yellow
     (d) green
     (e) red

21. The critical incident stress management (CISM) system is designed to assist EMS providers in dealing with stress following a critical incident. Critical incidents include calls that involve

     (1) multiple casualties.   (4) the injury or suicide of an EMS worker.
     (2) several dead victims.   (5) an EMS worker causing civilian injury or
     (3) unusual media attention.   death.
                                         (6) the injury or death of an infant or child.

     (a) 1 only.
     (b) 1 and 2.
     (c) 1, 2, and 4.
     (d) 1, 2, 4, and 6.
     (e) 1, 2, 3, 4, 5, and 6.

22. The CISM system consists of several components. A brief, informal meeting, provided within 2–4 hours post incident, is designed to give the involved EMS workers an opportunity to vent their feelings about the incident. This CISM intervention is called
     (a) a defusing.
     (b) a demobilization meeting.
     (c) a critical incident stress debriefing.
     (d) an individual consultation.
     (e) a follow-up service.

23. Within 24 to 72 hours post-event, a carefully planned, formal, and highly structured intervention is provided by a team of trained CISM providers. This CISM intervention is called
     (a) a defusing.
     (b) a demobilization meeting.

(c) a critical incident stress debriefing.

(d) an individual consultation.

(e) a follow-up service.

24. All paramedics are required to be trained and able to

(a) operate the Jaws of Life or similar types of equipment.

(b) recognize potential hazards and know how to summon personnel trained to handle them.

(c) don a self-contained breathing apparatus (SCBA).

(d) Answers (a) and (b).

(e) Answers (a), (b), and (c).

25. During the Arrival and Scene Size-Up phase of a rescue operation, you should

(a) determine the nature of the rescue situation and call for appropriate personnel, even if you merely suspect that they will be needed.

(b) survey the scene to determine the number of patients involved.

(c) begin providing treatment to the closest patient as soon as possible.

(d) Answers (a) and (b).

(e) Answers (a), (b), and (c).

26. Immediately following the Arrival and Scene Size-Up phase of any rescue operation, you should

(a) continue providing medical treatment to the closest patients.

(b) prepare for patient packaging and removal.

(c) begin to disentangle the patient(s).

(d) assure that all on-scene hazards are being dealt with by trained individuals prior to anyone entering the rescue area.

(e) call for a helicopter.

27. You arrive on the scene where a small passenger car has struck a utility pole. Your scene size-up reveals that there is a sparking electrical line lying across the hood of the car. The vehicle's driver is visibly unconscious, and gurgling respirations are audible. You should

(a) immediately access the patient to provide C-spine control and assess the airway.

(b) carefully remove the patient from the car without touching any metal surfaces.

(c) summon the utility company to immediately respond to the scene, and wait for their arrival before accessing the patient.

(d) attempt to rouse the patient without touching any metal.

(e) use rubber gloves to gain access to the gear shift, place it in neutral, and roll the vehicle away from the electrical line.

28. After accessing an entrapped patient, the amount of medical treatment you provide should be based upon
   (a) the safety conditions of the immediate environment for both you and the patient.
   (b) the number of other patients requiring assessment and treatment.
   (c) the patient's mental status.
   (d) Answers (a) and (b).
   (e) Answers (a), (b), and (c).

29. After skidding off an icy highway, a young female is alert but trapped inside her overturned car. The rescue team advises you that the car's position has been stabilized but that extricating her will be a lengthy process, requiring the Jaws of Life and cutting tools. You should perform all of the following measures, except
   (a) call medical control and obtain permission for limb removal in order to expedite rapid disentanglement, should the patient lose consciousness.
   (b) provide the patient with eye protection, preferably vented goggles held in place with an elastic band.
   (c) provide the patient with hearing protection (earplugs).
   (d) cover the patient, above and below her body, with warm wool blankets.
   (e) be prepared to periodically cover her head and face with a protective blanket.

30. Which of the following statements about rapid emergency removal of a patient to a safer location is false?
   (a) If danger of rescuer injury exists (in addition to increased patient injury risk), and the danger cannot be alleviated, patient stabilization may be delayed until after immediate removal.
   (b) If rapid removal will aggravate existing injuries (such as the C-spine), the patient cannot be moved, under any circumstances, until appropriately immobilized.
   (c) Patients in swift-running, rising water may require extrication before they are spinally immobilized.
   (d) Patients being assessed in a vehicle that suddenly catches fire may be rapidly removed before they are spinally immobilized.
   (e) Patients being assessed in a building that begins to collapse may be rapidly removed before they are spinally immobilized.

31. During periods of extended entrapment, the emotional needs of the patient are an important and vital aspect of patient care. Which of the following statements about providing emotional support is false?
   (a) Learning the patient's name as quickly as possible and using it as often as possible is important to improving any patient's sense of well-being.
   (b) Discuss all technical aspects of the rescue operation with the patient, using plain English, so that the patient will understand and be forewarned of what to expect.

    (c) If the situation is unstable and continues to be dangerous, and the patient voices a suspicion of this, maintain direct eye contact while you strongly assure the patient that the situation is completely safe.

    (d) If something you plan to do is going to hurt the patient, admit it to the patient.

    (e) If you don't know the answer to one or more of the patient's questions, admit it.

32. A 25-year-old man fell into an open manhole. He reports striking his head on something and being unconscious for an unknown period of time. He denies spine pain. Which of the following extrication devices is most appropriate for lifting him from the manhole?

    (a) A long spine board.

    (b) A KED.

    (c) A Sked® device.

    (d) A harness device with rope.

    (e) A stair chair.

33. The most dangerous water-rescue situations are those involving water that is

    (a) deep.

    (b) moving.

    (c) cold.

    (d) hot.

    (e) stagnant.

34. Which of the following statements regarding self-rescue techniques is false?

    (a) In moving water, float on your stomach with your feet pointing upstream, and your head raised to watch for upcoming obstacles, so that you can try to avoid them.

    (b) In cold water, float with your head out of the water and your body in a fetal tuck to conserve heat.

    (c) Do not attempt to stand up in rapidly moving water, no matter what the depth.

    (d) Even if you are only assisting trained individuals with a shore-based water rescue, you should wear a personal flotation device (PFD).

    (e) If you have fallen into cold water with other people, floating while huddled together (body contact) will delay hypothermia.

35. A 25-year-old male was drinking and "horse-playing" with his buddies on the end of a dock that extends approximately 25 feet out from a lake shore. He fell into the very cold water, and is panicking because he cannot reach the dock's edge to pull himself out (nor can his buddies grasp him and pull him out). The water is approximately 12 feet deep, and the ETA of the Water Rescue team is at least 20 minutes. You place a long stick in the water, but the panicking male will not grab hold of it. What should you do next?

    (a) Throw him your PFD to use as a floatation device, and remain on the dock to keep his buddies from jumping in after him.

    (b) If there is a row boat or raft in the area, use it to row out to the patient.

    (c) If there is any sort of flotation device in the area, throw it to him and counsel him to use it to stay afloat while you wait for the Water Rescue Team to arrive.

    (d) Cover your mouth and nose before jumping into the water (wearing your PFD), calm the patient and keep him afloat until the Water Rescue Team arrives.

    (e) Encourage the patient to swim toward shore or to vigorously tread water, so that he'll generate body heat and avoid hypothermia.

36. The mammalian diving reflex enhances the likelihood of drowning resuscitation, and occurs when

    (a) a person submerged in very cold water experiences laryngeal spasm, which prevents aspiration of water.

    (b) a person's face is plunged into very cold water.

    (c) a person's feet are plunged into very cold water.

    (d) Answer (a) or (b).

    (e) Answer (b) or (c).

37. Effects and vital signs that commonly accompany the mammalian diving reflex include all of the following, except

    (a) tachycardia.

    (b) decreased blood pressure and peripheral vasoconstriction.

    (c) bradycardia.

    (d) shunting of blood from less vital organs to the heart.

    (e) shunting of blood from less vital organs to the brain.

38. An elderly gentleman was playing with his grandson in a wading pool when bystanders suddenly noticed him submerged, face down, and unconscious. A bystander rolled him to his back and held his face out of the water, but he is still unconscious and is not breathing upon your arrival. You should do all of the following, except

    (a) recruit bystanders to assist you and your partner to maintain spinal immobilization as you rapidly slide the patient out of the water.

    (b) maintain an open airway.

    (c) begin pocket face mask rescue breathing while the patient is still in the water.

    (d) apply a cervical collar.

    (e) place a long spine board beneath the patient and secure him to it before removing him from the water.

39.   Which of the following statements regarding confined spaces is false?

    (a) Many confined spaces contain toxic fumes that can cause rescuer asphyxiation.

    (b) The greatest danger to rescuers entering confined spaces is that of limited visibility (darkness) contributing to rescuer head injury.

    (c) Oxygen-deficient atmospheres are a frequent cause of death for rescuers entering confined spaces.

    (d) Dust particles within grain silos (or the like) can cause explosions.

    (e) Electrical equipment within confined spaces may contain stored energy, and present electrical shock dangers even after power sources have been disconnected.

40.   The greatest hazard associated with EMS highway operations is that of

    (a) ignition of fuel spills causing a fire hazard.

    (b) traffic flow.

    (c) lacerations from sharp glass, metal, or fiberglass wreckage edges.

    (d) vehicle explosions.

    (e) downed power lines.

41.   Which of the following statements regarding EMS highway operations is true?

    (a) After parking your vehicle, leave headlights and all flashing lights on, to warn motorists away from the accident site.

    (b) Impaired drivers may actually drive toward flashing emergency lights.

    (c) If your vehicle is the first to arrive on scene, park it so that it blocks all lanes containing wrecked vehicles from oncoming traffic, to ensure a safety zone.

    (d) Do not place cones or flares about the scene; this is law enforcement's responsibility, and they have specific SOPs for their placement.

    (e) If bystanders have inappropriately placed flares about the scene, pick them up and extinguish them.

42.   Which of the following are also potential hazards when working a highway or motor vehicle accident scene?

    (a) Undeployed airbags.

    (b) Natural gas powered or electrically powered vehicles.

    (c) Energy-absorbing bumpers.

    (d) Answers (a) and (b).

    (e) Answers (a), (b), and (c)

43. Which of the following statements about cutting the battery cables of an extensively damaged vehicle is true?
    (a) Before cutting battery cables, move all electric seats back and lower all power windows.
    (b) You must first obtain the vehicle owner's permission to cut battery cables.
    (c) The fire hazard from connected battery cables is a myth, and they should never be cut.
    (d) Turning the ignition off eliminates all threat of electrical spark from the vehicle battery.
    (e) Cutting the battery cables may cause undeployed airbags to explode.

44. Which of the following statements about automobile glass is false?
    (a) Safety glass is composed of a layer of plastic laminate fused between two layers of glass.
    (b) When safety glass is shattered, it fractures into small beads of dull-edged glass so that there is little to no risk of glass lacerations or cuts.
    (c) When tempered glass is shattered, it fractures into small, sharp beads of glass.
    (d) Safety glass can produce glass dust when shattered, which can be damaging to the patient's eyes, nose, or mouth.
    (e) Although designed to remain intact when impacted, safety glass can fracture into long shards and cause lacerations or cuts.

45. Your patient's vehicle drove head-on into a light pole and remains there, with the light pole embedded approximately six inches into the engine compartment. The driver appears to be alone in the car, and is slumped over the wheel and unresponsive. All the doors are locked. You should
    (a) call for the Jaws of Life.
    (b) use a window punch to break the side window farthest from the patient.
    (c) use a window punch to break the front windshield; it is cheapest to replace.
    (d) use a window punch to break the rear windshield; it affords easiest access.
    (e) use a crowbar to pry open the driver's door.

46. Which of the following devices is most appropriate for carrying a nonambulatory patient over rough or hazardous terrain?
    (a) A long spine board.
    (b) A KED.
    (c) A Sked® device.
    (d) A Stokes (or similar basket-type) stretcher.
    (e) A stair chair.

47. Which of the following statements about Stokes or basket-type stretchers is false?
   (a) The older "military style" basket stretchers will not accept a long spine board.
   (b) A KED may be used in combination with a basket stretcher.
   (c) Plastic basket stretchers are stronger and more durable than wire-mesh ones.
   (d) A basket stretcher can provide spinal immobilization without the use of additional devices.
   (e) The most preferred type of basket stretchers are those with plastic bottoms and steel frames.

48. You respond to the scene of a train derailment. On your approach (from a distance), you see a green fog emitting from an overturned boxcar. Your first action is to
   (a) stop and avoid the scene until you can determine the safest place to park.
   (b) use extreme caution while quickly rescuing any obvious victims.
   (c) summon a hazmat team to the scene.
   (d) have your dispatcher notify the railroad commission of a derailment.
   (e) drive past the scene, counting potential victims as you size up the situation.

49. Continuing the previous scenario, your second action is to
   (a) stop and avoid the scene until you can determine the safest place to park.
   (b) use extreme caution and rescue any obvious patients as soon as possible.
   (c) summon a hazmat team to the scene.
   (d) have your dispatcher notify the railroad commission of a derailment.
   (e) drive past the scene, counting potential victims as you size up the situation.

50. An "open" hazmat incident is best defined as one in which
   (a) a potential exists for more patients to be generated by the event.
   (b) open containers of potentially hazardous materials are visible.
   (c) the hazardous material cannot be identified from a distance.
   (d) the scene is an open area, and the contamination zone may drift or move.
   (e) the final clean-up has not yet occurred, so the incident remains open.

51. A "closed" hazmat incident is best defined as one in which
   (a) the incident area has been evacuated, and no further patients are anticipated.
   (b) closed containers of potentially hazardous materials are visible.
   (c) the hazardous material cannot be identified from a distance.
   (d) the scene is in a closed area, such as a building, and the contamination zone cannot drift or move.
   (e) the final clean-up has occurred, so the incident is closed.

52. The best place to park when responding to a hazmat incident is
  (a) upwind from the incident scene.
  (b) uphill from the incident scene.
  (c) as close as possible to where the patients are.
  (d) Answers (a) and (b).
  (e) Answer (a), (b), or (c), depending upon the hazardous substance.

53. A tanker truck has overturned on the highway, and an unknown liquid substance
  is spilling from the ruptured tank. To determine the type of substance involved, you
  should
  (a) immediately find the vehicle's driver and question him as you assess him.
  (b) immediately search the cab of the wreck for the vehicle's shipping papers.
  (c) remain at a distance from the wreck and use binoculars to search for external
       placards identifying the vehicle's contents.
  (d) have your dispatcher call the trucking company, relaying a detailed
       description of the truck, so they can access their records and advise you
       of the cargo.
  (e) wait for the hazmat team to arrive and identify the substance.

54. Because of budget difficulties, emergency response providers are often ill equipped
  to ensure their personal safety. Every emergency response vehicle that may be first
  on scene ought to be equipped with all of the following items. However, not every
  EMS provider is trained to use
  (a) binoculars.
  (b) the North American Emergency Response Guidebook.
  (c) personal protective equipment.
  (d) SCBA equipment.
  (e) a bullhorn.

55. The first emergency response team to arrive at a hazmat scene should establish
  zones of operation. The most dangerous hazmat zone (the one that includes the
  immediate area of the hazmat incident) is the
  (a) "hot" (black) zone.
  (b) "hot" (red) zone.
  (c) "cold" (black) zone.
  (d) "cold" (red) zone.
  (e) "cold" (blue) zone.

56. Only highly trained and appropriately protected personnel are allowed in the
  hazmat danger zone. The best method for determining the appropriate distance
  required for any hazmat danger zone perimeter is to
  (a) ask the truck driver, train conductor, or plant manager.
  (b) observe how wide a distance the substance appears to be affecting and set
       your perimeter 10 feet beyond that.

(c) consult the North American Emergency Response Guidebook or call one of the 24-hour toll-free numbers for hazmat information.

(d) determine the wind direction and establish the danger zone immediately downwind from the site, establishing the "warm" zone immediately upwind.

(e) call your medical director.

57. All contaminated clothing or gear (worn by patients or responders) must be removed before they can be allowed to enter the

(a) black (safe treatment) zone.

(b) red (immediate treatment) zone.

(c) "warm" (control) zone.

(d) "cold" (safe treatment) zone.

(e) blue (removal) zone.

The following four questions involve this scenario: You are called to a farm by the wife of a 29-year-old male. She reports that "Jim just came in from the barn, and he's real sick!"

58. Jim is drooling, his nose is running, and his eyes are tearing. You notice that he appears to have been incontinent of urine, and you smell a fecal odor on approach. He is moaning and clutching his stomach and then begins to vomit. His heart rate is very slow (approximately 40). Jim is displaying all the classic _____ symptoms.

(a) RUNNY

(b) SLUDGE

(c) HARE

(d) BEE STING

(e) OD

59. Jim's signs and symptoms are classic for what kind of emergency?

(a) Street drug overdose.

(b) Hydrocarbon solvent exposure.

(c) Methane exposure.

(d) Pesticide exposure.

(e) Allergic reaction.

60. As soon as you've established Jim's IV, which of the following medications should you administer?

(a) Naloxone, 2 mg, repeated as needed to a max of 10 mg.

(b) Atropine, 0.5–1.0 mg IV, repeated every 5 minutes.

(c) Atropine, 2.0–4.0 mg IV, repeated every 2 minutes as needed.

(d) Epi 1:1, 0.5 mg SQ, once only.

(e) Epi 1:10, 1 mg IV, repeated every 5 minutes, as needed.

61. If Jim begins to exhibit seizure activity, you should administer

  (a) 1 mg of morphine, IV, to counteract the naloxone.
  (b) another dose of atropine, but increase the dose to 5 mg, IV.
  (c) one amp of $D_{50}W$, IV.
  (d) 5–10 mg of diazepam, IV.
  (e) another dose of epi 1:10, but decrease the dose to 0.5 mg, IV.

62. Carbon monoxide inhalation causes asphyxiation because it

  (a) binds to hemoglobin more strongly than oxygen, preventing oxygen from being delivered to the cells.
  (b) displaces all atmospheric oxygen, preventing oxygen from being inhaled.
  (c) alters an intracellular enzyme, preventing cells from being able to utilize oxygen.
  (d) Answers (a) and (c).
  (e) Answers (b) and (c).

63. Cyanide inhalation causes asphyxiation because it

  (a) binds to hemoglobin more strongly than oxygen, preventing oxygen from being delivered to the cells.
  (b) displaces all atmospheric oxygen, preventing oxygen from being inhaled.
  (c) alters an intracellular enzyme, preventing cells from being able to utilize oxygen.
  (d) Answers (a) and (c).
  (e) Answers (b) and (c).

64. Early S/Sx of carbon monoxide poisoning include all of the following, except

  (a) altered level of consciousness.
  (b) headache.
  (c) a classic "cherry-red" skin coloring.
  (d) chest pain.
  (e) unconsciousness or seizures.

65. Early S/Sx of cyanide poisoning include all of the following, except

  (a) gradual respiratory depression (slower and more shallow breathing patterns).
  (b) sudden loss of consciousness.
  (c) sudden onset of seizures.
  (d) sudden respiratory arrest.
  (e) sudden cardiac arrest.

66. For treatment of cyanide exposure, the first medication that should be administered from a cyanide kit is

  (a) sodium nitrite.
  (b) sodium nitrate.
  (c) sodium amytal.
  (d) sodium thiosulfate.
  (e) amyl nitrite.

67. Which medication contained in a cyanide kit is flammable when exposed to oxygen?
    (a) Sodium nitrite.
    (b) Sodium nitrate.
    (c) Sodium amytal.
    (d) Sodium thiosulfate.
    (e) Amyl nitrite.

68. You are dispatched to a shooting at a private residence. To ensure the safety of your approach to the scene, you should
    (a) rendezvous with a police escort and stay with it until the scene is reached.
    (b) stop your vehicle well before it can be observed from the scene and remain there until dispatch advises you that the police have secured the scene and the shooter is either in custody or absent.
    (c) slowly approach the scene with your siren sounding until you have arrived on the scene, so as to afford the shooter plenty of time to leave the area.
    (d) shut off all your lights and sirens (especially after sundown) and use a "stealth" approach to the scene's location, so that you do not become a target.
    (e) stop as soon as you can see the scene, and wait there, closely observing scene activities, until police report that the scene is secured.

69. In the late evening, you are dispatched to a "need an ambulance" call at a private residence; the caller hung up before dispatch could obtain more information. From a block away, you can see that the house is dark. You should park your vehicle
    (a) at the curb immediately in front of the residence you were called to, to afford you a rapid getaway if you need to flee the scene.
    (b) at least a block away and out of sight of the residence you were called to.
    (c) one or two houses away from the residence you were called to.
    (d) in the middle of the street, so that you will not be blocked in if you need to flee the scene.
    (e) backed into the driveway of the residence you were called to, to afford you a rapid, forward-moving getaway if you need to flee the scene.

70. When approaching any dark scene, you should
    (a) turn on the ambulance scene lights closest to the residence and use that illumination to guide your approach, instead of using a flash light and possibly becoming a target.
    (b) rest your flashlight on your shoulder, lessening the likelihood that your center mass will become a target.
    (c) hold your flashlight out to the side, away from the bodies of you and your partner.
    (d) Answer (a) or (b), according to personal preference.
    (e) Answer (a) or (c), according to personal preference.

71. Which of the following statements about safely approaching a dark house is false?
    (a) Avoid using the sidewalk or a path that leads directly to the door; approach across the lawn or follow the perimeter of the building.
    (b) Avoid looking into windows.
    (c) As you approach, look for an escape route that affords cover or concealment.
    (d) Avoid shining a light on, or from behind, your partner.
    (e) Avoid standing in front of a door, especially when knocking on it.

72. Which of the following statements about body armor (Kevlar® vests) is false?
    (a) Most vests will stop low-velocity weapons such as knives, ice picks, and small handgun bullets.
    (b) Most vests will not stop high-velocity rifle fire.
    (c) Kevlar® vests will not stop armor-piercing bullets.
    (d) The wearer's head, axillary areas, groin, and femurs are not protected when wearing body armor.
    (e) Most vests provide protection from blunt trauma to the torso.

73. Which of the following objects provides cover as well as concealment?
    (a) Trash cans.
    (b) Bushes.
    (c) Trees with large trunks.
    (d) Vehicle doors (especially ambulance rear compartment doors).
    (e) Large shrubs.

74. Which of the following objects does not provide cover as well as concealment?
    (a) Brick walls.
    (b) Large commercial dumpsters.
    (c) The engine block of an ambulance or other vehicles.
    (d) Vehicle doors, especially ambulance rear compartment doors.
    (e) Large rocks.

75. You and your partner are alone and performing CPR on a 50-year-old female when her 29-year-old husband approaches you with a gun and orders you to stop what you are doing or he will kill you both. You should
    (a) direct your partner to continue CPR while you try to calm the husband down; if you stop treatment that has begun, you can be charged with abandonment.
    (b) stall for time to allow backup to arrive, by slowly gathering your equipment.
    (c) leave your equipment behind and back out of the vicinity calmly but quickly.
    (d) cue your partner to join you in overpowering and disarming the husband.
    (e) continue providing care with your partner while attempting to calm the husband down; if you stop or diminish the level of treatment that has begun, you can be charged with abandonment.

76. If it becomes absolutely necessary for you to handle a weapon found on a scene (such as to prevent others from using it), you should

 (a) step on it, and remain standing on it until the police arrive to take possession.

 (b) put a blanket or pillow over it and sit on it until the police arrive.

 (c) pick it up by the barrel, without touching any other part, and secure it in the ambulance until the police arrive.

 (d) pick it up by sliding a pencil or a pen into the barrel, so that you do not touch any part of it, and secure it in the ambulance until the police arrive.

 (e) pick it up by the side grips of the handle, and secure it in the ambulance until the police arrive.

77. Which of the following statements regarding clandestine drug laboratories ("clan labs") is false?

 (a) Clan labs are often identified by neighbors complaining of foul odors.

 (b) Doors and walkways in and around clan labs may be "booby trapped."

 (c) Like automobiles, clan labs rarely "explode," except in the movies.

 (d) Disassembled clan labs may be mobile, and stored in the trunk of someone's car.

 (e) Clan labs are "hazmat scenes."

78. On a sunny afternoon, responding to an "unconscious diabetic, come to the side door" call, you arrive at a suburban house. There is a cat on the porch, and the home reeks of cat urine. The side door is open, and you can see a woman on the floor of the kitchen. You enter and are about to begin your assessment when you notice strange glassware on the table and counters. Next, you notice a large, scientific-looking container perched on the stove, with a gas flame burning beneath it. You should do all the following, except

 (a) turn off the burner.

 (b) notify the police.

 (c) leave the building immediately.

 (d) drive away from the building immediately.

 (e) consider initiating evacuation of nearby homes.

79. According to the United States Department of Justice, most acts of violence occur

 (a) in residences.

 (b) in schools.

 (c) on interstate highways.

 (d) in office buildings.

 (e) on the streets.

80. Which of the following statements regarding organized gang activity is false?

   (a) Danger from gang members is increased when wearing a uniform that associates you with the police.

   (b) Hand signals used by gang members often go unnoticed by those unfamiliar with them.

   (c) Gang activities are often funded by drug sales.

   (d) Gang clothing usually incorporates a common color.

   (e) Gang activity is limited to large cities and their closest suburbs.

81. All of the following techniques are suggested when fleeing danger, except

   (a) run in the straightest, fastest, most direct line to your ambulance.

   (b) consider throwing your equipment in the direction of your pursuer.

   (c) consider throwing your equipment in one direction as you run in another.

   (d) overturn objects in the path of your pursuer.

   (e) run in a different direction from your partner.

82. Which of the following statements regarding the concept of "contact" and "cover" partner activities at suspicious scenes is false?

   (a) The "contact" partner handles all interaction with the patient.

   (b) The "cover" partner maintains a global perspective of the surrounding scene, watching for signs of danger.

   (c) The "contact" partner focuses on assessment and care of the patient but maintains an awareness of what the "cover" partner is doing.

   (d) Until more than two providers are on the scene of a cardiac arrest that has occurred in a potentially dangerous location, the "contact" partner should perform one-person CPR.

   (e) The "cover" partner generally avoids performing patient-care activities that would distract from watching for signs of danger.

83. Crime scene evidence includes all of the following except

   (a) any fluids on the patient's body or clothing.

   (b) statements made by your partner on scene.

   (c) notes found within the patient's clothing.

   (d) blood spatters on furniture.

   (e) blood drops on floors.

84. Your victim is supine, with blood emanating from two holes in the right side of his T-shirt front. The best way to preserve this garment's evidence is to

   (a) log-roll the patient to his side, cut the T-shirt up the middle of the back, examine his back for exit or additional entry wounds, log-roll him back to a long spine board, remove the shirt from his body, and give it to the police.

   (b) avoid further damage to the shirt, by "treating around" it. This also preserves the "chain of evidence," because the shirt will remain with the victim.

(c) cut the center of the T-shirt open and treat the victim without entirely removing the shirt.

(d) cut through the holes to open the T-shirt and treat the victim without entirely removing the shirt.

(e) cut away the section of T-shirt containing the holes and give that shirt section to the police.

85. En route to the emergency department, you remove the blood-soaked shoe of a stabbing victim. To best preserve this garment's evidence, you should

(a) place it under the patient's foot, so that it remains with the patient.

(b) place it in a paper bag labeled with the victim's name on a piece of tape.

(c) place it in a "hazmat" or similar plastic bag labeled with the victim's name on a piece of tape.

(d) Answer (a) or (b), depending upon emergency department protocols.

(e) Answer (a) or (c), depending upon emergency department protocols.

86. When entering and working within a crime scene, wearing disposable surgical gloves will prevent you from

(a) leaving your fingerprints on the scene.

(b) smearing fingerprints that might be on surfaces you cannot avoid touching.

(c) transferring blood or other body fluids from one surface to another.

(d) Answers (a) and (b).

(e) Answers (a), (b), and (c).

87. Which of the following statements about walking through a crime scene is false?

(a) Avoid stepping in or on any wet substances on a hard floor.

(b) Avoid stepping in or on any wet substances on a carpet.

(c) Avoid stepping in or on any dried substances on hard floors or carpets.

(d) If the police have secured the scene and allowed you to enter it, the floor evidence has already been documented, and your concern should be to get to the patient as rapidly as possible, without worrying about the floors.

(e) Do not touch anything you do not absolutely have to touch.

88. Your patient was shot in the street and went directly from the street into your ambulance. Which of the following statements regarding safety and evidence preservation is incorrect?

(a) Be alert for sharp items that may be contained within the patient's clothing.

(b) To preserve body fluid evidence when a large piece of clothing is soaked with blood, place it as flat as possible inside a large paper bag and roll the bag to layer the sections of the garment with paper; then place the rolled bag into another large paper bag, before enclosing the double-paper-bagged item in a plastic bag.

(c) Separately bag every single item of clothing.

(d) Separately bag the contents of each pocket.

(e) Place all small items of clothing into one plastic bag and label it with the patient's name; similarly place all belongings into one plastic bag.

89.   All of the following observations you might make at a crime scene are important to police records of the crime, except

    (a) crime description related comments made by bystanders.

    (b) medications you administered to the patient during your care.

    (c) the patient's condition upon your arrival.

    (d) the position of the patient upon your arrival.

    (e) statements made by the patient regarding the crime.

90.   The stabbing victim you cared for was pronounced dead at the emergency department soon after your arrival. When writing your patient-care report for this crime victim, you should

    (a) document witness statements describing the stabber's direction of flight from the scene, but put witness statements inside quotation marks.

    (b) record any individual's name that you heard used at the scene.

    (c) record the number of individuals you saw on the scene, noting "colors" present, if gang activity is suspected.

    (d) document only the observations you made and the statements you heard that are medically pertinent to the record of your patient's physical condition and the medical care you rendered.

    (e) document everything you heard and everything you saw.

Whether or not you work in a system that allows prehospital administration of fibrinolytics, an understanding of the indications and contraindications for fibrinolytic administration may be required to make transportation decisions. The following 5 questions refer to fibrinolytic administration indications and contraindications. If you do not know this material, skip to Question 96 to continue your timed test-taking. Then, study the fibrinolytic administration indications and contraindications references before you take the test you are preparing for.

91.   Appropriate fibrinolytic administration can significantly limit or diminish the extent of

    (a) neurologic damage caused by a clot that travels to the brain.

    (b) myocardial tissue damage caused by a clot that travels to a coronary vessel.

    (c) neurologic deficits ultimately suffered by a CVA patient.

    (d) Answers (a) and (c).

    (e) Answers (a), (b), and (c).

92.   Fibrinolytic administration consideration is vitally important to the emergency care of any patient who complains of ischemic-sounding CP and has an ECG that exhibits

    (1) ST-segment elevation.   (3) an apparently "normal" ECG.

    (2) ST-segment depression.   (4) a new (or presumably new) LBBB.

        (a) 1, 2, 3, or 4.

        (b) 1, 2, or 4.

(c) 2 or 3.

(d) 1 or 4.

(e) 2 or 4.

93. Fibrinolytic administration is contraindicated for the emergency care of any patient who complains of ischemic-sounding CP and has an ECG that exhibits

(1) ST-segment elevation.

(2) ST-segment depression.

(3) an apparently "normal" ECG.

(4) a new (or presumably new) LBBB.

(a) 1, 2, 3, or 4.

(b) 1, 2, or 4.

(c) 2 or 3.

(d) 1 or 4.

(e) 2 or 4.

94. Which of the following criteria is not considered a potential contraindication for fibrinolytic administration consideration?

(a) The patient has a Hx of chronic bleeding or clotting problems, or is currently taking a prescribed blood thinner.

(b) The patient has a heart rate greater than 100 bpm, with a systolic B/P less than 100 mmHg.

(c) The patient has a persistent systolic B/P of greater than 180 mmHg.

(d) The patient has a persistent diastolic B/P of greater than 110 mmHg.

(e) The patient is pregnant.

95. Which of the following criteria is considered a potential contraindication for fibrinolytic administration consideration?

(a) The patient had a witnessed seizure when the current S/Sx began.

(b) The patient is already exhibiting neurological deficits.

(c) The patient was the victim of serious trauma 15 days ago.

(d) The patient had major surgery 20 days ago.

(e) The patient is only 20 years old.

Every paramedic is responsible for knowing how to perform BLS CPR according to the most current recommendations. Additionally, every paramedic should be aware of the manner in which lay rescuers are currently being trained to perform BLS CPR. Thus, the following 25 questions are designed to test your knowledge of the December 2005 American Heart Association BLS CPR performance recommendations—as well as some 2005 AHA ALS performance recommendations not covered elsewhere.

96.   To assess for breathing, every rescuer should position the unconscious victim so that he has an open airway, then look, listen, and feel for breathing for up to 10 seconds. Which of the following statements regarding initiation of rescue breaths (ventilation) is true?

   (a) If the lay rescuer or EMT is not confident that the victim is breathing within 10 seconds, rescue breathing should immediately be started.

   (b) If a lay rescuer is unwilling to give an apneic victim rescue breaths via mouth-to-mouth, she or he should immediately begin to perform chest compressions without pausing to check for a pulse.

   (c) If the victim is occasionally "gasping" for breath, rescue breathing (or mechanical ventilation) should not be initiated until after a pulse check is performed.

   (d) Answers (a) and (b) are true.

   (e) Answers (a) and (c) are true.

97.   For the first provision of rescue breathing or artificial ventilation, any rescuer should deliver

   (a) 1 long, deep and forceful ventilation (lasting 2 to 4 seconds), to fully inflate the respiratory arrest victim's lungs prior to providing subsequent ventilations.

   (b) 2 ventilations, each lasting no more or less than one full second.

   (c) 4 quick ventilations, each larger than the previous one (performed in a "stair-step" manner of increasing depth and forcefulness), to fully inflate the respiratory arrest victim's lungs.

   (d) Answer (a) or (b), depending upon local protocol.

   (e) Answer (a) or (c), depending upon local protocol.

98.   Which of the following statements regarding ventilation administration—via mouth-to-mouth or mouth-to-mask (MTM), or via bag-valve-mask (BVM)—is false?

   (a) Each MTM or BVM ventilation should last at least one full second.

   (b) Each MTM or BVM ventilation should produce a visible chest rise.

   (c) If a visible chest rise does not occur during a MTM ventilation, the victim's airway should be repositioned, and the ventilation immediately repeated.

   (d) If a visible chest rise does not occur during a BVM ventilation, the victim's airway should be repositioned, the mask resealed, and the ventilation immediately repeated.

   (e) When oxygen is attached and flowing at 15 LPM, and the BVM is equipped with an oxygen reservoir, the BVM is able to deliver a greater amount of oxygen than can MTM ventilation without oxygen. Thus, the recommended amount of time for each oxygenated BVM ventilation duration can be shortened by approximately one half.

99.   Which of the following statements regarding BVM operation before an advanced airway device is in place (ie, ET tube, Combitube, PtL, or LMA) is false?

(a) Regardless of any rescuer's hand size, strength, or experience, 1-person BVM ventilations are never as effective as 2-person BVM ventilations.

(b) Because the BVM produces substantial positive-pressure ventilation even before it is attached to an advanced airway device, 8 to 10 ventilations per minute can be delivered without requiring a chest compression pause.

(c) At the end of every cycle of chest compressions, the compressor should pause for 2 seconds to allow delivery of 2 BVM ventilations.

(d) Insertion of an oropharyngeal airway will significantly improve the effectiveness of BVM ventilations.

(e) Insertion of a nasopharyngeal airway can significantly improve the effectiveness of BVM ventilations.

100. Which of the following statements regarding BVM operation when an advanced airway device is in place (ie, ET tube, Combitube, PtL, or LMA) is true?

(a) Even after it is attached to an advanced airway device, 1-person BVM ventilations are never as effective as 2-person BVM ventilations.

(b) Once connected to an advanced airway device, 8 to 10 BVM ventilations per minute can be delivered without requiring a chest compression pause.

(c) To facilitate adequate lung ventilation, the compressor should still pause for 2 seconds at the end of every cycle of chest compressions, to allow delivery of 2 ventilations.

(d) To facilitate adequate lung ventilation, the compressor should still pause for 1 second at the end of every cycle of chest compressions, to allow delivery of 2 ventilations.

(e) To facilitate adequate lung ventilation, the compressor should still pause for 1 to 4 seconds at the end of every cycle of chest compressions, to allow delivery of 1 full ventilation.

*For the following nine questions (questions 101 to 110), **initial** ventilation has already been accomplished.*

101. The universal single-rescuer CPR compression-to-ventilation ratio for cardiopulmonary arrest victims of all ages (except infants) is

(a) 5:1 (1 ventilation for every 5 chest compressions).

(b) 15:2 (2 ventilations for every 15 chest compressions).

(c) 30:2 (2 ventilations for every 30 chest compressions).

(d) 15:1 (1 ventilation for every 15 chest compressions).

(e) 30:1 (1 ventilation for every 30 chest compressions).

102. The compression-to-ventilation ratio 2 lay rescuers should perform for adult CPR is

(a) 5:1 (1 ventilation for every 5 chest compressions).

(b) 15:2 (2 ventilations for every 15 chest compressions).

(c) 30:2 (2 ventilations for every 30 chest compressions).

(d) 15:1 (1 ventilation for every 15 chest compressions).

(e) 30:1 (1 ventilation for every 30 chest compressions).

103. The compression-to-ventilation ratio 2 trained rescuers should perform for adult CPR is

    (a) 5:1 (1 ventilation for every 5 chest compressions).

    (b) 15:2 (2 ventilations for every 15 chest compressions).

    (c) 30:2 (2 ventilations for every 30 chest compressions).

    (d) 15:1 (1 ventilation for every 15 chest compressions).

    (e) 30:1 (1 ventilation for every 30 chest compressions).

104. The compression-to-ventilation ratio 2 lay rescuers should perform for cardiopulmonary arrest victims aged 1 to 8 years old is

    (a) 5:1 (1 ventilation for every 5 chest compressions).

    (b) 15:2 (2 ventilations for every 15 chest compressions).

    (c) 30:2 (2 ventilations for every 30 chest compressions).

    (d) 15:1 (1 ventilation for every 15 chest compressions).

    (e) 30:1 (1 ventilation for every 30 chest compressions).

105. The compression-to-ventilation ratio 2 trained rescuers should perform for cardiopulmonary arrest victims aged 1 to 12 years old is

    (a) 5:1 (1 ventilation for every 5 chest compressions).

    (b) 15:2 (2 ventilations for every 15 chest compressions).

    (c) 30:2 (2 ventilations for every 30 chest compressions).

    (d) 15:1 (1 ventilation for every 15 chest compressions).

    (e) 30:1 (1 ventilation for every 30 chest compressions).

106. The compression-to-ventilation ratio a single lay or trained rescuer should perform for cardiopulmonary arrest victims younger than 1 year old (infants) is

    (a) 3:1 (1 ventilation for every 3 chest compressions).

    (b) 5:2 (2 ventilations for every 5 chest compressions).

    (c) 30:2 (2 ventilations for every 30 chest compressions).

    (d) 15:1 (1 ventilation for every 15 chest compressions).

    (e) 5:1 (1 ventilation for every 5 chest compressions).

107. The compression-to-ventilation ratio 2 trained rescuers should perform for infant cardiopulmonary arrest victims is

    (a) 3:1 (1 ventilation for every 3 chest compressions).

    (b) 5:2 (2 ventilations for every 5 chest compressions).

    (c) 30:2 (2 ventilations for every 30 chest compressions).

    (d) 15:1 (1 ventilation for every 15 chest compressions).

    (e) 5:1 (1 ventilation for every 5 chest compressions).

108. Which of the following statements regarding AHA recommendations for the performance of infant CPR by a single lay or trained rescuer is false?

    (a) Each ventilation should require only about ½ second.

    (b) Each chest compression should require only about ½ second.

(c) The infant's sternum should be compressed with 2 fingers placed at the intermammary line.

(d) Full chest relaxation (reexpansion) should be allowed, without removing the fingers from the sternum.

(e) The chest compression phase should be equal to or slightly longer than the allowance of chest relaxation (reexpansion) phase.

109. Which of the following statements regarding AHA recommendations for the performance of infant CPR by 2 trained rescuers is false?

(a) Each ventilation should require only about ½ second.

(b) Each chest compression should require only about ½ second.

(c) Place both thumbs together over the lower half of the sternum, spreading the fingers of both hands around the infant's thorax, and forcefully compress the sternum with your thumbs while squeezing the thorax with your fingers (providing counterpressure).

(d) The chest compression phase should be slightly shorter than the allowance of chest relaxation (reexpansion) phase.

(e) Anxious adults tend to maintain a degree of chest compression while "resting" their hand on the infant's chest. Thus, the hand performing chest compression should be lifted completely off of the infant's chest (by only ½ to 1 inch) between compressions, to ensure full chest reexpansion prior to performing the next compression.

110. Rescuer fatigue has been shown to result in inadequate performance of chest compression rates and/or depth. When 2 or more rescuers are available to perform adult CPR, the AHA has mandated that the chest-compressor should switch roles with the ventilator, or be replaced by another chest-compressor, after performing

(a) 5 cycles of chest compression.

(b) 5 minutes of chest compression.

(c) 10 cycles of chest compression.

(d) 10 minutes of chest compression.

(e) 15 minutes of chest compression.

111. When 2 or more trained rescuers are available to perform CPR for a child aged 8 to 13 years old, the AHA has mandated that the chest-compressor should switch roles with the ventilator, or be replaced by another chest-compressor, after performing

(a) 1 minute of chest compression.

(b) 2 minutes of chest compression.

(c) 5 minutes of chest compression.

(d) 10 minutes of chest compression.

(e) 15 minutes of chest compression.

112. When 2 or more trained rescuers are available to perform CPR for an infant, the AHA has mandated that the chest-compressor should switch roles with the ventilator, or be replaced by another chest-compressor, after performing
   (a) 1 minute of chest compression.
   (b) 2 minutes of chest compression.
   (c) 5 minutes of chest compression.
   (d) 10 minutes of chest compression.
   (e) 15 minutes of chest compression.

113. The AHA has mandated that the CPR pause required for a chest-compressor to switch roles with a ventilator, or to be replaced by another chest-compressor, should be limited to no more than
   (a) 1 minute.
   (b) 10 seconds.
   (c) 5 seconds.
   (d) 20 seconds.
   (e) 2 minutes.

114. The 2005 AHA-recommended chest compression rate per minute during single-rescuer CPR for victims of all ages (except infants)
   (a) refers only to the relative speed of compression performance. It does not indicate the actual number of compressions the AHA expects you to deliver each minute.
   (b) is accomplished by performing compressions in time to the mnemonic, "one-one-thousand, two-one-thousand, three-one-thousand," and so on.
   (c) is accomplished by performing compressions in time to the mnemonic, "one-and, two-and, three-and, four-and," and so on.
   (d) is accomplished by performing compressions in time to the mnemonic, "one-potato, two-potato, three-potato, four-potato," and so on.
   (e) Answer (c) or (d), depending upon the mnemonic you were originally trained to use.

115. The 2005 AHA-recommended chest compression rate per minute during 2-person CPR for victims of all ages (except infants)
   (a) refers only to the relative speed of compression performance. It does not indicate the actual number of compressions the AHA expects you to deliver each minute.
   (b) is accomplished by performing compressions in time to the mnemonic, "one-one-thousand, two-one-thousand, three-one-thousand," and so on.
   (c) is accomplished by performing compressions in time to the mnemonic, "one-and, two-and, three-and, four-and," and so on.
   (d) is accomplished by performing compressions in time to the mnemonic, "one-potato, two-potato, three-potato, four-potato," and so on.
   (e) Answer (c) or (d), depending upon the mnemonic you were originally trained to use.

116. Which of the following statements regarding effective chest-compression performance is false?

   (a) For chest compression to achieve effective artificial blood flow, the chest must be depressed at least 1½ to 2 inches (4 to 5 cm).

   (b) Many experienced EMS-providers fail to compress the chest deep enough to achieve maximally effective artificial blood flow.

   (c) The AHA-recommendation for effective chest compression is, "Push Hard and Push Fast."

   (d) Allowing the chest to "recoil" (to return to its original position) after each chest compression is not as important as ensuring that chest compressions be performed at least 1½ to 2 inches (4 to 5 cm) deep.

   (e) Unless the victim spontaneously begins to move, at least 5 cycles of chest compression should be performed before stopping to assess for pulse presence, or to defibrillate the cardiac arrest patient.

117. Which of the following statements regarding circulation checks during BLS CPR performance is false?

   (a) Lay rescuers should not interrupt chest compressions to check for circulation unless the victim begins moving.

   (b) The lay rescuer circulation check following initial ventilation delivery has been increased from a 10 second check to a 30 second check.

   (c) Lay rescuers should not interrupt chest compressions to check for circulation until an AED has arrived and is ready to be deployed.

   (d) Health-care providers should spend no more than 10 seconds when checking for a pulse.

   (e) Chest compressions should never be delayed more than 10 seconds to perform a pulse check.

118. Which of the following statements regarding paramedic defibrillation of an unwitnessed and atraumatic cardiac arrest victim is true?

   (a) Since you did not witness the arrest, your first act upon arrival should be to ensure that at least 2 minutes of effective CPR is performed prior to performing defibrillation.

   (b) Even though he's probably been "down" for more than 5 minutes, if the victim is still in V-fib his best chance for survival is to immediately be defibrillated. Upon your arrival, your first act should be to use your cardiac monitor's paddles to assess the victim's rhythm and provide defibrillation if he is still in V-fib.

   (c) Do not check the patient's rhythm or pulse immediately after shock delivery. Instead, immediately resume chest compressions.

   (d) Answers (a) and (b) are true.

   (e) Answers (a) and (c) are true.

119. When defibrillating a victim with V-fib or pulseless V-tach for the first time, the paramedic should

(a) administer up to 3 "stacked shocks": if the first shock fails to convert or change the rhythm, the next shock should immediately be performed, and should consist of a greater energy level; if the second shock does not convert or change the rhythm, the third shock should immediately be performed, and should consist of a greater energy level.

(b) check for spontaneous return of pulse immediately following each shock, before repeating the shock or initiating/resuming basic CPR performance.

(c) administer 1 shock, immediately followed by initiation or resumption of chest compressions without delaying to evaluate the rhythm produced, and without delaying to check for a pulse.

(d) Answers (a) and (b).

(e) None of the above. (Pulseless V-tach should be cardioverted, not defibrillated.)

120. The most vitally important 2005 AHA ACLS recommendation is this: Trained health-care providers must significantly

(a) minimize the amount of time that transpires between onset of cardiac arrest and the defibrillation of V-fib, especially when the arrest was unwitnessed.

(b) increase the amount of time that is spent ventilating (oxygenating) all cardiac arrest patients.

(c) minimize the amount of time that chest compressions are interrupted to perform any manner of evaluation or any type of procedure (including advanced airway insertion).

(d) Answers (a) and (b).

(e) Answers (a), (b), and (c).

The Answer Key to Test Section 12 is on page 460.

Test
Section
Thirteen

13

Test Section 13's subjects:

- Additional Anatomy and Physiology
- Additional Medical Terminology, Abbreviations, and Symbols

Test Section 13 consists of 126 questions and is allotted 2 hours and 6 minutes for completion.

1. The study of the body's structure is called
    (a) pathophysiology.
    (b) physiology.
    (c) physics.
    (d) anatomy.
    (e) biophysics.

2. The study of the body's body functions is called
    (a) pathophysiology.
    (b) physiology.
    (c) physics.
    (d) anatomy.
    (e) biophysics.

3. The study of the body's disease or injury processes is called
    (a) pathophysiology.
    (b) physiology.
    (c) physics.
    (d) anatomy.
    (e) biophysics.

4. The sacrum is formed by ___ vertebrae that are fused together.
    (a) 3–5
    (b) 5
    (c) 7
    (d) 12
    (e) 14

5. The coccyx is formed by ___ vertebrae that are fused together.
    (a) 3–5
    (b) 5
    (c) 7
    (d) 12
    (e) 14

6. Each vertebra consists of a vertebral body, a spinal foramen, and three processes:
    (a) two transverse and one spinous.
    (b) two spinous and one posterior.
    (c) two spinal and one transverse.
    (d) two lateral and one spinal.
    (e) two posterior and one spinal.

7.   The ears are sensory organs that normally function to facilitate
  (a) hearing.
  (b) positional sense (physical sense of balance).
  (c) an opening for intracranial pressure release.
  (d) Answers (a) and (b).
  (e) Answers (a), (b), and (c).

8.   The actual organ(s) of hearing, where vibrations are translated into nerve impulses, which are then transmitted to the brain by the auditory nerve,
  (a) is the tympanic membrane.
  (b) is the cochlea.
  (c) are the semicircular canals.
  (d) are the ossicles.
  (e) is the auricle.

9.   The _____ responsible for sensing body motion, position, and equilibrium.
  (a) tympanic membrane is
  (b) cochlea is
  (c) semicircular canals are
  (d) ossicles are
  (e) auricle is

10. The _____ the first structure(s) to perceive environmental sound wave vibrations.
  (a) tympanic membrane is
  (b) cochlea is
  (c) semicircular canals are
  (d) ossicles are
  (e) auricle is

11. Your patient complains that, "The room is spinning!" This indicates a disturbance of the patient's sense of balance that is most accurately called
  (a) "WADAO."
  (b) dizziness.
  (c) "the vapors."
  (d) syncope.
  (e) vertigo.

12. The medical term for an inner ear infection is
  (a) auditis.
  (b) rhinitis.
  (c) sinusitis.
  (d) cochlitis.
  (e) otitis.

13. Although not part of the ear, the _____ is composed of air-filled cells that are continuous with (connected to) the middle ear. Thus pain or tenderness in this area often accompanies an inner ear infection.

    (a) zygomatic bone

    (b) mandible

    (c) maxilla

    (d) maxillary bone's styloid process

    (e) temporal bone's mastoid process

14. The medical term for a "runny nose" is

    (a) rhinitis.

    (b) rhinorrhea.

    (c) sinusitis.

    (d) hematitis.

    (e) epistaxis.

15. The medical term for bleeding from the nose is

    (a) rhinitis.

    (b) rhinorrhea.

    (c) sinusitis.

    (d) hematorrhea.

    (e) epistaxis.

16. The colored area of the eye is called the

    (a) iris.

    (b) cornea.

    (c) sclera.

    (d) pupil.

    (e) conjunctiva.

17. The white area of the eye is called the

    (a) iris.

    (b) cornea.

    (c) sclera.

    (d) pupil.

    (e) conjunctiva.

18. The thin, external, membrane that covers the anterior eye, protecting the eye from foreign bodies, is called the

    (a) iris.

    (b) cornea.

    (c) sclera.

    (d) pupil.

    (e) conjunctiva.

19.   The clear structure that separates the anterior chamber of the eye from the external environment is called the
       (a) iris.
       (b) cornea.
       (c) sclera.
       (d) pupil.
       (e) conjunctiva.

20.   The part of the eye that transforms light rays into electrical impulses that are transmitted to the brain is called the
       (a) retina.
       (b) sclera.
       (c) cornea.
       (d) pupil.
       (e) lens.

21.   The thin membrane that lines the interior surface of the eyelids is called the
       (a) iris.
       (b) cornea.
       (c) sclera.
       (d) pupil.
       (e) conjunctiva.

22.   The _____ is the opening through which light waves travel to enter the interior eye.
       (a) iris
       (b) cornea
       (c) sclera
       (d) pupil
       (e) conjunctiva

23.   The _____ is a contractile muscle that constricts or dilates to control the amount of light that enters the eye.
       (a) iris
       (b) cornea
       (c) sclera
       (d) pupil
       (e) conjunctiva

24.   The _____ is a convex, transparent structure that allows images to focus onto the sensory structure of the eye.
       (a) retina
       (b) sclera
       (c) cornea
       (d) pupil
       (e) lens

25.   The human skeleton normally has _____ bones.
       (a) 106
       (b) 156
       (c) 206
       (d) 215
       (e) 226

26.   The only bone in the axial skeleton that does not connect to any other bone
       is the
       (a) patella.
       (b) nasal septum.
       (c) first cervical vertebra.
       (d) seventh sacral vertebra.
       (e) hyoid bone.

27.   The area of a long bone where growth occurs is called the
       (a) epiphysis.
       (b) metaphysis.
       (c) diaphysis.
       (d) apophysis.
       (e) aponeurosis.

28.   Each widened end of a long bone is called the
       (a) epiphysis.
       (b) metaphysis.
       (c) diaphysis.
       (d) apophysis.
       (e) aponeurosis.

29.   The central, cylinder-like shaft of a long bone is called the
       (a) epiphysis.
       (b) metaphysis.
       (c) diaphysis.
       (d) apophysis.
       (e) aponeurosis.

30.   The axial skeleton includes all of the following except
       (a) the skull.
       (b) the ribs and sternum.
       (c) the clavicles and scapulae.
       (d) the sacrum and coccyx.
       (e) the spinal column.

31.   The appendicular skeleton includes all of the following, except
       (a) the clavicles and scapulae.
       (b) the arms, forearms, wrists, hands, and fingers.
       (c) the legs, ankles, feet, and toes.
       (d) the sacrum and coccyx.
       (e) the pelvic bones.

32.   _____ is the medical term for the exaggerated inward curve of the lumbar
       spine, commonly called "swayback."
       (a) Osteoporosis
       (b) Kyphosis
       (c) Hypertrophy
       (d) Lordosis
       (e) Spondylosis

33.   Red blood cell production is a function of
       (a) yellow bone marrow.
       (b) the aponeurosis.
       (c) red bone marrow.
       (d) the metaphysis.
       (e) the periosteum.

34.   Fat storage is a function of
       (a) yellow bone marrow.
       (b) the aponeurosis.
       (c) red bone marrow.
       (d) the metaphysis.
       (e) the periosteum.

35.   A form of connective tissue that has the ability to stretch and connects bone to
       bone is called
       (a) cancellous tissue.
       (b) a ligament.
       (c) cartilage.
       (d) a tendon.
       (e) articulate tissue.

36.   A form of connective tissue that is inelastic and that connects muscle to bone is
       called
       (a) cancellous tissue.
       (b) a ligament.
       (c) cartilage.
       (d) a tendon.
       (e) articulate tissue.

37.   The medial malleolus is formed by the distal end of the
    (a)   humerus.
    (b)   radius.
    (c)   tibia.
    (d)   fibula.
    (e)   femur.

38.   The lateral malleolus is formed by the distal end of the
    (a)   humerus.
    (b)   radius.
    (c)   tibia.
    (d)   fibula.
    (e)   femur.

39.   The acetabulum is
    (a)   the medical term for the neck of the humerus.
    (b)   located between the neck and head of the femur.
    (c)   the hollow depression of the pelvis where the femoral head articulates.
    (d)   located between the shaft and neck of the femur.
    (e)   the hollow depression where the head of the humerus articulates.

40.   The most commonly fractured bone in the body is the
    (a)   mandible.
    (b)   clavicle.
    (c)   scapula.
    (d)   humerus.
    (e)   femur.

41.   In layman's terms, the _____ is called the shoulder blade.
    (a)   ulna
    (b)   clavicle
    (c)   humerus
    (d)   radius
    (e)   scapula

42.   The bone located on the lateral side of the forearm is the
    (a)   ulna.
    (b)   clavicle.
    (c)   humerus.
    (d)   radius.
    (e)   scapula.

43. In layman's terms, the _____ is called the collarbone.

    (a) ulna

    (b) clavicle

    (c) humerus

    (d) radius

    (e) scapula

44. The bone located on the medial side of the forearm is the

    (a) ulna.

    (b) clavicle.

    (c) humerus.

    (d) radius.

    (e) scapula.

45. The _____ are the bones of the wrist.

    (a) tarsals

    (b) metatarsals

    (c) carpals

    (d) metacarpals

    (e) phalanges

46. The _____ are the bones of the palm.

    (a) tarsals

    (b) metatarsals

    (c) carpals

    (d) metacarpals

    (e) phalanges

47. The _____ are the bones of the fingers.

    (a) tarsals

    (b) metatarsals

    (c) carpals

    (d) metacarpals

    (e) phalanges

48. The seven _____ include the talus and calcaneus bones that form the ankle.

    (a) tarsals

    (b) metatarsals

    (c) carpals

    (d) metacarpals

    (e) phalanges

49.   The _____ are the bones of the foot that articulate with the toes.
   - (a) tarsals
   - (b) metatarsals
   - (c) carpals
   - (d) metacarpals
   - (e) phalanges

50.   The _____ are the bones of the toes.
   - (a) tarsals
   - (b) metatarsals
   - (c) carpals
   - (d) metacarpals
   - (e) phalanges

51.   Each of the large winglike bones of the pelvis is called
   - (a) an ischium.
   - (b) an ilium.
   - (c) an iliac.
   - (d) a sacrum.
   - (e) a pubic bone.

52.   In layman's terms, the _____ is called the kneecap.
   - (a) patella
   - (b) tibia
   - (c) femur
   - (d) fibula
   - (e) malleolus

53.   The _____ is the bone that forms the anterior lower leg.
   - (a) patella
   - (b) tibia
   - (c) femur
   - (d) fibula
   - (e) malleolus

54.   The _____ is the bone within the posterior lower leg.
   - (a) patella
   - (b) tibia
   - (c) femur
   - (d) fibula
   - (e) malleolus

55. Paralysis limited to both lower extremities is called
    (a) paraplegia.
    (b) quadriplegia.
    (c) hemiplegia.
    (d) paresis.
    (e) paresthesia.

56. Paralysis limited to one side of the entire body is called
    (a) paraplegia.
    (b) quadriplegia.
    (c) hemiplegia.
    (d) paresis.
    (e) paresthesia.

57. Paralysis that includes the upper and lower extremities is called
    (a) paraplegia.
    (b) quadriplegia.
    (c) hemiplegia.
    (d) paresis.
    (e) paresthesia.

58. The right lung consists of
    (a) a single large lobe of lung tissue.
    (b) two lobes of lung tissue: the parietal and visceral lobes.
    (c) three lobes of lung tissue: the parietal, visceral, and mediastinal lobes.
    (d) two lobes of lung tissue: the upper and lower lobes.
    (e) three lobes of lung tissue: the upper, middle, and lower lobes.

59. The left lung consists of
    (a) a single large lobe of lung tissue.
    (b) two lobes of lung tissue: the parietal and visceral lobes.
    (c) three lobes of lung tissue: the parietal, visceral, and mediastinal lobes.
    (d) two lobes of lung tissue: the upper and lower lobes.
    (e) three lobes of lung tissue: the upper, middle, and lower lobes.

60. The medical term *medial* indicates a direction or area
    (a) on or toward the front of the body.
    (b) on or toward the back of the body.
    (c) away from the body's midline.
    (d) toward or at the body's midline.
    (e) on the surface of the body.

61.   The medical term *lateral* indicates a direction or area
   - (a) on or toward the front of the body.
   - (b) on or toward the back of the body.
   - (c) away from the body's midline.
   - (d) toward or at the body's midline.
   - (e) on the surface of the body.

62.   The medical term *ventral* indicates a direction or area
   - (a) on or toward the front of the body.
   - (b) on or toward the back of the body.
   - (c) away from the body's midline.
   - (d) toward or at the body's midline.
   - (e) on the surface of the body.

63.   The medical term *dorsal* indicates a direction or area
   - (a) on or toward the front of the body.
   - (b) on or toward the back of the body.
   - (c) away from the body's midline.
   - (d) toward or at the body's midline.
   - (e) on the surface of the body.

64.   The medical term *posterior* indicates a direction or area
   - (a) on or toward the front of the body or body part.
   - (b) on or toward the back of the body or body part.
   - (c) away from the body's midline.
   - (d) toward the top of the body or body part.
   - (e) toward the bottom of the body or body part.

65.   The medical term *anterior* indicates a direction or area
   - (a) on or toward the front of the body or body part.
   - (b) on or toward the back of the body or body part.
   - (c) away from the body's midline.
   - (d) toward the top of the body or body part.
   - (e) toward the bottom of the body or body part.

66.   The medical term *superior* indicates a direction or area
   - (a) on or toward the front of the body or body part.
   - (b) on or toward the back of the body or body part.
   - (c) away from the body's midline.
   - (d) toward the top of the body or body part.
   - (e) toward the bottom of the body or body part.

67.   The medical term *inferior* indicates a direction or area

    (a) on or toward the front of the body or body part.

    (b) on or toward the back of the body or body part.

    (c) away from the body's midline.

    (d) toward the top of the body or body part.

    (e) toward the bottom of the body or body part.

68.   When something is close to a given point of reference, it is considered to be

    (a) superficial.

    (b) palmar.

    (c) distal.

    (d) proximal.

    (e) deep.

69.   When something is away from a given point of reference, it is considered to be

    (a) superficial.

    (b) palmar.

    (c) distal.

    (d) proximal.

    (e) deep.

70.   The medical term _____ describes movement away from the body.

    (a) extension

    (b) flexion

    (c) aversion

    (d) adduction

    (e) abduction

71.   The medical term _____ describes the act of bending at a joint.

    (a) extension

    (b) flexion

    (c) aversion

    (d) adduction

    (e) abduction

72.   The medical term _____ describes the movement of a joint toward its "straight" position.

    (a) extension

    (b) flexion

    (c) aversion

    (d) adduction

    (e) abduction

73. The medical term _____ describes movement toward the body.
  (a) extension
  (b) flexion
  (c) aversion
  (d) adduction
  (e) abduction

74. A patient resting horizontally on the ventral surface of her/his body is in the
  (a) Trendelenburg position.
  (b) prone position.
  (c) laterally recumbent position.
  (d) semi-Fowler's position.
  (e) supine position.

75. A patient lying on the posterior surface of her/his body, with the lower body elevated higher than the level of her/his head, is in the
  (a) Trendelenburg position.
  (b) prone position.
  (c) laterally recumbent position.
  (d) semi-Fowler's position.
  (e) supine position.

76. A patient lying on her/his left or right side is in a
  (a) Trendelenburg position.
  (b) prone position.
  (c) laterally recumbent position.
  (d) semi-Fowler's position.
  (e) supine position.

77. A patient lying on the posterior surface of her/his body, with her/his head elevated less than 45 degrees is in the
  (a) Trendelenburg position.
  (b) prone position.
  (c) laterally recumbent position.
  (d) semi-Fowler's position.
  (e) supine position.

78. A patient horizontally resting on the dorsal surface of her/his body is in the
  (a) Trendelenburg position.
  (b) prone position.

    (c) laterally recumbent position.

    (d) semi-Fowler's position.

    (e) supine position.

79.   Another name for the hypopharynx is the
    (a) nasopharynx.
    (b) oropharynx.
    (c) laryngopharynx.
    (d) endopharynx.
    (e) hyperpharynx.

80.   Filtering, warming, and humidification of inspired air occurs in the
    (a) nasopharynx.
    (b) oropharynx.
    (c) laryngopharynx.
    (d) endopharynx.
    (e) hyperpharynx.

81.   The portion of the pharynx extending from the posterior soft palate to the base of the tongue is called the
    (a) nasopharynx.
    (b) oropharynx.
    (c) laryngopharynx.
    (d) endopharynx.
    (e) hyperpharynx.

82.   The portion of the pharynx extending from the base of the tongue to the lower border of the cricoid cartilage is called the
    (a) nasopharynx.
    (b) oropharynx.
    (c) laryngopharynx.
    (d) endopharynx.
    (e) hyperpharynx.

83.   A wall dividing two cavities is often called a
    (a) nare.
    (b) conchae.
    (c) ventricle.
    (d) cilia.
    (e) septum.

84. Hairlike processes projecting from epithelial cells, which propel mucus, pus, and dust particles toward areas where they can be removed, are called
    (a) nares.
    (b) conchae.
    (c) ganglia.
    (d) cilia.
    (e) septumnal fibers.

85. The tongue is attached to the
    (a) mandible.
    (b) hyoid bone.
    (c) epiglottis.
    (d) Answers (a) and (b).
    (e) Answers (a), (b), and (c).

86. The epiglottis
    (a) is a thin, leaf-shaped structure located immediately posterior to the base of the tongue.
    (b) covers the entrance of the larynx when an individual swallows, to prevent aspiration of food or liquids.
    (c) is connected to the hyoid bone.
    (d) Answers (a) and (b).
    (e) Answers (a), (b), and (c).

87. The structure located where the trachea bifurcates into the right and left main stem bronchus is the
    (a) corona.
    (b) septum.
    (c) carina.
    (d) cornea.
    (e) cilia.

88. The widest and straightest bronchus is
    (a) also called the "subtrachea."
    (b) the right main stem bronchus.
    (c) the left main stem bronchus.
    (d) something that differs with each individual patient.
    (e) None of the above; the bronchi are the same width and shape.

89. The more angled bronchus is
    (a) also called the "subtrachea."
    (b) the right main stem bronchus.

(c) the left main stem bronchus.

(d) something that differs with each individual patient.

(e) None of the above; the bronchi are the same width and shape.

90. The most important functional unit of the entire respiratory system is the

(a) alveoli.

(b) epiglottis.

(c) pulmonary arteries.

(d) trachea.

(e) bronchioles.

91. The exchange of oxygen and carbon dioxide occurs within the

(a) alveoli.

(b) bronchi.

(c) trachea.

(d) Answers (a) and (b).

(e) Answers (a), (b), and (c).

92. A woman who is pregnant for the first time is referred to as a

(a) unigravida.

(b) unipara.

(c) primagravida.

(d) primapara.

(e) nulligravida.

93. A woman who has given birth to her first child is referred to as a

(a) unigravida.

(b) unipara.

(c) primagravida.

(d) primapara.

(e) nulligravida.

94. A woman who has delivered more than one child is referred to as a

(a) bi-, tri- or quadragravida (and so on).

(b) bi-, tri- or quadrapara (and so on).

(c) megagravida.

(d) multipara.

(e) multigravida.

95. A woman who been pregnant more than once is referred to as a
    (a) bi-, tri- or quadragravida (and so on).
    (b) bi-, tri- or quadrapara (and so on).
    (c) megapara.
    (d) multipara.
    (e) multigravida.

96. The medical term for an abnormally large food intake or a complaint of excessive hunger is
    (a) polytrophia.
    (b) polyphagia.
    (c) polyopsia.
    (d) polydipsia.
    (e) polygastria.

97. The medical term for an excessive output of urine is
    (a) polyhydruria.
    (b) polysaccharose.
    (c) polygastria.
    (d) polydipsia.
    (e) polyuria.

98. The medical term for an abnormally large fluid intake or a complaint of excessive thirst is
    (a) polyhydrosis.
    (b) polygastria.
    (c) polyhydruria.
    (d) polydipsia.
    (e) polyuria.

99. The medical term for needle decompression of a tension pneumothorax is
    (a) thorocotomy.
    (b) thorectomy.
    (c) thoracentesis.
    (d) cricoidotomy.
    (e) thoracodicotomy.

100. The medical term for difficulty with swallowing, or the inability to swallow is
    (a) euthanasia.
    (b) eutronasia.
    (c) aphasia.
    (d) polyphagia.
    (e) dysphagia.

101. Vasovagal syncope occurs when strong coughing, forceful defecation, or other types of _____ maneuvers trigger bradycardia and loss of blood pressure.
- (a) Selick's
- (b) syncopal
- (c) Stokes-Adams
- (d) valsalva
- (e) perianal

102. Orthostatic syncope occurs when a person
- (a) coughs too strongly while standing up.
- (b) bears down too strongly when having a bowel movement.
- (c) rises too quickly from a supine or seated position.
- (d) Answer (a) or (b).
- (e) Answer (b) or (c).

103. _____ is a progressively degenerative disease of the brain that causes impaired memory, altered thinking, and atypical behaviors.
- (a) Alzheimer's disease
- (b) Amyotrophic lateral sclerosis
- (c) Shy-Drager syndrome
- (d) Lou Gehrig's disease
- (e) Parkinson's disease

104. _____ is a progressively degenerative nervous disease, characterized by increasing muscle tremors or weakness, loss of facial expression, and gait disturbances.
- (a) Alzheimer's disease
- (b) Amyotrophic lateral sclerosis
- (c) Shy-Drager syndrome
- (d) Lou Gehrig's disease
- (e) Parkinson's disease

105. The medical symbol that means "decreased" is ___.
- (a) <
- (b) =
- (c) >
- (d) ↑
- (e) ↓

106. The medical symbol that means "increased" is ___.
- (a) <
- (b) =
- (c) >
- (d) ↑
- (e) ↓

107. The medical symbol that means "greater than" or "more than" is ___.

    (a) $<$

    (b) $=$

    (c) $>$

    (d) ↑

    (e) ↓

108. The medical symbol that means "less than" is ___.

    (a) $<$

    (b) $=$

    (c) $>$

    (d) ↑

    (e) ↓

109. _____ is the medical abbreviation used to indicate that a patient's particular system, body area, or vital sign was found to be "normal" in function, appearance, or measurement.

    (a) WNL

    (b) WAP

    (c) NP

    (d) NFP

    (e) NNFT

110. The medical abbreviation/symbol that means "every" is ___.

    (a) $\bar{a}$

    (b) $\bar{p}$

    (c) $\bar{c}$

    (d) $\bar{s}$

    (e) $\bar{q}$

111. The medical abbreviation that indicates an activity should be performed, or a medication taken "as needed" is

    (a) PRN.

    (b) AN.

    (c) AMA.

    (d) PMA.

    (e) WN.

112. The medical abbreviation for "nothing by mouth" is

    (a) NBM.

    (b) NDM.

    (c) NPO.

    (d) NM.

    (e) N/M.

113. The medical abbreviation "Rx" means
- (a) prescription or therapy.
- (b) treatment.
- (c) regular.
- (d) right.
- (e) left (opposite of right).

114. The medical abbreviation "Tx" means
- (a) prescription or therapy.
- (b) treatment.
- (c) transportation.
- (d) total.
- (e) tearing.

115. The medical abbreviation "Fx" means
- (a) faulty.
- (b) fair.
- (c) fracture.
- (d) frequently.
- (e) fortunately.

116. The medical abbreviation "Hx" means
- (a) high.
- (b) low (opposite of high).
- (c) heterosexual.
- (d) homosexual.
- (e) history.

117. The medical abbreviation "Dx" means
- (a) delivery.
- (b) delivered.
- (c) discharged.
- (d) diagnosis.
- (e) dislocation.

118. The medical abbreviation "Sx" means
- (a) associated.
- (b) symptoms.
- (c) "says."
- (d) story.
- (e) sometimes.

119. The medical abbreviation that means "secondary to" ("because of" or "due to") is
    (a) ST.
    (b) 2T.
    (c) 2°.
    (d) DT.
    (e) BO.

120. The medical abbreviation indicating that the patient acted contrary to counsel provided by you and/or your medical advisor is
    (a) WOP.
    (b) AMA.
    (c) APA.
    (d) W/OP.
    (e) WNP.

121. The complete inability to move one side of the body is called
    (a) hemiphasia.
    (b) paresthesia.
    (c) hemiplegia.
    (d) quadriplegia.
    (e) hemiparesis.

122. The sensation of numbness, "prickling," or "tingling" of one or more body parts is called
    (a) hemiphasia.
    (b) paresthesia.
    (c) hemiplegia.
    (d) quadriplegia.
    (e) hemiparesis.

123. Profound weakness of one side of the body is called
    (a) hemiphasia.
    (b) paresthesia.
    (c) hemiplegia.
    (d) quadriplegia.
    (e) hemiparesis.

124. The medical term for generalized difficulty speaking is.
    (a) dysarthria.
    (b) aphasia.
    (c) quadriphasia.
    (d) parephasia.
    (e) dysphasia.

125. The medical term for inability to speak is
- (a) dysarthria.
- (b) aphasia.
- (c) quadriphasia.
- (d) parephasia.
- (e) dysphasia.

126. The medical term that specifically means impairment of the tongue or impairment of the muscles essential to speech is
- (a) dysarthria.
- (b) aphasia.
- (c) quadriphasia.
- (d) parephasia.
- (e) dysphasia.

The Answer Key for Test Section 13 is on page 464.

Test
Section
Fourteen

Test Section 14's subjects:

- Multiple-lead ECG analysis (4-lead and 12-lead ECGs)

Test Section 14 consists of 45 questions and is allotted 45 minutes for completion.

Author's Note: *Questions 1–21 (and 38–45) can be answered without having knowledge of "12-lead" ECG interpretation, and without having a 12-lead ECG machine. Any standard prehospital cardiac monitor with 3 cables can easily be used to view leads I, II, III, and MCL$_1$ (V$_1$). If you can view those four leads, you can make all of the important diagnostic determinations addressed by Questions 9–21. To learn how to use a standard prehospital cardiac monitor to easily view those four leads and make those diagnostic determinations, please refer to Brady's* Taigman's Advanced Cardiology (In Plain English).

1.   Bipolar limb leads provide ECG recordings of the direction of myocardial electrical impulse travel between two electrodes of opposite polarity (positive and negative). To achieve the most accurate prehospital bipolar limb lead ECG recordings, limb lead electrodes should be placed on the patient's
      (a) wrists and ankle, exactly as indicated by the lead identifiers. Although this placement may result in excessive limb-movement artifact interfering with the ECG recording clarity, it is the only acceptable placement for prehospital or inhospital electrodes.
      (b) biceps or outer shoulders, and the outer thigh area. This placement will diminish the likelihood of limb-movement artifact interfering with clarity of the ECG recording, yet be almost as accurate as wrist and ankle placement.
      (c) chest; in the upper right, upper left, and lower left areas. Chest placement of limb lead electrodes does not interfere in any way with the ability of bipolar limb lead recordings to be considered accurate.
      (d) right and left neck, and the patient's right foot.
      (e) Answer (a) or (d), depending on local protocol.

2.   When looking at lead I, the electrode button _____ is the positive electrode.
      (a) labeled "RA" for "right arm"
      (b) labeled "LA" for "left arm"
      (c) labeled "RL" for "right leg"
      (d) labeled "LL" for "left leg"
      (e) placed to the left of the sternum, in the 5th intercostal space

3.   When looking at lead I, the electrode button _____ is the negative electrode.
      (a) labeled "RA" for "right arm"
      (b) labeled "LA" for "left arm"
      (c) labeled "RL" for "right leg"
      (d) labeled "LL" for "left leg"
      (e) placed to the left of the sternum, in the 5th intercostal space

4. When looking at lead II, the electrode button _____ is the positive electrode.
   (a) labeled "RA" for "right arm"
   (b) labeled "LA" for "left arm"
   (c) labeled "RL" for "right leg"
   (d) labeled "LL" for "left leg"
   (e) placed to the left of the sternum, in the 5ᵗʰ intercostal space

5. When looking at lead II, the electrode button _____ is the negative electrode.
   (a) labeled "RA" for "right arm"
   (b) labeled "LA" for "left arm"
   (c) labeled "RL" for "right leg"
   (d) labeled "LL" for "left leg"
   (e) placed to the left of the sternum, in the 5ᵗʰ intercostal space

6. When looking at lead III, the electrode button _____ is the positive electrode.
   (a) labeled "RA" for "right arm"
   (b) labeled "LA" for "left arm"
   (c) labeled "RL" for "right leg"
   (d) labeled "LL" for "left leg"
   (e) placed to the left of the sternum, in the 5ᵗʰ intercostal space

7. When looking at lead III, the electrode button _____ is the negative electrode.
   (a) labeled "RA" for "right arm"
   (b) labeled "LA" for "left arm"
   (c) labeled "RL" for "right leg"
   (d) labeled "LL" for "left leg"
   (e) placed to the left of the sternum, in the 5ᵗʰ intercostal space

8. If a "grounding" electrode (a 4ᵗʰ electrode) is used, it is usually placed
   (a) on the right arm.
   (b) on the left arm.
   (c) on the right leg.
   (d) on the left leg.
   (e) to the left of the sternum, in the 5ᵗʰ intercostal space.

9. A normal QRS axis is demonstrated by which of the following ECG criteria?

(1) primarily upright QRS in lead I
(2) primarily upright QRS in lead II
(3) primarily upright QRS in lead III
(4) primarily negative QRS in lead I
(5) primarily negative QRS in lead II
(6) primarily negative QRS in lead III
(7) isoelectric (equally negative and positive) QRS in lead II

    (a) 1, 2, and 3.
    (b) 1, 2 or 7, and 6.
    (c) 1, 5, and 6.
    (d) 4 and 3, with 2, 5, or 7.
    (e) 4, 5, and 6.

10. A right shoulder axis deviation occurs when all electrical ventricular impulses are traveling toward the right shoulder. It is demonstrated by which of the ECG criteria offered in question #9?

    (a) 1, 2, and 3.
    (b) 1, 2 or 7, and 6.
    (c) 1, 5, and 6.
    (d) 4, 3, and 2, 5, or 7.
    (e) 4, 5, and 6.

11. A physiologic left axis deviation occurs when a full left axis deviation is not present, and does not conclusively indicate an abnormality. It is demonstrated by which of the ECG criteria offered in question #9?

    (a) 1, 2, and 3.
    (b) 1, 2 or 7, and 6.
    (c) 1, 5, and 6.
    (d) 4 and 3, with 2, 5, or 7.
    (e) 4, 5, and 6.

12. A right axis deviation is demonstrated by which of the ECG criteria offered in question #9?

    (a) 1, 2, and 3.
    (b) 1, 2 or 7, and 6.
    (c) 1, 5, and 6.
    (d) 4 and 3, with 2, 5, or 7.
    (e) 4, 5, and 6.

13. A pathologic left axis deviation is demonstrated by which of the ECG criteria offered in question #9?

    (a) 1, 2, and 3.
    (b) 1, 2 or 7, and 6.

(c) 1, 5, and 6.

(d) 4 and 3, with 2, 5, or 7.

(e) 4, 5, and 6.

14. A bundle branch block is recognized by observing a QRS complex that measures
    (a) 0.12 sec or greater in any lead (including the V leads), even if one or more of the other leads appear to show a QRS duration less than 0.12 sec.
    (b) 0.14 sec or greater in lead II. (Lead II is the "universal" QRS measurement lead.)
    (c) 0.14 sec or greater in lead III. (Lead III is the "universal" QRS measurement lead.)
    (d) 0.14 sec or greater in lead I. (Lead I is the "universal" QRS measurement lead.)
    (e) 0.12 sec or greater in all leads. (If one or more leads show a QRS duration less than 0.12 sec, the shortest QRS duration is considered the "true" QRS measurement, and the patient does not have a bundle branch block.)

15. A right bundle branch block (RBBB) is recognized by observing
    (a) a "rabbit ears" QRS configuration in lead II.
    (b) an RSR' QRS configuration in V_1 (or MCL_1).
    (c) a downward deflection of the last portion of the QRS complex in V_1 (or MCL_1).
    (d) Answer (a) or (c).
    (e) Answer (b) or (c).

16. A left bundle branch block (LBBB) is recognized by observing
    (a) a "rabbit ears" QRS configuration in lead II.
    (b) an RSR' QRS configuration in V_1 (or MCL_1).
    (c) a downward deflection of the last portion of the QRS complex in V_1 (or MCL_1).
    (d) Answer (a) or (c).
    (e) Answer (b) or (c).

17. When a hemiblock is present, the QRS complex will
    (a) be greater than or equal to 0.12 sec only if the larger (posterior) fascicle is blocked.
    (b) be greater than or equal to 0.12 sec only if the larger (anterior) fascicle is blocked.
    (c) be within normal limits, unless a RBBB is present.
    (d) always be within normal limits. If it is not, a hemiblock cannot be detected.
    (e) always be greater than or equal to 0.12 sec (no matter which fascicle is blocked).

18. Which of the following criteria are (or may be) present when the patient has an anterior hemiblock?

   (1) a normal axis
   (2) a pathologic left axis deviation
   (3) a physiologic left axis deviation
   (4) a right axis deviation
   (5) a right shoulder axis deviation
   (6) a RBBB
   (7) a LBBB
   (8) a QRS of 0.12 sec or greater
   (9) a QRS of less than 0.12 sec.

   (a) 1 and 8.
   (b) 2 and 6 or 9.
   (c) 4 and 6 or 9.
   (d) 3 or 4, and 9.
   (e) 8, with 1, 2, 3, 4, or 5.

19. Which of the criteria offered in Question 18 are (or may be) present when the patient has a posterior hemiblock?

   (a) 1 and 8.
   (b) 2 and 6 or 9.
   (c) 4 and 6 or 9.
   (d) 3 or 4, and 9.
   (e) 8, with 1, 2, 3, 4, or 5.

20. Which of the following statements regarding a patient with a LBBB is true?

   (a) If the patient has a LBBB, localization of an acute area of ischemia or infarction is impossible, even with a 12-lead ECG.
   (b) Because of its size, a LBBB is rather common. Thus, by itself, a LBBB should not cause immediate concern that significant or widespread myocardial disease exists.
   (c) A LBBB almost always indicates significant or widespread myocardial disease. Thus, any patient with a LBBB should be considered to have an acute area of ischemia or infarction until proven otherwise.
   (d) Answers (a) and (b) are true.
   (e) Answers (a) and (c) are true.

21. Which of the following statements regarding a patient with a RBBB is true?

   (a) If the patient has a RBBB, localization of an acute area of ischemia or infarction is impossible, even with a 12-lead ECG.
   (b) Because of its size, a RBBB is rather common. Thus, by itself, a RBBB should not cause immediate concern that significant or widespread myocardial disease exists.

(c) A RBBB almost always indicates significant or widespread myocardial disease. Thus, any patient with a RBBB should be considered to have an acute area of ischemia or infarction until proven otherwise.

(d) Answers (a) and (b) are true.

(e) Answers (a) and (c) are true.

22. Palpating the angle of Louis assists chest lead electrode placement because its location corresponds with the

(a) 1st intercostal space.

(b) 2nd intercostal space.

(c) 3rd intercostal space.

(d) 4th intercostal space.

(e) 5th intercostal space.

23. The electrode for chest lead V_1 should be placed

(1) midway between the previous and next lead locations.

(2) at the same level as the electrode for V_4.

(3) next to the sternum, on the right.

(4) in the center of the sternum.

(5) next to the sternum, on the left.

(6) at the midaxillary line.

(7) at the midclavicular line.

(8) at the anterior axillary line.

(9) in the 3rd intercostal space.

(10) in the 4th intercostal space.

(11) in the 5th intercostal space.

(12) in the 6th intercostal space.

(13) in the 7th intercostal space.

   (a) 3 and 9.

   (b) 3 and 10.

   (c) 3 and 11.

   (d) 1 and 4.

   (e) 2 and 4.

24. Using the location indicators offered in question 23, indicate where the electrode for chest lead V_2 should be placed.

(a) 1 and 4.

(b) 2 and 4.

(c) 5 and 10.

(d) 5 and 11.

(e) 5 and 12.

25. Using the location indicators offered in question 23, indicate where the electrode for chest lead V_3 should be placed.

(a) 1.

(b) 1 and 2.

(c) 5 and 10.

(d) 5 and 11.

(e) 5 and 12.

26.   Using the location indicators offered in question 23, indicate where the electrode for chest lead V_4 should be placed.

    (a) 7, 10.
    (b) 7, 11.
    (c) 7, 12.
    (d) 6, 10.
    (e) 6, 11.

27.   Using the location indicators offered in question 23, indicate where the electrode for chest lead V_5 should be placed.

    (a) 1.
    (b) 7 and 11.
    (c) 2 and 6.
    (d) 2 and 7.
    (e) 2 and 8.

28.   Using the location indicators offered in question 23, indicate where the electrode for chest lead V_6 should be placed.

    (a) 2 and 6.
    (b) 6 and 11.
    (c) 8 and 11.
    (d) 8 and 12.
    (e) 8 and 13.

29.   Which of the following statements regarding ECG signs that identify acute myocardial ischemia or infarct is false?

    (a) The degree of a paramedic's index of suspicion that a patient is suffering from myocardial ischemia or infarct should primarily be based on the patient's complaints and medical Hx, not on the physical exam or ECG findings.
    (b) If the paramedic is able to run a 12-lead ECG, and all leads demonstrate absolutely no "abnormal" findings, the patient is not experiencing acute myocardial ischemia or infarct. The paramedic must look elsewhere for the cause of the patient's complaints.
    (c) Without a recent previous ECG for comparison, the paramedic cannot know whether any abnormal ECG findings are old or new.
    (d) A patient suffering from an acute area of myocardial ischemia or infarct may not experience any kind of chest pain or chest discomfort.
    (e) ECG changes caused by an acute area of myocardial infarct may take hours or days to develop.

30.   T wave inversion is one of the earliest signs of myocardial ischemia. However, T waves may normally be inverted in leads

    (a) I, II, or III.
    (b) V_4, V_5, or V_6.

(c) aVR, aVL, or V_1.

(d) Answer (a), (b), or (c). (Each patient is different.)

(e) None of the above. A normal 12-lead has upright T waves in all leads.

31. Another common and early sign of myocardial ischemia is

(a) S-T segment elevation in leads opposite to the area of ischemia.

(b) S-T segment depression in leads over the area of ischemia.

(c) S-T segment elevation in leads over the area of ischemia.

(d) Answer (a) or (b).

(e) Answer (a), (b), or (c). (Each patient is different.)

32. An early sign that an area of ischemia is developing into an area of injury (but not yet infarction) is

(a) T waves that have returned to normal direction in leads over the area of injury.

(b) S-T segment depression developing in leads over the area of injury.

(c) S-T segment elevation developing in leads over the area of injury.

(d) Answer (a) or (b).

(e) Answer (a) or (c).

33. Which of the following statements regarding Q waves is true?

(a) There are many noninfarction conditions that may produce Q waves.

(b) A myocardial infarction may occur in an area of the heart without producing Q waves.

(c) Instead of producing Q waves, a posterior wall infarction may produce unusually large R waves ("reciprocal" changes) in leads V_1 and V_2.

(d) Answers (b) and (c) are true.

(e) Answers (a), (b), and (c) are true.

34. The heart doesn't have clearly defined boundaries between anterior, posterior, lateral, and inferior areas. Additionally, an infarction may overlap two areas. Generally speaking, however, an anterior infarction is commonly recognized by observing abnormal ECG changes in leads

(a) V_1 and V_2.

(b) II, III, and aVF.

(c) I, V_2, V_3, and V_4.

(d) V_5 and V_6.

(e) I, II, V_5, and V_6.

35. Generally speaking, an inferior infarction is most often recognized by observing abnormal ECG changes in leads

(a) V_1 and V_2.

(b) II, III, and aVF.

(c) I, V_2, V_3, and V_4.

(d) V_5 and V_6.

(e) I, II, V_5, and V_6.

36. Generally speaking, a lateral infarction is most often recognized by observing abnormal ECG changes in leads

   (a) V_1 and V_2.
   (b) II, III, and aVF.
   (c) I, V_2, V_3, and V_4.
   (d) V_5 and V_6.
   (e) I, II, V_5, and V_6.

37. Which of the following statements regarding a right ventricular infarction (RVI) is true?

   (a) Any patient with inferior infarct ECG changes and bradycardia should be considered to have an RVI until proven otherwise.
   (b) The classic triad of clinical signs that accompany RVI are JVD, bilateral pulmonary edema, and hypotension.
   (c) Any patient with inferior infarct ECG changes, hypotension, and bilaterally clear lung fields should be considered to have an RVI until proven otherwise.
   (d) Answers (a) and (b) are true.
   (e) Answers (a) and (c) are true.

Author's Note: *Questions 38–45 can be answered without having knowledge of "12-lead" ECG interpretation, and without having a 12-lead ECG machine. If you are familiar with using leads I, II, III, and MCL$_1$ (V$_1$), you can make all of the important determinations addressed by Questions 38–45.*

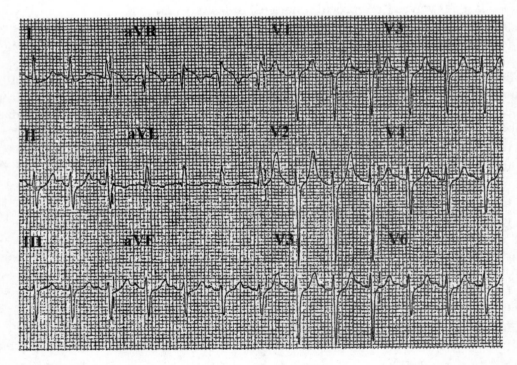

Figure 14–1

38. Figure 14-1 is the ECG of an unconscious 35 y/o male named David, who has a carotid pulse of 142 beats per minute. Which of the following statements regarding his ECG is true?

    (a) David is in sinus tachycardia.

    (b) David has a physiologic left axis deviation.

    (c) David has a pathologic left axis deviation.

    (d) Answers (a) and (b) are true.

    (e) Answers (a) and (c) are true.

39. According to the ECG in Figure 14-1, David has a

    (a) left posterior hemiblock.

    (b) left anterior hemiblock.

    (c) left bundle branch block.

    (d) right anterior hemiblock.

    (e) right bundle branch block.

40. According to the ECG in Figure 14-1, David has a

    (a) prolonged PRI and his QRS is WNL.

    (b) normal PRI and his QRS is WNL.

    (c) normal PRI and his QRS is wider than normal.

    (d) prolonged PRI and his QRS is wider than normal, indicating (at least) a bifascicular block.

    (e) Either answer (a) or (c), depending on the lead that is observed.

41. Figure 14-2 is the ECG of an unconscious 35 y/o female named Gloria, who has a carotid pulse of 62 beats per minute. Which of the following statements regarding her ECG is true?

    (a) Gloria has a sinus rhythm at a rate WNL for an unconscious individual.

    (b) Gloria has a right shoulder axis deviation.

    (c) Gloria has a right axis deviation.

    (d) Answers (a) and (b) are true.

    (e) Answers (a) and (c) are true.

42. According to the ECG in Figure 14-2, Gloria has a

    (a) left posterior hemiblock.

    (b) left bundle branch block.

    (c) right posterior hemiblock.

    (d) right anterior hemiblock.

    (e) right bundle branch block.

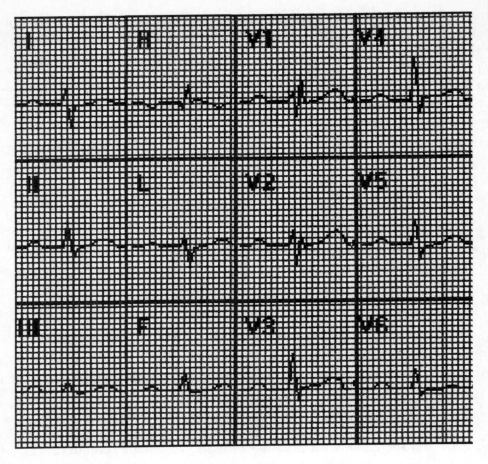

Figure 14–2

43. According to the ECG in Figure 14-2, Gloria has a
  (a) prolonged PRI and her QRS is WNL.
  (b) normal PRI and her QRS is WNL.
  (c) normal PRI and her QRS is wider than normal.
  (d) prolonged PRI and her QRS is wider than normal, indicating (at least) a bifascicular block.
  (e) Either answer (a) or (c), depending on the lead that is observed.

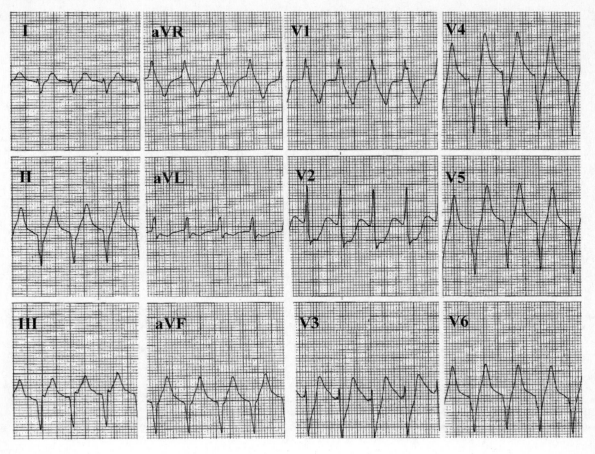

Figure 14–3

44.   Figure 14-3 is the ECG of a 53 y/o male who just walked up to your ambulance. He complains of feeling a "fluttering" sensation in his chest. He seems tired and confused—he doesn't know his own name when you meet him. But, when you get him seated on your wheeled stretcher, he exclaims, "John! My name is John!" John has a weak radial pulse of 120 beats per minute, and a blood pressure of 100/60. Which of the following statements regarding his ECG is true?

    (a) John is in V-tach and synchronized cardioversion should be performed ASAP.

    (b) John is in a supraventricular tachycardia causing rate-related aberrant ventricular conduction.

    (c) An IV should be started immediately and adenosine administered rapid IVP.

    (d) An IV should be started immediately and verapamil administered slow IVP.

    (e) Answers (b) and (c) or (d), depending upon local protocol.

45. Which of the following clues lead you to the correct assessment of (and treatment for) John's condition?

(1) John has a right shoulder axis deviation.

(2) John has a left axis deviation.

(3) John has a normal QRS axis.

(4) The duration of John's QRS in V_2 is clearly 0.12 sec or greater, which means that all of his QRS complexes are 0.12 sec or greater in duration.

(5) In the majority of leads (all but 1), John's QRS complexes appear less than 0.12 sec in duration, so John's QRS complexes are WNL.

(6) John has a pulse, and walked up to your ambulance.

(7) John's QRS shows "left rabbit ear taller than the right" in V_1, a morphology that strongly indicates a ventricular tachycardia.

(8) In V_1, John's QRS shows a version of the classic RSR' morphology, indicating right bundle branch block aberration due to the rapid rate of his supraventricular tachycardia.

(a) 1, 4, and 7.

(b) 3, 5, 6, and 8.

(c) 6.

(d) 2, 5, 6, and 8.

(e) 3, 5, and 6.

The Answer Key for Test Section 14 is on page 469.

Appendix
Contents

DIRECTIONS FOR COPYING THE MASTER BLANK ANSWER SHEETS

To use this Self Test text, photocopy the Blank Answer Sheet form we've created and included in this text. The first page of this 2-page form provides answer sets for 120 questions. The second page provides another 120 answer sets (a total of 240 answer sets, if both pages are copied). But, only 5 of the 14 Test Sections contain more than 120 questions.

To make a *full set* of Blank Answer Sheets (those needed to use all 14 Test Sections):

- Make 14 photocopies of Blank Answer Sheet page 1.
- Make 5 photocopies of Blank Answer Sheet page 2
  (for Test Sections 2, 4, 6, 7, and 13).
- Make 1 photocopy of the ECG Identification Blank Answer Sheet pages, for Test Section 7's Name-That-Strip exercise.

Paramedic National Standards Self Test, 5ᵗʰ Edition

BLANK ANSWER SHEET Page 1 for Test Section _____.

1. ⓐ ⓑ ⓒ ⓓ ⓔ
2. ⓐ ⓑ ⓒ ⓓ ⓔ
3. ⓐ ⓑ ⓒ ⓓ ⓔ
4. ⓐ ⓑ ⓒ ⓓ ⓔ
5. ⓐ ⓑ ⓒ ⓓ ⓔ
6. ⓐ ⓑ ⓒ ⓓ ⓔ
7. ⓐ ⓑ ⓒ ⓓ ⓔ
8. ⓐ ⓑ ⓒ ⓓ ⓔ
9. ⓐ ⓑ ⓒ ⓓ ⓔ
10. ⓐ ⓑ ⓒ ⓓ ⓔ
11. ⓐ ⓑ ⓒ ⓓ ⓔ
12. ⓐ ⓑ ⓒ ⓓ ⓔ
13. ⓐ ⓑ ⓒ ⓓ ⓔ
14. ⓐ ⓑ ⓒ ⓓ ⓔ
15. ⓐ ⓑ ⓒ ⓓ ⓔ
16. ⓐ ⓑ ⓒ ⓓ ⓔ
17. ⓐ ⓑ ⓒ ⓓ ⓔ
18. ⓐ ⓑ ⓒ ⓓ ⓔ
19. ⓐ ⓑ ⓒ ⓓ ⓔ
20. ⓐ ⓑ ⓒ ⓓ ⓔ
21. ⓐ ⓑ ⓒ ⓓ ⓔ
22. ⓐ ⓑ ⓒ ⓓ ⓔ
23. ⓐ ⓑ ⓒ ⓓ ⓔ
24. ⓐ ⓑ ⓒ ⓓ ⓔ

25. ⓐ ⓑ ⓒ ⓓ ⓔ
26. ⓐ ⓑ ⓒ ⓓ ⓔ
27. ⓐ ⓑ ⓒ ⓓ ⓔ
28. ⓐ ⓑ ⓒ ⓓ ⓔ
29. ⓐ ⓑ ⓒ ⓓ ⓔ
30. ⓐ ⓑ ⓒ ⓓ ⓔ
31. ⓐ ⓑ ⓒ ⓓ ⓔ
32. ⓐ ⓑ ⓒ ⓓ ⓔ
33. ⓐ ⓑ ⓒ ⓓ ⓔ
34. ⓐ ⓑ ⓒ ⓓ ⓔ
35. ⓐ ⓑ ⓒ ⓓ ⓔ
36. ⓐ ⓑ ⓒ ⓓ ⓔ
37. ⓐ ⓑ ⓒ ⓓ ⓔ
38. ⓐ ⓑ ⓒ ⓓ ⓔ
39. ⓐ ⓑ ⓒ ⓓ ⓔ
40. ⓐ ⓑ ⓒ ⓓ ⓔ
41. ⓐ ⓑ ⓒ ⓓ ⓔ
42. ⓐ ⓑ ⓒ ⓓ ⓔ
43. ⓐ ⓑ ⓒ ⓓ ⓔ
44. ⓐ ⓑ ⓒ ⓓ ⓔ
45. ⓐ ⓑ ⓒ ⓓ ⓔ
46. ⓐ ⓑ ⓒ ⓓ ⓔ
47. ⓐ ⓑ ⓒ ⓓ ⓔ
48. ⓐ ⓑ ⓒ ⓓ ⓔ

49. ⓐ ⓑ ⓒ ⓓ ⓔ
50. ⓐ ⓑ ⓒ ⓓ ⓔ
51. ⓐ ⓑ ⓒ ⓓ ⓔ
52. ⓐ ⓑ ⓒ ⓓ ⓔ
53. ⓐ ⓑ ⓒ ⓓ ⓔ
54. ⓐ ⓑ ⓒ ⓓ ⓔ
55. ⓐ ⓑ ⓒ ⓓ ⓔ
56. ⓐ ⓑ ⓒ ⓓ ⓔ
57. ⓐ ⓑ ⓒ ⓓ ⓔ
58. ⓐ ⓑ ⓒ ⓓ ⓔ
59. ⓐ ⓑ ⓒ ⓓ ⓔ
60. ⓐ ⓑ ⓒ ⓓ ⓔ
61. ⓐ ⓑ ⓒ ⓓ ⓔ
62. ⓐ ⓑ ⓒ ⓓ ⓔ
63. ⓐ ⓑ ⓒ ⓓ ⓔ
64. ⓐ ⓑ ⓒ ⓓ ⓔ
65. ⓐ ⓑ ⓒ ⓓ ⓔ
66. ⓐ ⓑ ⓒ ⓓ ⓔ
67. ⓐ ⓑ ⓒ ⓓ ⓔ
68. ⓐ ⓑ ⓒ ⓓ ⓔ
69. ⓐ ⓑ ⓒ ⓓ ⓔ
70. ⓐ ⓑ ⓒ ⓓ ⓔ
71. ⓐ ⓑ ⓒ ⓓ ⓔ
72. ⓐ ⓑ ⓒ ⓓ ⓔ

73. ⓐ ⓑ ⓒ ⓓ ⓔ
74. ⓐ ⓑ ⓒ ⓓ ⓔ
75. ⓐ ⓑ ⓒ ⓓ ⓔ
76. ⓐ ⓑ ⓒ ⓓ ⓔ
77. ⓐ ⓑ ⓒ ⓓ ⓔ
78. ⓐ ⓑ ⓒ ⓓ ⓔ
79. ⓐ ⓑ ⓒ ⓓ ⓔ
80. ⓐ ⓑ ⓒ ⓓ ⓔ
81. ⓐ ⓑ ⓒ ⓓ ⓔ
82. ⓐ ⓑ ⓒ ⓓ ⓔ
83. ⓐ ⓑ ⓒ ⓓ ⓔ
84. ⓐ ⓑ ⓒ ⓓ ⓔ
85. ⓐ ⓑ ⓒ ⓓ ⓔ
86. ⓐ ⓑ ⓒ ⓓ ⓔ
87. ⓐ ⓑ ⓒ ⓓ ⓔ
88. ⓐ ⓑ ⓒ ⓓ ⓔ
89. ⓐ ⓑ ⓒ ⓓ ⓔ
90. ⓐ ⓑ ⓒ ⓓ ⓔ
91. ⓐ ⓑ ⓒ ⓓ ⓔ
92. ⓐ ⓑ ⓒ ⓓ ⓔ
93. ⓐ ⓑ ⓒ ⓓ ⓔ
94. ⓐ ⓑ ⓒ ⓓ ⓔ
95. ⓐ ⓑ ⓒ ⓓ ⓔ
96. ⓐ ⓑ ⓒ ⓓ ⓔ

97. ⓐ ⓑ ⓒ ⓓ ⓔ
98. ⓐ ⓑ ⓒ ⓓ ⓔ
99. ⓐ ⓑ ⓒ ⓓ ⓔ
100. ⓐ ⓑ ⓒ ⓓ ⓔ
101. ⓐ ⓑ ⓒ ⓓ ⓔ
102. ⓐ ⓑ ⓒ ⓓ ⓔ
103. ⓐ ⓑ ⓒ ⓓ ⓔ
104. ⓐ ⓑ ⓒ ⓓ ⓔ
105. ⓐ ⓑ ⓒ ⓓ ⓔ
106. ⓐ ⓑ ⓒ ⓓ ⓔ
107. ⓐ ⓑ ⓒ ⓓ ⓔ
108. ⓐ ⓑ ⓒ ⓓ ⓔ
109. ⓐ ⓑ ⓒ ⓓ ⓔ
110. ⓐ ⓑ ⓒ ⓓ ⓔ
111. ⓐ ⓑ ⓒ ⓓ ⓔ
112. ⓐ ⓑ ⓒ ⓓ ⓔ
113. ⓐ ⓑ ⓒ ⓓ ⓔ
114. ⓐ ⓑ ⓒ ⓓ ⓔ
115. ⓐ ⓑ ⓒ ⓓ ⓔ
116. ⓐ ⓑ ⓒ ⓓ ⓔ
117. ⓐ ⓑ ⓒ ⓓ ⓔ
118. ⓐ ⓑ ⓒ ⓓ ⓔ
119. ⓐ ⓑ ⓒ ⓓ ⓔ
120. ⓐ ⓑ ⓒ ⓓ ⓔ

Paramedic National Standards Self Test, 5th Edition
BLANK ANSWER SHEET Page 2 for Test Section _____.

121. ⓐ ⓑ ⓒ ⓓ ⓔ	145. ⓐ ⓑ ⓒ ⓓ ⓔ	169. ⓐ ⓑ ⓒ ⓓ ⓔ	193. ⓐ ⓑ ⓒ ⓓ ⓔ	217. ⓐ ⓑ ⓒ ⓓ ⓔ
122. ⓐ ⓑ ⓒ ⓓ ⓔ	146. ⓐ ⓑ ⓒ ⓓ ⓔ	170. ⓐ ⓑ ⓒ ⓓ ⓔ	194. ⓐ ⓑ ⓒ ⓓ ⓔ	218. ⓐ ⓑ ⓒ ⓓ ⓔ
123. ⓐ ⓑ ⓒ ⓓ ⓔ	147. ⓐ ⓑ ⓒ ⓓ ⓔ	171. ⓐ ⓑ ⓒ ⓓ ⓔ	195. ⓐ ⓑ ⓒ ⓓ ⓔ	219. ⓐ ⓑ ⓒ ⓓ ⓔ
124. ⓐ ⓑ ⓒ ⓓ ⓔ	148. ⓐ ⓑ ⓒ ⓓ ⓔ	172. ⓐ ⓑ ⓒ ⓓ ⓔ	196. ⓐ ⓑ ⓒ ⓓ ⓔ	220. ⓐ ⓑ ⓒ ⓓ ⓔ
125. ⓐ ⓑ ⓒ ⓓ ⓔ	149. ⓐ ⓑ ⓒ ⓓ ⓔ	173. ⓐ ⓑ ⓒ ⓓ ⓔ	197. ⓐ ⓑ ⓒ ⓓ ⓔ	221. ⓐ ⓑ ⓒ ⓓ ⓔ
126. ⓐ ⓑ ⓒ ⓓ ⓔ	150. ⓐ ⓑ ⓒ ⓓ ⓔ	174. ⓐ ⓑ ⓒ ⓓ ⓔ	198. ⓐ ⓑ ⓒ ⓓ ⓔ	222. ⓐ ⓑ ⓒ ⓓ ⓔ
127. ⓐ ⓑ ⓒ ⓓ ⓔ	151. ⓐ ⓑ ⓒ ⓓ ⓔ	175. ⓐ ⓑ ⓒ ⓓ ⓔ	199. ⓐ ⓑ ⓒ ⓓ ⓔ	223. ⓐ ⓑ ⓒ ⓓ ⓔ
128. ⓐ ⓑ ⓒ ⓓ ⓔ	152. ⓐ ⓑ ⓒ ⓓ ⓔ	176. ⓐ ⓑ ⓒ ⓓ ⓔ	200. ⓐ ⓑ ⓒ ⓓ ⓔ	224. ⓐ ⓑ ⓒ ⓓ ⓔ
129. ⓐ ⓑ ⓒ ⓓ ⓔ	153. ⓐ ⓑ ⓒ ⓓ ⓔ	177. ⓐ ⓑ ⓒ ⓓ ⓔ	201. ⓐ ⓑ ⓒ ⓓ ⓔ	225. ⓐ ⓑ ⓒ ⓓ ⓔ
130. ⓐ ⓑ ⓒ ⓓ ⓔ	154. ⓐ ⓑ ⓒ ⓓ ⓔ	178. ⓐ ⓑ ⓒ ⓓ ⓔ	202. ⓐ ⓑ ⓒ ⓓ ⓔ	226. ⓐ ⓑ ⓒ ⓓ ⓔ
131. ⓐ ⓑ ⓒ ⓓ ⓔ	155. ⓐ ⓑ ⓒ ⓓ ⓔ	179. ⓐ ⓑ ⓒ ⓓ ⓔ	203. ⓐ ⓑ ⓒ ⓓ ⓔ	227. ⓐ ⓑ ⓒ ⓓ ⓔ
132. ⓐ ⓑ ⓒ ⓓ ⓔ	156. ⓐ ⓑ ⓒ ⓓ ⓔ	180. ⓐ ⓑ ⓒ ⓓ ⓔ	204. ⓐ ⓑ ⓒ ⓓ ⓔ	228. ⓐ ⓑ ⓒ ⓓ ⓔ
133. ⓐ ⓑ ⓒ ⓓ ⓔ	157. ⓐ ⓑ ⓒ ⓓ ⓔ	181. ⓐ ⓑ ⓒ ⓓ ⓔ	205. ⓐ ⓑ ⓒ ⓓ ⓔ	229. ⓐ ⓑ ⓒ ⓓ ⓔ
134. ⓐ ⓑ ⓒ ⓓ ⓔ	158. ⓐ ⓑ ⓒ ⓓ ⓔ	182. ⓐ ⓑ ⓒ ⓓ ⓔ	206. ⓐ ⓑ ⓒ ⓓ ⓔ	230. ⓐ ⓑ ⓒ ⓓ ⓔ
135. ⓐ ⓑ ⓒ ⓓ ⓔ	159. ⓐ ⓑ ⓒ ⓓ ⓔ	183. ⓐ ⓑ ⓒ ⓓ ⓔ	207. ⓐ ⓑ ⓒ ⓓ ⓔ	231. ⓐ ⓑ ⓒ ⓓ ⓔ
136. ⓐ ⓑ ⓒ ⓓ ⓔ	160. ⓐ ⓑ ⓒ ⓓ ⓔ	184. ⓐ ⓑ ⓒ ⓓ ⓔ	208. ⓐ ⓑ ⓒ ⓓ ⓔ	232. ⓐ ⓑ ⓒ ⓓ ⓔ
137. ⓐ ⓑ ⓒ ⓓ ⓔ	161. ⓐ ⓑ ⓒ ⓓ ⓔ	185. ⓐ ⓑ ⓒ ⓓ ⓔ	209. ⓐ ⓑ ⓒ ⓓ ⓔ	233. ⓐ ⓑ ⓒ ⓓ ⓔ
138. ⓐ ⓑ ⓒ ⓓ ⓔ	162. ⓐ ⓑ ⓒ ⓓ ⓔ	186. ⓐ ⓑ ⓒ ⓓ ⓔ	210. ⓐ ⓑ ⓒ ⓓ ⓔ	234. ⓐ ⓑ ⓒ ⓓ ⓔ
139. ⓐ ⓑ ⓒ ⓓ ⓔ	163. ⓐ ⓑ ⓒ ⓓ ⓔ	187. ⓐ ⓑ ⓒ ⓓ ⓔ	211. ⓐ ⓑ ⓒ ⓓ ⓔ	235. ⓐ ⓑ ⓒ ⓓ ⓔ
140. ⓐ ⓑ ⓒ ⓓ ⓔ	164. ⓐ ⓑ ⓒ ⓓ ⓔ	188. ⓐ ⓑ ⓒ ⓓ ⓔ	212. ⓐ ⓑ ⓒ ⓓ ⓔ	236. ⓐ ⓑ ⓒ ⓓ ⓔ
141. ⓐ ⓑ ⓒ ⓓ ⓔ	165. ⓐ ⓑ ⓒ ⓓ ⓔ	189. ⓐ ⓑ ⓒ ⓓ ⓔ	213. ⓐ ⓑ ⓒ ⓓ ⓔ	237. ⓐ ⓑ ⓒ ⓓ ⓔ
142. ⓐ ⓑ ⓒ ⓓ ⓔ	166. ⓐ ⓑ ⓒ ⓓ ⓔ	190. ⓐ ⓑ ⓒ ⓓ ⓔ	214. ⓐ ⓑ ⓒ ⓓ ⓔ	238. ⓐ ⓑ ⓒ ⓓ ⓔ
143. ⓐ ⓑ ⓒ ⓓ ⓔ	167. ⓐ ⓑ ⓒ ⓓ ⓔ	191. ⓐ ⓑ ⓒ ⓓ ⓔ	215. ⓐ ⓑ ⓒ ⓓ ⓔ	239. ⓐ ⓑ ⓒ ⓓ ⓔ
144. ⓐ ⓑ ⓒ ⓓ ⓔ	168. ⓐ ⓑ ⓒ ⓓ ⓔ	192. ⓐ ⓑ ⓒ ⓓ ⓔ	216. ⓐ ⓑ ⓒ ⓓ ⓔ	240. ⓐ ⓑ ⓒ ⓓ ⓔ

Paramedic National Standards Self Test, 5th Edition

BLANK ANSWER SHEET page 1 for Test Section _____.

1. ⓐ ⓑ ⓒ ⓓ ⓔ
2. ⓐ ⓑ ⓒ ⓓ ⓔ
3. ⓐ ⓑ ⓒ ⓓ ⓔ
4. ⓐ ⓑ ⓒ ⓓ ⓔ
5. ⓐ ⓑ ⓒ ⓓ ⓔ
6. ⓐ ⓑ ⓒ ⓓ ⓔ
7. ⓐ ⓑ ⓒ ⓓ ⓔ
8. ⓐ ⓑ ⓒ ⓓ ⓔ
9. ⓐ ⓑ ⓒ ⓓ ⓔ
10. ⓐ ⓑ ⓒ ⓓ ⓔ
11. ⓐ ⓑ ⓒ ⓓ ⓔ
12. ⓐ ⓑ ⓒ ⓓ ⓔ
13. ⓐ ⓑ ⓒ ⓓ ⓔ
14. ⓐ ⓑ ⓒ ⓓ ⓔ
15. ⓐ ⓑ ⓒ ⓓ ⓔ
16. ⓐ ⓑ ⓒ ⓓ ⓔ
17. ⓐ ⓑ ⓒ ⓓ ⓔ
18. ⓐ ⓑ ⓒ ⓓ ⓔ
19. ⓐ ⓑ ⓒ ⓓ ⓔ
20. ⓐ ⓑ ⓒ ⓓ ⓔ
21. ⓐ ⓑ ⓒ ⓓ ⓔ
22. ⓐ ⓑ ⓒ ⓓ ⓔ
23. ⓐ ⓑ ⓒ ⓓ ⓔ
24. ⓐ ⓑ ⓒ ⓓ ⓔ

25. ⓐ ⓑ ⓒ ⓓ ⓔ
26. ⓐ ⓑ ⓒ ⓓ ⓔ
27. ⓐ ⓑ ⓒ ⓓ ⓔ
28. ⓐ ⓑ ⓒ ⓓ ⓔ
29. ⓐ ⓑ ⓒ ⓓ ⓔ
30. ⓐ ⓑ ⓒ ⓓ ⓔ
31. ⓐ ⓑ ⓒ ⓓ ⓔ
32. ⓐ ⓑ ⓒ ⓓ ⓔ
33. ⓐ ⓑ ⓒ ⓓ ⓔ
34. ⓐ ⓑ ⓒ ⓓ ⓔ
35. ⓐ ⓑ ⓒ ⓓ ⓔ
36. ⓐ ⓑ ⓒ ⓓ ⓔ
37. ⓐ ⓑ ⓒ ⓓ ⓔ
38. ⓐ ⓑ ⓒ ⓓ ⓔ
39. ⓐ ⓑ ⓒ ⓓ ⓔ
40. ⓐ ⓑ ⓒ ⓓ ⓔ
41. ⓐ ⓑ ⓒ ⓓ ⓔ
42. ⓐ ⓑ ⓒ ⓓ ⓔ
43. ⓐ ⓑ ⓒ ⓓ ⓔ
44. ⓐ ⓑ ⓒ ⓓ ⓔ
45. ⓐ ⓑ ⓒ ⓓ ⓔ
46. ⓐ ⓑ ⓒ ⓓ ⓔ
47. ⓐ ⓑ ⓒ ⓓ ⓔ
48. ⓐ ⓑ ⓒ ⓓ ⓔ

49. ⓐ ⓑ ⓒ ⓓ ⓔ
50. ⓐ ⓑ ⓒ ⓓ ⓔ
51. ⓐ ⓑ ⓒ ⓓ ⓔ
52. ⓐ ⓑ ⓒ ⓓ ⓔ
53. ⓐ ⓑ ⓒ ⓓ ⓔ
54. ⓐ ⓑ ⓒ ⓓ ⓔ
55. ⓐ ⓑ ⓒ ⓓ ⓔ
56. ⓐ ⓑ ⓒ ⓓ ⓔ
57. ⓐ ⓑ ⓒ ⓓ ⓔ
58. ⓐ ⓑ ⓒ ⓓ ⓔ
59. ⓐ ⓑ ⓒ ⓓ ⓔ
60. ⓐ ⓑ ⓒ ⓓ ⓔ
61. ⓐ ⓑ ⓒ ⓓ ⓔ
62. ⓐ ⓑ ⓒ ⓓ ⓔ
63. ⓐ ⓑ ⓒ ⓓ ⓔ
64. ⓐ ⓑ ⓒ ⓓ ⓔ
65. ⓐ ⓑ ⓒ ⓓ ⓔ
66. ⓐ ⓑ ⓒ ⓓ ⓔ
67. ⓐ ⓑ ⓒ ⓓ ⓔ
68. ⓐ ⓑ ⓒ ⓓ ⓔ
69. ⓐ ⓑ ⓒ ⓓ ⓔ
70. ⓐ ⓑ ⓒ ⓓ ⓔ
71. ⓐ ⓑ ⓒ ⓓ ⓔ
72. ⓐ ⓑ ⓒ ⓓ ⓔ

73. ⓐ ⓑ ⓒ ⓓ ⓔ
74. ⓐ ⓑ ⓒ ⓓ ⓔ
75. ⓐ ⓑ ⓒ ⓓ ⓔ
76. ⓐ ⓑ ⓒ ⓓ ⓔ
77. ⓐ ⓑ ⓒ ⓓ ⓔ
78. ⓐ ⓑ ⓒ ⓓ ⓔ
79. ⓐ ⓑ ⓒ ⓓ ⓔ
80. ⓐ ⓑ ⓒ ⓓ ⓔ
81. ⓐ ⓑ ⓒ ⓓ ⓔ
82. ⓐ ⓑ ⓒ ⓓ ⓔ
83. ⓐ ⓑ ⓒ ⓓ ⓔ
84. ⓐ ⓑ ⓒ ⓓ ⓔ
85. ⓐ ⓑ ⓒ ⓓ ⓔ
86. ⓐ ⓑ ⓒ ⓓ ⓔ
87. ⓐ ⓑ ⓒ ⓓ ⓔ
88. ⓐ ⓑ ⓒ ⓓ ⓔ
89. ⓐ ⓑ ⓒ ⓓ ⓔ
90. ⓐ ⓑ ⓒ ⓓ ⓔ
91. ⓐ ⓑ ⓒ ⓓ ⓔ
92. ⓐ ⓑ ⓒ ⓓ ⓔ
93. ⓐ ⓑ ⓒ ⓓ ⓔ
94. ⓐ ⓑ ⓒ ⓓ ⓔ
95. ⓐ ⓑ ⓒ ⓓ ⓔ
96. ⓐ ⓑ ⓒ ⓓ ⓔ

97. ⓐ ⓑ ⓒ ⓓ ⓔ
98. ⓐ ⓑ ⓒ ⓓ ⓔ
99. ⓐ ⓑ ⓒ ⓓ ⓔ
100. ⓐ ⓑ ⓒ ⓓ ⓔ
101. ⓐ ⓑ ⓒ ⓓ ⓔ
102. ⓐ ⓑ ⓒ ⓓ ⓔ
103. ⓐ ⓑ ⓒ ⓓ ⓔ
104. ⓐ ⓑ ⓒ ⓓ ⓔ
105. ⓐ ⓑ ⓒ ⓓ ⓔ
106. ⓐ ⓑ ⓒ ⓓ ⓔ
107. ⓐ ⓑ ⓒ ⓓ ⓔ
108. ⓐ ⓑ ⓒ ⓓ ⓔ
109. ⓐ ⓑ ⓒ ⓓ ⓔ
110. ⓐ ⓑ ⓒ ⓓ ⓔ
111. ⓐ ⓑ ⓒ ⓓ ⓔ
112. ⓐ ⓑ ⓒ ⓓ ⓔ
113. ⓐ ⓑ ⓒ ⓓ ⓔ
114. ⓐ ⓑ ⓒ ⓓ ⓔ
115. ⓐ ⓑ ⓒ ⓓ ⓔ
116. ⓐ ⓑ ⓒ ⓓ ⓔ
117. ⓐ ⓑ ⓒ ⓓ ⓔ
118. ⓐ ⓑ ⓒ ⓓ ⓔ
119. ⓐ ⓑ ⓒ ⓓ ⓔ
120. ⓐ ⓑ ⓒ ⓓ ⓔ

Paramedic National Standards Self Test, 5th Edition

BLANK ANSWER SHEET Page 2 for Test Section _____.

121. Ⓐ Ⓑ Ⓒ Ⓓ Ⓔ	145. Ⓐ Ⓑ Ⓒ Ⓓ Ⓔ	169. Ⓐ Ⓑ Ⓒ Ⓓ Ⓔ	193. Ⓐ Ⓑ Ⓒ Ⓓ Ⓔ	217. Ⓐ Ⓑ Ⓒ Ⓓ Ⓔ
122. Ⓐ Ⓑ Ⓒ Ⓓ Ⓔ	146. Ⓐ Ⓑ Ⓒ Ⓓ Ⓔ	170. Ⓐ Ⓑ Ⓒ Ⓓ Ⓔ	194. Ⓐ Ⓑ Ⓒ Ⓓ Ⓔ	218. Ⓐ Ⓑ Ⓒ Ⓓ Ⓔ
123. Ⓐ Ⓑ Ⓒ Ⓓ Ⓔ	147. Ⓐ Ⓑ Ⓒ Ⓓ Ⓔ	171. Ⓐ Ⓑ Ⓒ Ⓓ Ⓔ	195. Ⓐ Ⓑ Ⓒ Ⓓ Ⓔ	219. Ⓐ Ⓑ Ⓒ Ⓓ Ⓔ
124. Ⓐ Ⓑ Ⓒ Ⓓ Ⓔ	148. Ⓐ Ⓑ Ⓒ Ⓓ Ⓔ	172. Ⓐ Ⓑ Ⓒ Ⓓ Ⓔ	196. Ⓐ Ⓑ Ⓒ Ⓓ Ⓔ	220. Ⓐ Ⓑ Ⓒ Ⓓ Ⓔ
125. Ⓐ Ⓑ Ⓒ Ⓓ Ⓔ	149. Ⓐ Ⓑ Ⓒ Ⓓ Ⓔ	173. Ⓐ Ⓑ Ⓒ Ⓓ Ⓔ	197. Ⓐ Ⓑ Ⓒ Ⓓ Ⓔ	221. Ⓐ Ⓑ Ⓒ Ⓓ Ⓔ
126. Ⓐ Ⓑ Ⓒ Ⓓ Ⓔ	150. Ⓐ Ⓑ Ⓒ Ⓓ Ⓔ	174. Ⓐ Ⓑ Ⓒ Ⓓ Ⓔ	198. Ⓐ Ⓑ Ⓒ Ⓓ Ⓔ	222. Ⓐ Ⓑ Ⓒ Ⓓ Ⓔ
127. Ⓐ Ⓑ Ⓒ Ⓓ Ⓔ	151. Ⓐ Ⓑ Ⓒ Ⓓ Ⓔ	175. Ⓐ Ⓑ Ⓒ Ⓓ Ⓔ	199. Ⓐ Ⓑ Ⓒ Ⓓ Ⓔ	223. Ⓐ Ⓑ Ⓒ Ⓓ Ⓔ
128. Ⓐ Ⓑ Ⓒ Ⓓ Ⓔ	152. Ⓐ Ⓑ Ⓒ Ⓓ Ⓔ	176. Ⓐ Ⓑ Ⓒ Ⓓ Ⓔ	200. Ⓐ Ⓑ Ⓒ Ⓓ Ⓔ	224. Ⓐ Ⓑ Ⓒ Ⓓ Ⓔ
129. Ⓐ Ⓑ Ⓒ Ⓓ Ⓔ	153. Ⓐ Ⓑ Ⓒ Ⓓ Ⓔ	177. Ⓐ Ⓑ Ⓒ Ⓓ Ⓔ	201. Ⓐ Ⓑ Ⓒ Ⓓ Ⓔ	225. Ⓐ Ⓑ Ⓒ Ⓓ Ⓔ
130. Ⓐ Ⓑ Ⓒ Ⓓ Ⓔ	154. Ⓐ Ⓑ Ⓒ Ⓓ Ⓔ	178. Ⓐ Ⓑ Ⓒ Ⓓ Ⓔ	202. Ⓐ Ⓑ Ⓒ Ⓓ Ⓔ	226. Ⓐ Ⓑ Ⓒ Ⓓ Ⓔ
131. Ⓐ Ⓑ Ⓒ Ⓓ Ⓔ	155. Ⓐ Ⓑ Ⓒ Ⓓ Ⓔ	179. Ⓐ Ⓑ Ⓒ Ⓓ Ⓔ	203. Ⓐ Ⓑ Ⓒ Ⓓ Ⓔ	227. Ⓐ Ⓑ Ⓒ Ⓓ Ⓔ
132. Ⓐ Ⓑ Ⓒ Ⓓ Ⓔ	156. Ⓐ Ⓑ Ⓒ Ⓓ Ⓔ	180. Ⓐ Ⓑ Ⓒ Ⓓ Ⓔ	204. Ⓐ Ⓑ Ⓒ Ⓓ Ⓔ	228. Ⓐ Ⓑ Ⓒ Ⓓ Ⓔ
133. Ⓐ Ⓑ Ⓒ Ⓓ Ⓔ	157. Ⓐ Ⓑ Ⓒ Ⓓ Ⓔ	181. Ⓐ Ⓑ Ⓒ Ⓓ Ⓔ	205. Ⓐ Ⓑ Ⓒ Ⓓ Ⓔ	229. Ⓐ Ⓑ Ⓒ Ⓓ Ⓔ
134. Ⓐ Ⓑ Ⓒ Ⓓ Ⓔ	158. Ⓐ Ⓑ Ⓒ Ⓓ Ⓔ	182. Ⓐ Ⓑ Ⓒ Ⓓ Ⓔ	206. Ⓐ Ⓑ Ⓒ Ⓓ Ⓔ	230. Ⓐ Ⓑ Ⓒ Ⓓ Ⓔ
135. Ⓐ Ⓑ Ⓒ Ⓓ Ⓔ	159. Ⓐ Ⓑ Ⓒ Ⓓ Ⓔ	183. Ⓐ Ⓑ Ⓒ Ⓓ Ⓔ	207. Ⓐ Ⓑ Ⓒ Ⓓ Ⓔ	231. Ⓐ Ⓑ Ⓒ Ⓓ Ⓔ
136. Ⓐ Ⓑ Ⓒ Ⓓ Ⓔ	160. Ⓐ Ⓑ Ⓒ Ⓓ Ⓔ	184. Ⓐ Ⓑ Ⓒ Ⓓ Ⓔ	208. Ⓐ Ⓑ Ⓒ Ⓓ Ⓔ	232. Ⓐ Ⓑ Ⓒ Ⓓ Ⓔ
137. Ⓐ Ⓑ Ⓒ Ⓓ Ⓔ	161. Ⓐ Ⓑ Ⓒ Ⓓ Ⓔ	185. Ⓐ Ⓑ Ⓒ Ⓓ Ⓔ	209. Ⓐ Ⓑ Ⓒ Ⓓ Ⓔ	233. Ⓐ Ⓑ Ⓒ Ⓓ Ⓔ
138. Ⓐ Ⓑ Ⓒ Ⓓ Ⓔ	162. Ⓐ Ⓑ Ⓒ Ⓓ Ⓔ	186. Ⓐ Ⓑ Ⓒ Ⓓ Ⓔ	210. Ⓐ Ⓑ Ⓒ Ⓓ Ⓔ	234. Ⓐ Ⓑ Ⓒ Ⓓ Ⓔ
139. Ⓐ Ⓑ Ⓒ Ⓓ Ⓔ	163. Ⓐ Ⓑ Ⓒ Ⓓ Ⓔ	187. Ⓐ Ⓑ Ⓒ Ⓓ Ⓔ	211. Ⓐ Ⓑ Ⓒ Ⓓ Ⓔ	235. Ⓐ Ⓑ Ⓒ Ⓓ Ⓔ
140. Ⓐ Ⓑ Ⓒ Ⓓ Ⓔ	164. Ⓐ Ⓑ Ⓒ Ⓓ Ⓔ	188. Ⓐ Ⓑ Ⓒ Ⓓ Ⓔ	212. Ⓐ Ⓑ Ⓒ Ⓓ Ⓔ	236. Ⓐ Ⓑ Ⓒ Ⓓ Ⓔ
141. Ⓐ Ⓑ Ⓒ Ⓓ Ⓔ	165. Ⓐ Ⓑ Ⓒ Ⓓ Ⓔ	189. Ⓐ Ⓑ Ⓒ Ⓓ Ⓔ	213. Ⓐ Ⓑ Ⓒ Ⓓ Ⓔ	237. Ⓐ Ⓑ Ⓒ Ⓓ Ⓔ
142. Ⓐ Ⓑ Ⓒ Ⓓ Ⓔ	166. Ⓐ Ⓑ Ⓒ Ⓓ Ⓔ	190. Ⓐ Ⓑ Ⓒ Ⓓ Ⓔ	214. Ⓐ Ⓑ Ⓒ Ⓓ Ⓔ	238. Ⓐ Ⓑ Ⓒ Ⓓ Ⓔ
143. Ⓐ Ⓑ Ⓒ Ⓓ Ⓔ	167. Ⓐ Ⓑ Ⓒ Ⓓ Ⓔ	191. Ⓐ Ⓑ Ⓒ Ⓓ Ⓔ	215. Ⓐ Ⓑ Ⓒ Ⓓ Ⓔ	239. Ⓐ Ⓑ Ⓒ Ⓓ Ⓔ
144. Ⓐ Ⓑ Ⓒ Ⓓ Ⓔ	168. Ⓐ Ⓑ Ⓒ Ⓓ Ⓔ	192. Ⓐ Ⓑ Ⓒ Ⓓ Ⓔ	216. Ⓐ Ⓑ Ⓒ Ⓓ Ⓔ	240. Ⓐ Ⓑ Ⓒ Ⓓ Ⓔ

Test Section 7's Name-That-Strip ECG Figure Write-In Descriptions

Write in your description of each Figure's ECG. Then compare your description to the answer (the author's description) provided in Test Section 7's Answer Key.

Figure 7–15: _____

_____

_____

Figure 7–16: _____

_____

_____

Figure 7–17: _____

_____

_____

Figure 7–18: _____

_____

_____

Figure 7–19: _____

_____

_____

Figure 7–20: _____

_____

_____

Figure 7–21: _____

_____

_____

Figure 7–22: _____

_____

_____

Figure 7–23: _____

_____

_____

Figure 7–24: _____

_____

_____

Figure 7–25: _____

_____

_____

Figure 7–26: _____

_____

_____

Figure 7–27: _____

_____

_____

Figure 7–28: _____

_____

_____

Figure 7–29: _____

_____

_____

Figure 7–30: _____

_____

_____

Figure 7–31: _____

_____

_____

Figure 7–32: _____

_____

_____

Figure 7–33: _____

_____

_____

Figure 7–34: _____

_____

_____

Figure 7–35: _____

_____

_____

Figure 7–36: _____

_____

_____

Figure 7–37: _____

_____

_____

Figure 7–38: _____

_____

_____

Figure 7–39: _____

_____

_____

Figure 7–40: _____

_____

_____

Figure 7–41: _____

_____

_____

Figure 7–42: _____

_____

_____

Figure 7–43: _____

_____

_____

Figure 7–44: _____

_____

_____

Figure 7–45: _____

_____

_____

Figure 7–46: _____

_____

_____

Figure 7–47: _____

_____

_____

Figure 7–48: _____

_____

_____

Figure 7–49: _____

_____

_____

Figure 7–50: _____

_____

_____

Figure 7–51: _____

_____

_____

Figure 7–52: _____

_____

_____

Figure 7–53: _____

_____

_____

Figure 7–54: _____

_____

_____

Figure 7–55: _____

_____

_____

Figure 7–56: _____

_____

_____

Figure 7–57: _____

_____

_____

Figure 7–58: _____

_____

_____

Figure 7–59: _____

_____

_____

Figure 7–60: _____

_____

_____

Figure 7–61: _____

_____

Figure 7–62: _____

_____

_____

Figure 7–63: _____

_____

_____

Figure 7–64: _____

_____

_____

In the following answer keys:

- The first column contains the question number.
- The second column indicates each question's correct answer letter.
- The Subject column identifies each question's basic subject.

TEST SECTION 1 ANSWER KEY

Page Key texts:

Brady's *Paramedic Care: Principles & Practice*, AHA updated 2nd edition, 5-volume series of texts:

V1 indicates Volume 1: Introduction to Advanced Prehospital Care
V2 indicates Volume 2: Patient Assessment

EPC indicates Brady's *Essentials of Paramedic Care,* 2nd edition.

		PAGE KEY	SUBJECT
1.	(e)	V1: 86–87, 142–161; EPC: 53–63	ethics
2.	(a)	V1: 86–92; EPC: 27–30	professionalism
3.	(c)	V1: 78–86; EPC: 23–27	roles of the paramedic
4.	(b)	V1: 59, 117; EPC: 16, 68, 69	certification
5.	(c)	V1: 59, 117; EPC: 16, 68	licensure
6.	(a)	V1: 59; EPC: 16	reciprocity
7.	(b)	V1: 55; EPC: 14	protocols
8.	(e)	V1: 55; EPC: 14	protocols
9.	(e)	V1: 44–75; EPC: 8	the EMS system
10.	(b)	V1: 55; EPC: 14	off-line (indirect) medical control
11.	(a)	V1: 54–55; EPC: 13	on-line (direct) medical control
12.	(b)	V1: 55; EPC: 14	off-line (indirect) medical control
13.	(c)	V1: 55; EPC: 13	on-scene control of emergency care
14.	(c)	V1: 115–116; EPC: 67	civil (tort) law
15.	(a)	V1: 115; EPC: 66	criminal law
16.	(c)	V1: 118–121; EPC: 70–72	negligence, medical malpractice
17.	(a)	V1: 116–117; EPC: 8, 68	scope of practice – refusing orders
18.	(c)	V1: 122; EPC: 72–73	Borrowed Servant Doctrine
19.	(b)	V1: 122; EPC: 73	civil rights
20.	(a)	V1: 118, 121; EPC: 69, 72	Good Samaritan laws
21.	(d)	V1: 117; EPC: 69	mandatory reporting requirements
22.	(e)	V1: 132–135; EPC: 80–81, 1756, 1757	Advanced Directives, Living Wills, DNR orders

		PAGE KEY	SUBJECT
23.	(e)	V1: 132–135; EPC: 80–81, 1756, 1757	Advanced Directives, Living Wills, DNR orders
24.	(b)	V1: 118–120; EPC: 70–72	negligence
25.	(a)	V1: 118–120; EPC: 70	malpractice
26.	(b)	V1: 118–120; EPC: 70	misfeasance
27.	(c)	V1: 125, 155; EPC: 75	implied consent
28.	(a)	V1: 125; EPC: 75	informed and expressed consent
29.	(c)	V1: 130; EPC: 78	abandonment
30.	(a)	V1: 127; EPC: 77, 753–754	refusal of treatment
31.	(b)	V1: 130; EPC: 78	battery
32.	(a)	V1: 130; EPC: 78, 1877	assault
33.	(c)	V1: 130; EPC: 78	false imprisonment
34.	(e)	V1: 123–124, 153–155; EPC: 58–59, 73–74, 756	patient confidentiality
35.	(b)	V1: 121; EPC: 72	"best protection" in malpractice litigation
36.	(e)	V1: 33; EPC: 282	stress definitions
37.	(c)	V1: 33, 261; EPC: 41, 283	stressor
38.	(e)	V1: 33, 261; EPC: 41, 283	stressors
39.	(a)	V1: 36; EPC: 286	physiological effects of stress
40.	(d)	V1: 263–266; EPC: 284–286	hormone-response to stress
41.	(d)	V1: 266; EPC: 286–288	stress-/immune-related diseases/ conditions
42.	(b)	V1: 261; EPC: 42, 282	resistance/adaptation stage of stress
43.	(e)	V1: 34, 261; EPC: 42	exhaustion stage of stress
44.	(a)	V1: 34, 261; EPC: 42, 283	alarm stage of stress
45.	(a)	V1: 35–36; EPC: 43, 285, 287	physical signs of excessive stress
46.	(c)	V1: 36; EPC: 43	cognitive signs of excessive stress
47.	(a)	V1: 36; EPC: 43	emotional signs of excessive stress
48.	(e)	V1: 36; EPC: 43	behavioral signs of excessive stress
49.	(d)	V1: 35; EPC: 42	circadian rhythms
50.	(c)	V1: 35; EPC: 43	sleep deprivation
51.	(b)	V1: 35; EPC: 43	sleep techniques that minimize stress
52.	(e)	V1: 37; EPC: 44	critical incidents
53.	(b)	V1: 38; EPC: 45, 1829	critical incident stress debriefing (CISD)

		PAGE KEY	SUBJECT
54.	(d)	V1: 487; EPC: 294	startle or Moro reflex
55.	(b)	V1: 487; EPC: 294	rooting reflex
56.	(a)	V1: 488–489; EPC: 295, 296	infant parental-separation reactions
57.	(b)	V1: 489–491; EPC: 297, 298	Toddler life-span development stage
58.	(d)	V1: 493; EPC: 298, 299	School-age development stages, vital signs
59.	(c)	V1: 467; EPC: 533–534	trusting and therapeutic patient rapport
60.	(c)	V1: 473; EPC: 539	communication failure
61.	(a)	V1: 469–470; EPC: 535, 536	nonverbal communication (body language)
62.	(c)	V1: 471–472; EPC: 537	leading questions
63.	(d)	V1: 471–472; EPC: 537	open-ended questions
64.	(c)	V1: 473–474; EPC: 539–540	listening to patients
65.	(e)	V1: 30; EPC: 38	denial stage of loss/death and dying
66.	(a)	V1: 30; EPC: 39	anger stage of loss/death and dying
67.	(e)	V1: 30; EPC: 39	bargaining stage of loss/death and dying
68.	(d)	V1: 30; EPC: 41	paramedic responses to death and dying
69.	(a)	V: 131–32; EPC: 39–41	notifying family/friends of death
70.	(e)	V2: 257; EPC: 718	UHF
71.	(a)	V2: 266; EPC: 727	simplex transmission systems
72.	(b)	V2: 266; EPC: 727	duplex transmission systems
73.	(e)	V2: 267; EPC: 727–728	multiplex transmission systems
74.	(c)	V2: 262; EPC: 15, 724	EMS dispatcher
75.	(d)	V2: 258; EPC: 719–720	radio codes
76.	(b)	V2: 270–271; EPC: 725, 730–732	prehospital report information
77.	(a)	V2: 270–271; EPC: 725, 730–732	prehospital report information
78.	(d)	V1: 270–271; V2: ; EPC: 725, 730–731	paramedic/physician communications
79.	(c)	V2: 279, 289; EPC: 735–757	prehospital patient care reports
80.	(e)	V1: 136–137; V2: 290; EPC: 747	altering/amending a patient care report
81.	(c)	V1: 136–137; V2: 290; EPC: 747	altering/amending a patient care report
82.	(b)	V2: 284; EPC: 742	medical abbreviation/symbol for, "after"
83.	(a)	V2: 284; EPC: 742	medical abbreviation/symbol for, "before"
84.	(c)	V2: 284; EPC: 742	medical abbreviation/symbol for, "with"
85.	(d)	V2: 284; EPC: 742	medical abbreviation/symbol for, "without"

		PAGE KEY	SUBJECT
86.	(c)	V2: 282; EPC: 740	medical symbol for, "woman" or "female"
87.	(e)	V2: 282; EPC: 740	medical symbol for, "man" or "male"
88.	(b)	V2: 284; EPC: 742	medical symbol for, "change"
89.	(a)	V2: 284; EPC: 742	medical symbol for, "approximately"
90.	(d)	V2: 284; EPC: 741	the medical abbreviation, "BS"

TEST SECTION 2 ANSWER KEY

Page Key texts:

Brady's *Paramedic Care: Principles & Practice*, AHA updated 2nd edition, 5-volume series of texts:

   V1 indicates Volume 1: Introduction to Advanced Prehospital Care
   V2 indicates Volume 2: Patient Assessment
   V3 indicates Volume 3: Medical Emergencies
   V4 indicates Volume 4: Trauma Emergencies
   V5 indicates Volume 5: Special Considerations/Operations

EPC indicates Brady's *Essentials of Paramedic Care*, 2nd edition.

		PAGE KEY	SUBJECT
1.	(e)	V2: 49–54; EPC: 105–106, 585–588	the largest organ of the body
2.	(b)	V1: 171; EPC: 92	homeostasis
3.	(e)	V1: 168; EPC: 89	cytoplasm or protoplasm
4.	(c)	V1: 168; EPC: 89–90	organelles
5.	(d)	V1: 168; EPC: 88, 89	mitochondria
6.	(e)	V1: 168; EPC: 88, 89	lysosomes
7.	(a)	V1: 168; EPC: 89	nucleus
8.	(e)	V1: 168; EPC: 89, 258	adenosine triphosphate (ATP)
9.	(a)	V1: 169; EPC: 90	seven major cell functions
10.	(e)	V1: 169; EPC: 90	epithelial tissue
11.	(b)	V1: 169; EPC: 92	connective tissue
12.	(e)	V1: 169; EPC: 91	muscle tissue
13.	(c)	V1: 169; EPC: 134, 137	skeletal muscle tissue
14.	(a)	V1: 169; EPC: 134, 137	cardiac muscle tissue
15.	(d)	V1: 172; EPC: 1281	endocrine glands
16.	(b)	V1: 172; EPC: 1281	exocrine glands
17.	(c)	V1: 176, 555; EPC: 243, 991	hypoxia
18.	(b)	V1: 23, 177; EPC: 244, 1296, 1443	pathogen
19.	(c)	V1: 176; EPC: 243, 838	ischemia
20.	(c)	V1: 180–181; EPC: 94	average water percentage of total body weight
21.	(c)	V1: 180–181; EPC: 94	average water percentage of total body weight
22.	(a)	V1: 181; EPC: 95	intracellular fluid percentage
23.	(a)	V1: 183; EPC: 96	dehydration

		PAGE KEY	**SUBJECT**
24.	(a)	V1: 185; EPC: 98–99, 195	electrolytes
25.	(d)	V1: 185; EPC: 98	cations
26.	(d)	V1: 185; EPC: 98	anions
27.	(d)	V1: 186; EPC: 99	electrolyte measurement
28.	(b)	V2: 7; EPC: 537–538	open-ended questions
29.	(c)	V2: 7; EPC: 537–538	open-ended questions
30.	(d)	V2: 12; EPC: 551–552	OPQRST-ASPN mnemonic
31.	(b)	V2: 12; EPC: 551–552	OPQRST-ASPN mnemonic
32.	(c)	V2: 30–31; EPC: 567	palpation
33.	(c)	V2: 35–37; EPC: 570	blood pressure and radial pulse
34.	(b)	V2: 35–37; EPC: 570	blood pressure and carotid pulse
35.	(b)	V2: 35–37; EPC: 571	systolic blood pressure
36.	(e)	V2: 35–37; EPC: 571	diastolic blood pressure
37.	(a)	V2: 35–37; EPC: 572	pulse pressure
38.	(a)	V2: 35–37; EPC: 572, 1533, 1534, 1691	hypertension
39.	(c)	V2: 36; EPC: 572, 992, 1578	"tilt test" and hypovolemia
40.	(d)	V2: 36; EPC: 572	orthostatic vital sign changes
41.	(c)	V2: 37; EPC: 572–573	normal body temperature
42.	(d)	V2: 37–38; EPC: 573	stethoscope "bell"
43.	(e)	V2: 38; EPC: 574	accurate blood pressure measurement
44.	(c)	V2: 38; EPC: 574	accurate blood pressure measurement
45.	(e)	V2: 45–46; EPC: 579	normal oxygen saturation at sea level
46.	(b)	V2: 45–46; EPC: 582	pulse oximetry and carbon monoxide
47.	(b)	V2: 48; EPC: 584	blood glucose measurement ("glucometers")
48.	(e)	V2: 51; EPC: 586	spider angioma (skin lesion)
49.	(d)	V2: 51; EPC: 586	venous star (skin lesion)
50.	(b)	V2: 51; EPC: 585, 586, 1395	petechiae (skin lesion)
51.	(e)	V2: 57–58; EPC: 139, 144	facial anatomy
52.	(b)	V2: 57–58; EPC: 139, 144	cranial anatomy
53.	(e)	V2: 57–58; EPC: 143–145	orbits
54.	(b)	V4: 283; EPC: 146	aqueous humor of the eye
55.	(c)	V4: 283; EPC: 146	vitreous humor of the eye

		PAGE KEY	SUBJECT
56.	(e)	V4: 282–284; EPC: 146–147	the lens of the eye
57.	(b)	V4: 282–284; EPC: 147–147	the cornea
58.	(b)	V2: 124, 126; EPC: 153–154	cervical spine
59.	(d)	V2: 124, 126; EPC: 154	thoracic spine
60.	(a)	V2: 124, 126; EPC: 154	lumbar spine
61.	(c)	V2: 126; EPC: 625	lordosis
62.	(a)	V2: 126; EPC: 625	kyphosis
63.	(e)	V2: 126; EPC 625	scoliosis
64.	(d)	V2: 138–146; EPC: 172, 632–635, 688	cranial nerves
65.	(a)	V2: 138–146; EPC: 172, 632–635, 688	cranial nerves
66.	(e)	V2: 178–186; EPC: 654	Scene Size-Up
67.	(c)	V2: 186–200; EPC: 663	order of Initial Assessment
68.	(e)	V2: 189–190; EPC: 665–666, 1248	AVPU
69.	(b)	V2: 189–190; EPC: 665–666, 1248	AVPU
70.	(a)	V2: 189–190; EPC: 665–666, 1248	AVPU
71.	(d)	V2: 189–190; EPC: 665–666, 1248	AVPU
72.	(c)	V2: 85; EPC: 462, 608, 1092	stridor/upper airway obstruction
73.	(c)	V2: 85; EPC: 462, 608, 1092	stridor/upper airway obstruction
74.	(d)	V2: 191; EPC: 469–471	manual airway maneuvers-atraumatic patient
75.	(c)	V2: 191; EPC: 469–471	manual airway maneuvers-trauma patient
76.	(a)	V2: 33–34; EPC: 570	weak, thready pulse
77.	(b)	V2: 50; EPC: 585	capillary refill
78.	(e)	V2: 197–200; EPC: 672	Top Priority patients
79.	(e)	V2: 201–202; EPC: 673	predictors of serious internal injury
80.	(e)	V2: 203; EPC: 676	DECAP-BTLS
81.	(b)	V2: 203; EPC: 676	DECAP-BTLS
82.	(c)	V2: 203; EPC: 676	DECAP-BTLS
83.	(a)	V2: 203; EPC: 676	DECAP-BTLS
84.	(a)	V2: 223; EPC: 1040, 1090, 1191, 1251	jugular vein distention

		PAGE KEY	SUBJECT
85.	(e)	V2: 223; EPC: 1040, 1090, 1191, 1251	jugular vein distention
86.	(c)	V2: 206; EPC: 1038	pneumothorax ("simple")
87.	(b)	V2: 206; EPC: 1040–1041	tension pneumothorax
88.	(d)	V2: 96, 208; EPC: 613, 678, 1310	Cullen's sign
89.	(e)	V2: 96, 208; EPC: 613, 678, 1310	Grey-Turner's sign
90.	(c)	V2: 210, 211; EPC: 682	SAMPLE history
91.	(b)	V2: 210, 211; EPC: 682	SAMPLE history
92.	(a)	V2: 210, 211; EPC: 682	SAMPLE history
93.	(d)	V2: 210, 211; EPC: 682	SAMPLE history
94.	(b)	V2: 175; EPC: 653	order of patient assessment components
95.	(e)	V2: 228–232; EPC: 698–699, 701–702	Ongoing Assessment
96.	(a)	V2: 246–248; V5: 300–308; EPC: 711–714	Clinical Decision Making and Assessment Based Management skills
97.	(d)	V4: 6–7; EPC: 763	Trauma Center designations
98.	(a)	V4: 4–11; EPC: 765	immediate transport trauma triage criteria
99.	(d)	V4: 19; EPC: 771	Newton's First Law of Motion/Inertia
100.	(a)	V4: 20; EPC: 772	Conservation of Energy law
101.	(d)	V4: 19–21; EPC: 771	kinetics
102.	(b)	V4: 21; EPC: 774	ligamentum teres and deceleration trauma
103.	(a)	V4: 48; EPC: 797	ligamentum arteriosum and deceleration trauma
104.	(c)	V4: 30; EPC: 781	"down and under" frontal impact injuries
105.	(e)	V4: 28; EPC: 779, 781	"up and over" frontal impact injuries
106.	(e)	V4: 28; EPC: 779, 781	"up and over" frontal impact injuries
107.	(c)	V4: 30; EPC: 784	ejection injuries
108.	(e)	V4: 31; EPC: 782	rotational impact injuries
109.	(d)	V4: 25–27; EPC: 777–778	restraint systems
110.	(c)	V4: 35; EPC: 786	alcohol/drug intoxication and index of suspicion
111.	(e)	V4: 36–37; EPC: 787	motorcycle trauma
112.	(a)	V4: 36–37; EPC: 788	motorcycle trauma

		PAGE KEY	SUBJECT
113.	(a)	V4: 28; EPC: 779	frontal impact most common
114.	(b)	V4: 40–41; EPC: 790–793	blast injury mechanisms
115.	(c)	V4: 45–46; EPC: 795	pulmonary blast injuries
116.	(e)	V4: 47–48; EPC: 796–797	falls and deceleration trauma
117.	(a)	V4: 47–48; EPC: 796	fall mechanism of injury evaluation
118.	(e)	V4: 70; EPC: 812	bullet wound characteristics
119.	(e)	V1: 205–221; V4: 101–123; EPC: 822	shock
120.	(c)	V1: 207; V3: 79–80; EPC: 1128	stroke volume
121.	(d)	V3: 79–80; V4: 81–84; EPC: 1128	normal adult resting stroke volume
122.	(a)	V1: 207; V3: 80; EPC: 1129	cardiac output
123.	(c)	V1: 207; V3: 80; EPC: 1128	Frank-Starling mechanism
124.	(d)	V4: 83; EPC: 198	lumen
125.	(b)	V4: 83, 132, 133; EPC: 198	tunica adventitia
126.	(d)	V4: 83, 132, 133; EPC: 115, 198	tunica intima
127.	(c)	V4: 83, 132, 133; EPC: 198	tunica media
128.	(c)	V4: 83, 132, 133; EPC: 114, 198	tunica media
129.	(d)	V4: 83; EPC: 198	arterioles
130.	(a)	V1: 208; V4: 83; EPC: 198–199	arteries
131.	(c)	V1: 208; V4: 83–84; EPC: 198	veins
132.	(a)	V4: 83; EPC: 203	peripheral vascular resistance
133.	(b)	V4: 83–84; EPC: 201	venous system blood volume
134.	(b)	V4: 84; EPC: 108	erythrocyte
135.	(b)	V4: 84; EPC: 108, 248	erythrocytes and hemoglobin
136.	(e)	V4: 84; EPC: 114	platelets
137.	(d)	V1: 177, 210; V4: 102; EPC: 258	anaerobic metabolism
138.	(c)	V4: 106; EPC: 92	baroreceptors
139.	(c)	V4: 110; EPC: 260–261	decompensated shock
140.	(d)	V1: 215–216; V4: 111; EPC: 262	causes of hypovolemic shock
141.	(a)	V3: 357; V4: 91; EPC: 828	melena
142.	(d)	V3: 360; V4: 97; EPC: 834	hematochezia
143.	(b)	V1: 218–219; V4: 112; EPC: 265	septic shock
144.	(d)	V1: 216–217; V4: 113, 350, 373–374; EPC: 263	neurogenic shock

		PAGE KEY	SUBJECT
145.	(c)	V4: 111–112; EPC: 261, 264	distributive shock, anaphylaxis, sepsis
146.	(c)	V4: 111–112, 190; EPC: 582, 901, 1359, 1362	distributive shock, carbon monoxide poisoning
147.	(d)	V4: 112; EPC: 261	causes of obstructive shock
148.	(a)	V4: 120–122, 255, 501–502; EPC: 263	PASG complications
149.	(b)	V4: 10–11; EPC: 1808	helicopter transportation considerations
150.	(a)	V4: 101–123; EPC: 257–267	anticipation of "shock"

TEST SECTION 3 ANSWER KEY

Page Key texts:
Brady's *Paramedic Care: Principles & Practice*, AHA updated 2nd edition, 5-volume series
of texts:

> V1 indicates Volume 1: Introduction to Advanced Prehospital Care
> V3 indicates Volume 3: Medical Emergencies
> V4 indicates Volume 4: Trauma Emergencies

EPC indicates Brady's *Essentials of Paramedic Care*, 2nd edition.
PEP indicates Brady's *Prehospital Emergency Pharmacology*, 6th edition.
AHA indicates the 2005 American Heart Association Guidelines (*Circulation*, December
2005 Supplement), and is followed by the PART number of the section to reference.

		PAGE KEY	SUBJECT
1.	(a)	V1: 185; EPC: 98	chief extracellular ion
2.	(b)	V1: 185; EPC: 98; PEP: 88	chief intracellular ion
3.	(d)	V1: 186; EPC: 98; PEP: 88	sodium
4.	(c)	V1: 186; EPC: 98	chloride
5.	(e)	V1: 186; EPC: 98: PEP: 98	magnesium
6.	(e)	V1: 186; EPC: 98	bicarbonate
7.	(b)	V1: 186; EPC: 98; PEP: 88	calcium
8.	(c)	V1: 186; EPC: 98; PEP: 88	potassium
9.	(a)	V1: 186; EPC: 98	phosphate
10.	(b)	V1: 186; EPC: 98	bicarbonate
11.	(c)	V1: 186; EPC: 99; PEP: 98	isotonic fluids
12.	(d)	V1: 186; EPC: 99; PEP: 98	hypotonic fluids
13.	(b)	V1: 186; EPC: 99; PEP: 98	hypertonic fluids
14.	(e)	V1: 186–187; EPC: 99–100; PEP: 23	solute/solvent movement
15.	(d)	V1: 186–187; EPC: 100; PEP: 23	solute/solvent movement
16.	(d)	V1: 186–188, 289, 519; EPC: 99–100; PEP: 26, 90	diffusion
17.	(a)	V1: 187–188, 289; EPC: 100; PEP: 26, 91, 117	osmosis
18.	(a)	V1: 187–188, 289; EPC: 100; PEP: 26, 91, 117	osmosis
19.	(d)	V1: 186–188, 289, 519; EPC: 99 PEP: 26, 90	diffusion
20.	(a)	V1: 187–188, 289; EPC: 100; PEP: 26, 91, 117	osmosis
21.	(b)	V1: 188, 289; EPC: 100; PEP: 25, 91	active transport
22.	(d)	V1: 188, 289; EPC: 100; PEP: 90	facilitated diffusion

		PAGE KEY	SUBJECT
23.	(b)	V1: 193; EPC: 249; PEP: 94	colloid IV solutions
24.	(d)	V1: 193–194; EPC: 249; PEP: 95	crystalloid IV solutions
25.	(d)	V1: 193–194; EPC: 249; PEP: 95	crystalloid IV solutions
26.	(b)	V1: 193; EPC: 249; PEP: 94	colloid IV solutions
27.	(d)	V1: 193–194; EPC: 250; PEP: 95, 104, 106	crystalloid IV solutions
28.	(b)	V1: 193; EPC: 249; PEP: 94–100	colloid IV solutions
29.	(d)	V1: 193–194; EPC: 250; PEP: 95, 103	crystalloid IV solutions
30.	(e)	V1: 194; EPC: 250; PEP: 95	hypertonic solutions
31.	(a)	V1: 194; EPC: 249; PEP: 95	isotonic solutions
32.	(c)	V1: 194; EPC: 250; PEP: 95	hypotonic solutions
33.	(e)	V1: 194; EPC: 250; PEP: 95	hypertonic solutions
34.	(c)	V1: 194; EPC: 250; PEP: 95	hypotonic solutions
35.	(c)	V1: 195–200; EPC: 101; PEP: 43	acid-base balance
36.	(d)	V1: 195, 196; EPC: 101–102	pH/acid-base balance
37.	(a)	V1: 195; EPC: 102	acidosis
38.	(b)	V1: 196; EPC: 102	alkalosis
39.	(c)	V1: 195; EPC: 102	normal pH range
40.	(b)	V1: 197; EPC: 102–104	pH/acid-base balance regulation
41.	(a)	V1: 196; EPC: 102	bicarbonate buffer system
42.	(c)	V1: 197; EPC: 104	renal system
43.	(b)	V1: 196; EPC: 104	bicarbonate buffer system
44.	(e)	V1: 96; EPC: 103	carbonic acid
45.	(d)	V1: 197, 516; EPC: 103–104	respirations and pH balance
46.	(c)	V1: 197, 516; EPC: 103–104	respirations and pH balance
47.	(c)	V1: 198; EPC: 103–104	respiratory acidosis
48.	(d)	V1: 198–199; EPC: 251	respiratory alkalosis
49.	(e)	V1: 199; EPC: 251	metabolic acidosis
50.	(a)	V1: 199; EPC: 251	metabolic alkalosis
51.	(c)	V1: 279; EPC: 310; PEP: 4	mineral drug sources
52.	(a)	V1: 279; EPC: 310; PEP: 4	animal drug sources
53.	(b)	V1: 279; EPC: 310; PEP: 3–4	vegetable drug sources
54.	(d)	V1: 279; EPC: 310; PEP: 4–5	synthetic drug sources
55.	(c)	V1: 281; EPC: 311–312; PEP: 14	controlled substance act
56.	(d)	V1: 279; EPC: 310; PEP: 17	official drug names

		PAGE KEY	SUBJECT
57.	(e)	V1: 278; EPC: 310; PEP: 17	chemical drug names
58.	(b)	V1: 278; EPC: 310; PEP: 17	generic drug names
59.	(c)	V1: 278; EPC: 310; PEP: 17	trade or proprietary drug names
60.	(c)	V1: 294; EPC: 321; PEP: 5	drug solutions
61.	(b)	V1: 294; EPC: 321; PEP: 6	tinctures
62.	(d)	V1: 294; EPC: 321; PEP: 6	suspensions
63.	(a)	V1: 294; EPC: 321; PEP: 6	solid drug forms
64.	(a)	V1: 298; EPC: 324; PEP: 22	synergism
65.	(c)	V1: 298; EPC: 324; PEP: 21	cumulative effect
66.	(d)	V1: 298; EPC: 323; PEP: 21	antagonism
67.	(b)	V1: 298; EPC: 324; PEP: 22	potentiation
68.	(a)	V1: 297; EPC: 324; PEP: 22	side effects
69.	(e)	V1: 253, 298; EPC: 324; PEP: 21	hypersensitivity/allergic reaction
70.	(d)	V1: 298; EPC: 324; PEP: 21	idiosyncrasy
71.	(c)	V1: 280; EPC: 311; PEP: 21	contraindications
72.	(a)	PEP: 22; MedD	refractory
73.	(c)	V1: 293; EPC: 320; PEP: 23	enteral drug routes
74.	(d)	V1: 293–294; EPC: 320–321; PEP: 23	parenteral drug routes
75.	(d)	V1: 293–294; EPC: 320–321; PEP: 23	parenteral drug routes
76.	(d)	V1: 293–294; EPC: 320–321; PEP: 23	parenteral drug routes
77.	(a)	V1: 298–299; EPC: 325–326; PEP: 38–39	drug response factors
78.	(b)	AHA 7.2; V1: 293–294; EPC: 317–318; PEP: 26–28	drug-route absorption factors
79.	(a)	V1: 292–293; EPC: 319–320; PEP: 32	drug elimination
80.	(e)	V1: 295–300; EPC: 322; PEP: 34	pharmacodynamics
81.	(b)	V1: 296; EPC: 323; PEP: 35	agonist drugs
82.	(c)	V3: 80, 82, 249, 269; EPC: 176	parasympathetic nervous system
83.	(d)	V3: 81–82, 249, 261; EPC: 172–175	sympathetic nervous system
84.	(c)	V1: 311–314; V3: 82, 251; EPC: 176; PEP: 120	acetylcholine
85.	(d)	V1: 263, 311, 324, 343; V3: 39, 81, 158, 251, 342, 344; EPC: 174–175; PEP: 120, 122	epinephrine and norepinephrine
86.	(a)	V1: 311–314; EPC: 176; PEP: 125	acetylcholine stimulation
87.	(d)	V1: 264, 321, 314; EPC: 175; PEP: 124	beta 2 receptor stimulation

		PAGE KEY	SUBJECT
88.	(e)	V1: 323; EPC: 176; PEP: 124	dopaminergic stimulation
89.	(c)	V1: 264, 321, 322; EPC: 340; PEP: 124	beta 1 receptor stimulation
90.	(b)	V1: 264, 321, 322; EPC: 340; PEP: 124	alpha 1 receptor stimulation
91.	(c)	V1: 264, 321, 322; EPC: 340; PEP: 124	beta 1 receptor stimulation
92.	(a)	V1: 311; EPC: 340; PEP: 135	adrenergic drugs
93.	(b)	V1: 304; EPC: 340; PEP: 124	antiadrenergic drugs
94.	(b)	V1: 321; EPC: 340; PEP: 124	sympatholytic drugs
95.	(a)	V1: 321; EPC: 340; PEP: 124	sympathomimetic drugs
96.	(d)	V1: 314–316; EPC: 337; PEP: 127	parasympathomimetic/cholinergic drugs
97.	(e)	V1: 314–316; EPC: 338; PEP: 127	parasympathetic blockers
98.	(a)	V1: 323; EPC: 194; PEP: 128	inotrope/inotropic
99.	(e)	V1: 323; EPC: 194; PEP: 128	chronotrope/chronotropic
100.	(a)	V1: 451; EPC: 443; PEP: 72	pounds to kilograms conversion
101.	(e)	V1: 451; EPC: 443; PEP: 72	kilograms to pounds conversion
102.	(b)	V1: 451–457; EPC: 445; PEP: 74	drug dose calculations
103.	(c)	V1: 451–457; EPC: 445; PEP: 73	drug dose calculations
104.	(b)	V1: 451–457; EPC: 446; PEP: 75	drug dose calculations
105.	(a)	V1: 451–457; EPC: 448; PEP: 80	drug dose calculations
106.	(d)	V1: 451–457; EPC: 448; PEP: 80	drug dose calculations
107.	(c)	V1: 451–457; EPC: 448; PEP: 79	drug dose calculations
108.	(a)	V1: 451–457; EPC: 445, 447; PEP: 72	drug dose calculations
109.	(e)	V1: 193, 409; EPC: 407; PEP: 94	colloidal IV solutions
110.	(a)	V1: 193, 409; EPC: 407–408; PEP: 95	crystalloidal IV solutions
111.	(c)	V1: 194, 410; EPC: 408; PEP: 96	Lactated Ringer's solution
112.	(e)	V1: 411; EPC: 409; PEP: 112	macrodrip administration sets
113.	(b)	V1: 411; EPC: 409; PEP: 112	microdrip administration sets
114.	(a)	V1: 415–417; EPC: 413; PEP: 111	IV catheter sizes
115.	(b)	V4: 413; EPC: 111	IV catheter sizes
116.	(a)	V1: 425; EPC: 420	extravasation
117.	(d)	V1: 426; EPC: 421	pyrogenic reaction
118.	(a)	V1: 427; EPC: 422;	thrombophlebitis
119.	(c)	V1: 426; EPC: 421	pyrogenic reaction S/Sx
120.	(d)	V1: 427; EPC: 422	thrombophlebitis S/Sx

TEST SECTION 4 ANSWER KEY

Page Key texts:

Brady's *Paramedic Care: Principles & Practice*, AHA updated 2nd edition, 5-volume series of texts:

> V1 indicates Volume 1: Introduction to Advanced Prehospital Care
> V3 indicates Volume 3: Medical Emergencies
> V5 indicates Volume 5: Special Considerations/Operations

EPC indicates Brady's *Essentials of Paramedic Care*, 2nd edition
PEP indicates Brady's *Prehospital Emergency Pharmacology*, 6th edition.
AHA indicates the 2005 American Heart Association Guidelines (*Circulation*, December 2005 Supplement), and is followed by the PART number of the section to reference.

		PAGE KEY	SUBJECT
1.	(d)	AHA: 1	Class III ACLS interventions
2.	(a)	AHA: 1	Class I ACLS interventions
3.	(c)	AHA: 1	Class IIb ACLS interventions
4.	(c)	AHA: 7.2, 7.4; V1: 263–264; EPC: 284–285; PEP: 138–139	epinephrine
5.	(c)	V1: 324; V3: 158, 359; EPC: 1183, 1231; PEP: 138–139	epinephrine 1:10,000
6.	(c)	AHA: 12; V3: 359; PEP: 138–139	epinephrine 1:10,000
7.	(d)	AHA: 7.2; V3: 359; PEP: 138–139	epinephrine 1:10,000
8.	(e)	AHA: 7.2; V3: 359; PEP: 138–139	epinephrine 1:10,000
9.	(c)	V3: 159, 308, 453; EPC: 360; PEP: 149–150, 463	vasopressin
10.	(a)	AHA: 7.2, 7.4; V3: 159, 308, 453; EPC: 360; PEP: 149–150, 463	vasopressin
11.	(d)	AHA: 7.2; V3: 159, 308, 453; EPC: 360; PEP: 149–150, 463	vasopressin
12.	(c)	V3: 158; EPC: 342; PEP: 141–143, 436–437	isoproterenol
13.	(c)	V3: 158–159; EPC: 1027; PEP: 143–145	dopamine
14.	(d)	AHA: 7.4; V3: 158–159; EPC: 1027 PEP: 143–145	dopamine
15.	(c)	AHA: 7.4; V3: 158–159; EPC: 1027 PEP: 143–145	dopamine
16.	(e)	V3: 158–159; EPC: 1027; PEP: 143–145	dopamine
17.	(b)	AHA: 7.4; V3: 158–159; EPC: 1027; PEP: 143–145	dopamine

		PAGE KEY	SUBJECT
18.	(a)	AHA: 7.4; V3: 159; PEP: 145–146	dobutamine
19.	(b)	AHA: 7.2, 7.3; V3: 158; EPC: 344; PEP: 172–173	amiodarone
20.	(b)	AHA: 7.2, 7.3; V3: 158; EPC 344; PEP: 172–173	amiodarone
21.	(c)	AHA: 7.2, 7.3; V3: 158; EPC: 344; PEP: 172–173	amiodarone
22.	(a)	AHA: 7.3; V3: 158; EPC: 344; PEP: 172–173	amiodarone
23.	(d)	AHA: 7.3; V3: 158; EPC: 344; PEP: 172–173	amiodarone
24.	(a)	AHA: 7.2, 7.3; V3: 157; EPC: 343; PEP: 160–163	lidocaine
25.	(e)	AHA: 7.2, 7.3; V3: 157; EPC: 343; PEP: 160–163	lidocaine
26.	(e)	AHA: 7.3; V3: 157; EPC: 343; PEP: 160–163	lidocaine
27.	(e)	AHA: 7.3; V3: 157; EPC: 343; PEP: 160–163	lidocaine
28.	(c)	V3: 157; EPC 343; PEP: 160–163	lidocaine
29.	(e)	AHA: 7.2; V3: 157; EPC: 343; PEP: 160–163	lidocaine
30.	(d)	AHA: 7.2, 7.3; V3: 157; EPC: 343; PEP: 160–163	lidocaine
31.	(b)	V3: 157; EPC: 343; PEP: 163–165	procainamide
32.	(a)	AHA: 7.2, 7.3; V3: 157; EPC: 343; PEP: 163–165	procainamide
33.	(c)	AHA: 7.3; V3: 157; EPC: 343; PEP: 163–165	procainamide
34.	(b)	AHA: 7.3; V3: 157; EPC: 343; PEP: 163–165	procainamide
35.	(e)	AHA: 7.3; V3: 157; EPC: 343; PEP: 163–165	procainamide
36.	(c)	AHA: 7.3; V3: 157; EPC: 343; PEP: 163–165	procainamide
37.	(a)	V3: 112, 115, 158, 162; EPC: 344, 350; PEP: 170–171	verapamil
38.	(d)	AHA: 7.3; V3: 112, 115, 158, 162; EPC: 344, 350; PEP: 170–171	verapamil

		PAGE KEY	SUBJECT
39.	(e)	AHA: 7.3; V3: 112, 115, 158, 162; EPC: 344, 350; PEP: 170–171	verapamil
40.	(e)	AHA: 7.3; V3: 112, 115, 158, 162; EPC: 344, 350; PEP: 170–171	verapamil
41.	(b)	AHA: 7.3; V3: 157; EPC: 344, 1155; PEP: 167, 170	adenosine
42.	(a)	AHA: 7.3; V3: 157; EPC: 344, 1155; PEP: 167.170	adenosine
43.	(e)	AHA: 7.3, 10.2; V3: 157; EPC: 344, 1155; PEP: 167, 170	adenosine
44.	(b)	AHA: 7.3; V3: 157; EPC: 344, 1155; PEP: 167, 170	adenosine
45.	(c)	AHA: 7.3; V3: 157; EPC: 344, 1155; PEP: 167, 170	adenosine
46.	(b)	AHA: 7.3; V3: 157; EPC: 344, 1155; PEP: 167, 170	adenosine
47.	(d)	AHA: 7.3; V3: 157; EPC: 344, 1155; PEP: 167, 170	adenosine
48.	(e)	AHA: 7.2, 12; V3: 157; EPC: 994, 1027; PEP: 419	atropine
49.	(b)	AHA: 7.2, 7.3, 10.2; V3: 157; EPC: 994, 1027; PEP: 419	atropine
50.	(e)	AHA: 7.2; V3: 157; EPC: 994, 1027; PEP: 419	atropine
51.	(d)	AHA: 7.2, 7.3; V3: 157; EPC: 994, 1027; PEP: 178–181	atropine
52.	(c)	AHA: 7.2, 7.3; V3: 157; EPC: 994, 1027; PEP: 342–343	atropine
53.	(b)	AHA: 10.2, 12; V3: 157; EPC: 994, 1027; PEP: 419	atropine
54.	(c)	AHA: 7.4; V3: 162; EPC: 98, 1572, 1622; PEP: 195–196	sodium bicarbonate
55.	(a)	AHA: 7.4; V3: 162; EPC: 98, 1572, 1622; PEP: 195–196	sodium bicarbonate
56.	(c)	AHA: 12; V3: 162; EPC: 98, 1572, 1622; PEP: 195–196	sodium bicarbonate
57.	(c)	AHA: 7.4; V3: 162; EPC: 98, 1572, 1622; PEP: 195–196	sodium bicarbonate
58.	(a)	AHA: 8; V3: 160; EPC: 328, 953, 994; PEP: 196–199	morphine sulfate

		PAGE KEY	SUBJECT
59.	(d)	AHA: 8; V3: 160; EPC: 328, 953, 994; PEP: 196–199	morphine sulfate
60.	(d)	AHA: 8; V3: 160; EPC: 328, 953, 994; PEP: 196–199	morphine sulfate
61.	(e)	V3: 160; EPC: 328, 953, 994; PEP: 196–199	morphine sulfate
62.	(c)	V3: 160; EPC: 328, 953, 994; PEP: 196–199	morphine sulfate
63.	(e)	AHA: 8; V3: 160; EPC: 328, 953, 994; PEP: 196–199	morphine sulfate
64.	(c)	AHA: 7.2, 8; V3: 160; EPC: 328, 953, 994; PEP: 196–199	morphine sulfate
65.	(c)	AHA: 7.4; V3: 161; EPC: 346, 992, 1392	furosemide
66.	(d)	AHA: 7.4; V3: 161; EPC: 346, 992, 1392; PEP: 201–202	furosemide
67.	(a)	V3: 161; EPC: 346, 992, 1392; PEP: 201–202	furosemide
68.	(e)	AHA: 7.4, 10.1; V3: 161; EPC: 346, 992, 1392; PEP: 201–202	furosemide
69.	(b)	AHA: 7.2; V3: 161; EPC: 346, 992, 1392; PEP: 201–202	furosemide
70.	(b)	AHA: 8; V3: 159; EPC: 350; PEP: 207, 210	nitroglycerin
71.	(d)	AHA: 8; V3: 159; EPC: 350; PEP: 207, 210	nitroglycerin
72.	(b)	AHA: 8; V3: 159; EPC: 350; PEP: 207, 210	nitroglycerin
73.	(c)	AHA: 8; V3: 159; EPC: 350; PEP: 207, 210	nitroglycerin
74.	(c)	AHA: 8; V3: 159; EPC: 350; PEP: 207, 210	nitroglycerin
75.	(a)	AHA: 7.4; V5: 85; EPC: 1622; PEP: 216–217	calcium chloride
76.	(c)	AHA: 7.4, 10.1; V5: 85; EPC: 1622; PEP: 216–217	calcium chloride
77.	(a)	AHA: 10.1; V5: 85; EPC: 1622; PEP: 216–217	calcium chloride
78.	(c)	AHA: 10.5, 10.6; V3: 344; EPC: 1301–1303; PEP: 427–428	epinephrine 1:1000

		PAGE KEY	SUBJECT
79.	(d)	AHA: 10.5, 10.6; V3: 344; EPC: 1301–1303; PEP: 427–428	epinephrine 1:1000
80.	(b)	AHA: 10.5, 10.6; V3: 344; EPC: 1301–1303; PEP: 427–428	epinephrine 1:1000
81.	(c)	V3: 343; EPC: 332, 1302; PEP: 236–237	aminophylline
82.	(d)	V3: 343; EPC: 332, 1302; PEP: 236–237	aminophylline
83.	(e)	V3: 343; EPC: 332, 1302; PEP: 236–237	aminophylline
84.	(c)	PEP: 231–232	racemic epinephrine
85.	(a)	V3: 39, 46, 343; EPC: 353, 1302; PEP: 229–230	albuterol
86.	(c)	V1: 355; EPC: 995; PEP: 276–277	dextrose
87.	(e)	V1: 355; EPC: 995; PEP: 276–277	$D_{50}W$
88.	(d)	V1: 355; EPC: 995; PEP: 276–277	$D_{50}W$
89.	(b)	V1: 355; EPC: 995; PEP: 276–277	$D_{50}W$
90.	(b)	V1: 355; EPC: 995; PEP: 276–277	$D_{50}W$
91.	(b)	V3: 274; EPC: 396, 995; PEP: 277, 281	thiamine
92.	(e)	V3: 274; EPC: 396, 995; PEP: 277, 281	thiamine
93.	(e)	V3: 274; EPC: 396, 995; PEP: 277, 281	thiamine
94.	(d)	V3: 274; EPC: 396, 995; PEP: 277, 281	thiamine
95.	(a)	V3: 274; EPC: 396, 995; PEP: 277, 281	thiamine
96.	(a)	V3: 343; PEP: 284–285	dexamethasone
97.	(c)	PEP: 287	steroid preparations
98.	(e)	V3: 343; PEP: 241–242	methylprednisolone
99.	(a)	V3: 161; EPC: 329, 496, 953, 993–994; PEP: 289	diazepam
100.	(d)	V3: 161; EPC: 329, 496, 953, 993–994; PEP: 289–290	diazepam
101.	(b)	V3: 161; EPC: 329, 496, 953, 993–994; PEP: 289–290	diazepam
102.	(e)	V3: 161; EPC: 329, 496, 953, 993–994; PEP: 289–290	diazepam

		PAGE KEY	SUBJECT
103.	(c)	V3: 308; EPC: 364; PEP: 302–303	oxytocin
104.	(a)	AHA: 7.2, 10.1, 12; EPC: 345; PEP: 176–177	magnesium sulfate
105.	(a)	AHA: 7.2, 10.1, 12; EPC: 345; PEP: 176–177	magnesium sulfate
106.	(e)	EPC: 345; PEP: 176–177	magnesium sulfate
107.	(b)	AHA: 7.2; EPC: 345; PEP: 176–177	magnesium sulfate
108.	(d)	AHA: 7.2; EPC: 345; PEP: 176–177	magnesium sulfate
109.	(a)	EPC: 345; PEP: 176–177	magnesium sulfate
110.	(b)	V3: 342; EPC: 335–336, 356, 1302; PEP: 425, 499	diphenhydramine
111.	(b)	V3: 342; EPC: 335–336, 356, 1302; PEP: 425, 499	diphenhydramine
112.	(c)	V3: 342; EPC: 335–336, 356, 1302; PEP: 425, 499	diphenhydramine
113.	(e)	V3: 421; EPC: 1354; PEP: 313	ipecac
114.	(d)	V3: 421; EPC: 1354; PEP: 413–414	activated charcoal
115.	(c)	AHA: 10.2; V3: 450; EPC: 323, 329, 994, 1573; PEP: 338–340	naloxone
116.	(a)	V3: 450; EPC: 323, 329, 994, 1573; PEP: 338–340	naloxone
117.	(c)	V3: 450; EPC: 323, 329, 994, 1573; PEP: 338–340	naloxone
118.	(b)	V3: 450; EPC: 323, 329, 994, 1573; PEP: 338–340	naloxone
119.	(e)	AHA: 7.2; V3: 450; EPC 323, 329, 994, 1573; PEP: 338–340	naloxone
120.	(c)	V3: 285; EPC: 330, 994; PEP: 318–320	flumazenil
121.	(c)	V3: 285; EPC: 330, 994; PEP: 318–320	flumazenil
122.	(d)	V3: 285; EPC: 330, 994; PEP: 318–320	flumazenil
123.	(b)	V3: 285; EPC: 330, 994; PEP: 318–320	flumazenil
124.	(c)	V3: 285; EPC: 330, 994; PEP: 318–320	flumazenil
125.	(d)	V3: 285; EPC: 330, 994; PEP: 318–320	flumazenil

		PAGE KEY	SUBJECT
126.	(a)	V3: 315; EPC: 186, 362, 364; PEP: 274–276	glucagon
127.	(c)	V3: 315; EPC: 186, 362, 364; PEP: 274–276	glucagon
128.	(d)	AHA: 7.3; V3: 162; EPC: 116, 351; PEP: 151–152	propranolol
129.	(a)	AHA: 7.3, 10.2; V3: 162; EPC: 116, 351; PEP: 151–152	propranolol
130.	(e)	AHA: 7.3; V3: 162; EPC: 344, 346; PEP: 151–152	propranolol
131.	(c)	AHA: 8; V3: 160; EPC: 116, 351; PEP: 186–187	aspirin
132.	(d)	AHA: 8; V3: 160; EPC: 116, 351; PEP: 186–187	aspirin
133.	(d)	AHA: 4, 8, 9; V3: 160; EPC: 116, 351; PEP: 186–187	aspirin
134.	(d)	AHA: 4, 8, 9; V3: 160; EPC: 116, 351; PEP: 186–187	aspirin
135.	(d)	AHA: 4, 8; V3: 160; EPC: 116, 351; PEP: 186–187	aspirin

TEST SECTION 5 ANSWER KEY

Page Key texts:

Brady's *Paramedic Care: Principles & Practice*, AHA updated 2nd edition, 5-volume series of texts:

 V1 indicates Volume 1: Introduction to Advanced Prehospital Care
 V3 indicates Volume 3: Medical Emergencies

EPC indicates Brady's *Essentials of Paramedic Care*, 2nd edition.

PEP indicates Brady's *Prehospital Emergency Pharmacology*, 6th edition.

AHA indicates the 2005 American Heart Association Guidelines (*Circulation*, December 2005 Supplement), and is followed by the PART number of the section to reference.

		PAGE KEY	SUBJECT
1.	(a)	V1: 511–512; EPC: 207–208	structures of the larynx
2.	(d)	V1: 514; EPC: 1424	surfactant
3.	(c)	V1: 514; EPC: 209	atelectasis
4.	(e)	V1: 515; EPC: 1595	pediatric cricoid cartilage
5.	(b)	V1: 485, 495; EPC: 571	DOT normal adult resting respiratory rate (some systems recognize 12–20, as do Brady texts)
6.	(c)	V1: 485; EPC: 1601	child respiratory rate
7.	(d)	V1: 485; EPC: 1601	infant respiratory rate
8.	(c)	V1: 517–518; EPC: 213	room air oxygen percent
9.	(a)	V1: 519; EPC: 214	gas diffusion
10.	(b)	V1: 519; EPC: 214	oxygen diffusion/transportation
11.	(a)	V1: 518; EPC: 214	room air nitrogen
12.	(e)	V1: 519; EPC: 214	arterial PO_2
13.	(c)	V1: 519; EPC: 214	arterial PCO_2
14.	(a)	V1: 519–520; EPC: 214–215	blood oxygen concentration factors
15.	(c)	V1: 521; EPC: 216	respiratory mechanics
16.	(d)	V1: 521; EPC: 216	respiratory mechanics
17.	(e)	V1: 521; EPC: 216	respiratory mechanics
18.	(b)	V1: 523–524; EPC: 457	airway obstruction
19.	(d)	V1: 523–534; EPC: 457	airway obstruction
20.	(e)	V1: 523–524; EPC: 458	aspiration of vomitus
21.	(a)	V1: 524; EPC: 457–458	laryngeal spasm
22.	(e)	V1: 524; EPC: 457–458	laryngeal spasm
23.	(b)	V1: 524; EPC: 457–458	laryngeal spasm

		PAGE KEY	SUBJECT
24.	(c)	V1: 528; EPC: 1088	cyanosis
25.	(c)	V1: 528; EPC: 460	anoxia
26.	(a)	V1: 521; EPC: 216	hypoxemia
27.	(b)	V1: 176, 555; EPC: 243	hypoxia
28.	(a)	V1: 525–526; EPC: 571	earliest sign of respiratory distress
29.	(d)	V1: 528; EPC: 157	accessory muscles of respiration
30.	(b)	V1: 529; EPC: 461	central neurogenic hyperventilation
31.	(c)	V1: 529; EPC: 461	Biot's (ataxic) respirations
32.	(d)	V1: 529; EPC: 461	Cheyne-Stokes respirations
33.	(e)	V1: 529; EPC: 461	Kussmaul's respirations
34.	(a)	V1: 529; EPC: 461	agonal respirations
35.	(e)	V1: 529; EPC: 461	preferred area for breath sound auscultation
36.	(d)	V1: 530; EPC: 462	causes of difficult ventilation
37.	(c)	V1: 531–532; EPC: 463	pulse oximetry
38.	(c)	V1: 531–532; EPC: 463	pulse oximetry
39.	(b)	V1: 531–532; EPC: 463	pulse oximetry
40.	(d)	V1: 531–532; EPC: 463	pulse oximetry
41.	(a)	V1: 512, 541–542; EPC: 470	Sellick's maneuver
42.	(c)	V1: 544–546; EPC: 473	oropharyngeal airway
43.	(e)	V1: 544–546; EPC: 471	nasopharyngeal airway
44.	(b)	V1: 548–550; EPC: 478	ET tube inflatable cuff size
45.	(a)	AHA 7.2, 7.3; V1: 384; EPC: 384; PEP: 522	ET tube medication administration
46.	(c)	V1: 547–550; EPC: 476	straight/Miller laryngoscope blade
47.	(a)	V1: 547–550; EPC: 476	curved/Macintosh laryngoscope blade
48.	(e)	V1: 547–550; EPC: 480	forceps for intubation
49.	(e)	V1: 547–550; EPC: 476	straight/Miller laryngoscope blade
50.	(d)	V1: 547–550; EPC: 476	curved/Macintosh laryngoscope blade
51.	(a)	V1: 547–550; EPC: 476	curved/Macintosh laryngoscope blade
52.	(b)	V1: 547–550; EPC: 476	straight/Miller laryngoscope blade
53.	(a)	V1: 547–550; EPC: 476	laryngoscope blade preferences
54.	(b)	V1: 572–573; EPC: 497	pediatric intubation
55.	(d)	V1: 547–550; EPC: 478	flexible/malleable stylet use in intubation
56.	(a)	V1: 555; EPC: 481–82	hyperventilation and intubation

		PAGE KEY	SUBJECT
57.	(e)	AHA: 7.1; V1: 554–556; EPC: 480–81	ET intubation side effects
58.	(c)	V1: 555; EPC: 482	esophageal intubation
59.	(b)	V1: 555–556; EPC: 482	endobronchial intubation
60.	(d)	V1: 555–556; EPC: 482	endobronchial intubation
61.	(d)	AHA: 7.1; V1: 559–560; EPC: 486	endotracheal intubation confirmation
62.	(c)	V1: 555–556; EPC: 483	endotracheal intubation
63.	(b)	V1: 576–579; EPC: 501–503	nasotracheal intubation
64.	(a)	V1: 576–579; EPC: 501–503	nasotracheal intubation
65.	(c)	V1: 576–579; EPC: 501–503	nasotracheal intubation
66.	(b)	V1: 589; EPC: 513	cricoid cartilage
67.	(a)	V1: 589; EPC: 513	thyroid cartilage
68.	(d)	V1: 589; EPC: 513	cricothyroid membrane
69.	(d)	V1: 588–592; EPC: 512	transtracheal jet insufflation
70.	(e)	V1: 589; EPC: 512	barotrauma
71.	(b)	V1: 587; EPC: 516	open cricothyrotomy
72.	(d)	V1: 597–600; EPC: 521–522	suction
73.	(c)	V1: 597–600; EPC: 521–522	suction
74.	(b)	V1: 601–602; EPC: 524	nasal cannula
75.	(b)	V1: 601–602; EPC: 524	nasal cannula
76.	(d)	V1: 601–602; EPC: 524	simple face mask
77.	(c)	V1: 601–602; EPC: 524	simple face mask
78.	(a)	V1: 601–602; EPC: 524	nonrebreather mask
79.	(e)	V1: 601–602; EPC: 524	nonrebreather mask
80.	(c)	V1: 601–602; EPC: 524	Venturi mask
81.	(c)	V1: 601–602; EPC: 524	Venturi mask
82.	(a)	V1: 603; EPC: 525	mouth-to-mouth/nose/stoma ventilation
83.	(b)	V1: 603; EPC: 525–526	mouth-to-mask ventilation
84.	(b)	V1: 603–604; EPC: 526–527	bag-valve-mask ventilation
85.	(c)	V1: 603–604; EPC: 526–527	bag-valve-mask ventilation
86.	(e)	V1: 603–604; EPC: 526–527	bag-valve-mask ventilation
87.	(e)	V1: 603–604; EPC: 526–527	bag-valve-mask ventilation
88.	(d)	V1: 605–606; EPC: 527–528	demand-valve device
89.	(e)	V1: 605–606; EPC: 527–528	demand-valve device

		PAGE KEY	SUBJECT
90.	(e)	V1: 699; EPC: 522–523	gastric distention
91.	(d)	AHA: 7.1, 7.4; V1: 532–535; EPC 464–467, 583	capnometry
92	(c)	AHA: 7.1, 7.4; V1: 532–535; EPC 464–467, 583	capnometry
93.	(e)	AHA: 7.1; V1: 532–535; EPC 464–467, 583	ETCO$_2$ monitoring
94.	(a)	V1: 533; EPC 465	changes in ETCO$_2$
95.	(b)	V1: 533; EPC 465	changes in ETCO$_2$
96.	(e)	V1: 533; EPC 465	changes in ETCO$_2$
97.	(d)	V1: 533; EPC 465	changes in ETCO$_2$
98.	(d)	V1: 533; EPC 465	changes in ETCO$_2$
99.	(c)	V1: 533; EPC 465	changes in ETCO$_2$
100.	(e)	V1: 533; EPC 465	changes in ETCO$_2$
101.	(e)	V1: 533; EPC 465	changes in ETCO$_2$
102.	(b)	V1: 533; EPC 465	changes in ETCO$_2$

TEST SECTION 6 ANSWER KEY

Page Key texts:
Brady's *Paramedic Care: Principles & Practice*, AHA updated 2nd edition, 5-volume series of texts:

- V2 indicates Volume 2: Patient Assessment
- V3 indicates Volume 3: Medical Emergencies
- V4 indicates Volume 4: Trauma Emergencies

EPC indicates Brady's *Essentials of Paramedic Care*, 2nd edition.
PEP indicates Brady's *Prehospital Emergency Pharmacology*, 6th edition.

		PAGE KEY	SUBJECT
1.	(a)	V4: 11; EPC: 766–767	trauma triage rapid transfer criteria
2.	(c)	V4: 288, 290; EPC: 964	retroauricular ecchymosis
3.	(a)	V4: 288–290; EPC: 964	bilateral periorbital ecchymosis
4.	(e)	V4: 288–290; EPC: 964	signs of basilar skull fracture
5.	(c)	V4: 348; EPC: 1004–1005	spinal fracture
6.	(a)	V4: 347; EPC: 1003	axial loading injury
7.	(b)	V4: 347; EPC: 1003	distraction injury
8.	(b)	V4: 300; EPC: 973	zygomatic fracture
9.	(e)	V4: 292; EPC: 966	contrecoup injury
10.	(d)	V4: 275; EPC: 140	the meninges
11.	(c)	V4: 275; EPC: 140	dura mater
12.	(a)	V4: 275; EPC: 140	pia mater
13.	(b)	V4: 276; EPC: 140	arachnoid membrane
14.	(b)	V4: 293; EPC: 967	subdural hematoma
15.	(d)	V4: 294; EPC: 968	cerebral concussion and spine injury
16.	(b)	V4: 297; EPC: 970	anterograde amnesia
17.	(a)	V4: 292; EPC: 970	retrograde amnesia
18.	(d)	V4: 294; EPC: 967–968	concussion and epidural hematoma
19.	(c)	V4: 297; EPC: 971	Cushing's reflex
20.	(d)	V4: 298; EPC: 1252	decorticate posturing
21.	(b)	V4: 298; EPC: 1252	decerebrate posturing
22.	(d)	V4: 298; EPC: 968	brainstem injury and posturing
23.	(c)	V4: 294; EPC: 968	intracerebral hemorrhage
24.	(b)	V4: 293; EPC: 967	subdural hematoma
25.	(c)	V4: 295–296; EPC: 969–970	increased intracranial pressure

		PAGE KEY	SUBJECT
26.	(d)	V3: 271, 275; V4: 295–296; EPC: 969–970	increased intracranial pressure
27.	(e)	V4: 302; EPC: 974–975	inner ear injuries
28.	(e)	V4: 342–343; EPC: 169	dermatomes
29.	(a)	V4: 342–343; EPC: 168–169	collar region dermatome
30.	(c)	V4: 342–343; EPC: 168–169	nipple-line dermatome
31.	(d)	V4: 342–343; EPC: 168–169	umbilical-area dermatome
32.	(b)	V4: 342–343; EPC: 168–169	phrenic nerve origin
33.	(d)	V4: 113, 350, 373–374; EPC: 263–264	neurogenic shock
34.	(e)	V4: 349, 355; EPC: 1010	spine injury signs
35.	(a)	V4: 373, 374; EPC: 1026–1027	spine injury management
36.	(c)	V4: 303; EPC: 976	retinal detachment
37.	(e)	V4: 303; EPC: 975	corneal abrasion
38.	(b)	V4: 303; EPC: 976	acute retinal artery occlusion
39.	(a)	V4: 303; EPC: 975	hyphema
40.	(d)	V4: 303; EPC: 976	conjunctival hemorrhage
41.	(a)	V4: 303; EPC: 975	hyphema
42.	(d)	V4: 302; EPC: 975	vitreous humor loss
43.	(d)	V4: 315; EPC: 987	parasympathetic stimulation of intubation
44.	(c)	V4: 296–298; EPC: 970–971	increased ICP S/Sx
45.	(c)	V4: 319; EPC: 990	increased ICP management
46.	(d)	V4: 321–324; EPC: 992	increased ICP management
47.	(e)	V4: 310–311; EPC: 982	Glasgow Coma Scale
48.	(b)	V4: 310–311; EPC: 982	Glasgow Coma Scale
49.	(c)	V4: 310–311; EPC: 982	Glasgow Coma Scale
50.	(e)	V4: 309–310, 320; EPC: 982, 991	open neck wound management
51.	(c)	V4: 309–310, 320; EPC: 982, 991	open neck wound management
52.	(c)	V4: 120–121, 320; EPC: 991	PASG and head trauma
53.	(a)	V4: 321; EPC: 992; PEP: 283–296	head trauma management
54.	(e)	V2: 65; EPC: 594	aniscoria
55.	(a)	V2: 65; EPC: 594	dysconjugate gaze
56.	(a)	V4: 299; EPC: 594	unilateral pupillary changes
57.	(d)	V4: 299; EPC: 594	pupillary changes

		PAGE KEY	SUBJECT
58.	(b)	EPC: 323; PEP: 285–286	mannitol
59.	(e)	V4: 321; EPC: 992; PEP: 201–202	furosemide
60.	(d)	V3: 281; EPC: 970	seizures and head trauma
61.	(e)	V4: 403–405; EPC: 1042–1043	lung contusion
62.	(d)	V4: 405; EPC: 1043	hemoptysis
63.	(e)	V4: 396–397; EPC: 1035–1036	rib fracture
64.	(c)	V4: 398–399; EPC: 1037–1038	flail chest segment
65.	(b)	V4: 398–399; EPC: 1038	paradoxical movement
66.	(e)	V4: 407; EPC: 1045	pulsus paradoxus
67.	(d)	V4: 407; EPC: 1045	electrical alternans
68.	(b)	V4: 410; EPC: 1048	traumatic asphyxia
69.	(c)	V4410: ; EPC: 1048	traumatic asphyxia
70.	(e)	V4: 399; V3: 57–58; EPC: 1038	pneumothorax
71.	(c)	V4: 400; EPC: 1038	paper bag syndrome
72.	(c)	V4: 401–402; EPC: 1040–1041	tension pneumothorax
73.	(c)	V4: 401–402; EPC: 1040–1041	tension pneumothorax S/Sx
74.	(b)	V4: 401–402; EPC: 1040–1041	tension pneumothorax S/Sx
75.	(d)	V4: 401–402; EPC: 1040–1041	tension pneumothorax S/Sx
76.	(a)	V4: 403; EPC: 1041–1042	hemothorax S/Sx
77.	(b)	V4: 405–406; EPC: 1043–1044	myocardial contusion S/Sx
78.	(a)	V4: 405–406; EPC: 1043–1044	myocardial contusion
79.	(a)	V4: 391; EPC: 188–189	epicardium or visceral pericardium
80.	(b)	V4: 391; EPC: 188–189	parietal pericardium
81.	(c)	V4: 391; EPC: 188	pericardial sac
82.	(e)	V4: 406–407, 421; EPC: 1044–1045	pericardial tamponade
83.	(c)	V4: 406–407, 421; EPC: 1044–1045	pericardial tamponade S/Sx
84.	(d)	V4: 406–407, 421; EPC: 1058	pericardial tamponade management
85.	(a)	V4: 406–407, 421; EPC: 1046	thoracic aneurysm S/Sx
86.	(b)	V4: 263–264; EPC: 953; PEP: 199–201	nitrous oxide indications
87.	(a)	V4: 263–264; EPC: 953; PEP: 199–201	nitrous oxide contraindications
88.	(d)	V4: 263–264; EPC: 953; PEP: 199–201	nitrous oxide administration
89.	(a)	V4: 150, 418–419; EPC: 1055–1056	occlusive chest dressing
90.	(d)	V4: 419–420; EPC: 1056–1057	needle decompression of the chest

		PAGE KEY	SUBJECT
91.	(b)	V4: 419–420; EPC: 1056–1057	needle decompression of the chest
92.	(a)	V4: 421; EPC: 1057–1058	myocardial contusion management
93.	(e)	V4: 171; EPC: 818	impaled chest management
94.	(e)	V4: 429; EPC: 218	retroperitoneal structures
95.	(b)	V4: 429; EPC: 218	abdominal organs
96.	(c)	V4: 435–436; EPC: 218	common iliac arteries
97.	(c)	V4: 443; EPC: 1067	peritonitis
98.	(b)	V2: 208; V4: 443; EPC: 1067	rebound tenderness
99.	(d)	V4: 443; EPC: 1067	abdominal guarding
100.	(a)	V4: 452, 454; EPC: 1075	abdominal injury IV management
101.	(d)	V4: 452, 454; EPC: 1075	prehospital IV infusion limit
102.	(d)	V4: 452, 454; EPC: 1075	evisceration management
103.	(d)	V4: 452, 454; EPC: 1075	impaled abdomen management
104.	(b)	V4: 235; EPC: 927	strain injury
105.	(a)	V4: 234; EPC: 926–927	compartment syndrome
106.	(c)	V4: 235; EPC: 928	sprain injury
107.	(e)	V4: 232; EPC: 121	tendons
108.	(d)	V4: 223; EPC: 120	ligaments
109.	(e)	V4: 259; EPC: 948–949	hip injury S/Sx
110.	(a)	V4: 259; EPC: 948–949	anterior hip dislocation
111.	(b)	V4: 259; EPC: 948–949	posterior hip dislocation
112.	(c)	V4: 261; EPC: 951	inferior shoulder dislocation
113.	(a)	V4: 261; EPC: 951	anterior shoulder dislocation
114.	(b)	V4: 261; EPC: 951	posterior shoulder dislocation
115.	(a)	V2: 129; EPC: 199	popliteal pulse
116.	(e)	V2: 129; EPC: 199	posterior tibial pulse
117.	(b)	V2: 129; EPC: 199	dorsalis pedis pulse
118.	(d)	V4: 235–236, 239; EPC: 938–955	musculoskeletal management
119.	(c)	V4: 239; EPC: 938–940	bandage and splint immobilization
120.	(b)	V4: 443; EPC: 945	pelvic fracture management
121.	(a)	V4: 251–252; EPC: 942	traction splints
122.	(b)	V4: 129–130; EPC: 105	skin functions
123.	(d)	V4: 131, 179; EPC: 106	dermis

		PAGE KEY	SUBJECT
124.	(a)	V4: 131, 180; EPC: 106	subcutaneous tissue
125.	(e)	V4: 131, 179; EPC: 105	epidermis
126.	(a)	V4: 132, 180; EPC: 106	subcutaneous tissue
127.	(d)	V4: 131; EPC: 106	sebum
128.	(e)	V4: 131; EPC: 106	sudoriferous glands
129.	(b)	V4: 135–140; EPC: 855–861	ecchymosis
130.	(a)	V4: 135–140; EPC: 855–861	erythema
131.	(e)	V4: 135–140; EPC: 855–861	contusion
132.	(a)	V4: 135–140; EPC: 855–861	hematoma
133.	(c)	V4: 135–140; EPC: 855–861	abrasion
134.	(d)	V4: 135–140; EPC: 855–861	incision
135.	(b)	V4: 135–140; EPC: 855–861	laceration
136.	(b)	V4: 135–140; EPC: 855–861	avulsion
137.	(c)	V4: 135–140: ; EPC: 855–861	amputation
138.	(b)	V4: 190–191; EPC: 902	partial-thickness burn
139.	(c)	V4: 190–191; EPC: 902	full-thickness burn
140.	(a)	V4: 192; EPC: 903	adult rule of nines
141.	(e)	V4: 192; EPC: 903	adult rule of nines
142.	(a)	V4: 192; EPC: 903	adult rule of nines
143.	(b)	V4: 192; EPC: 903	adult rule of nines
144.	(c)	V4: 192; EPC: 903	adult rule of nines
145.	(e)	V4: 193; EPC: 903–904	palmar surface area (rule of palms)
146.	(d)	V4: 193; EPC: 903–904	palmar surface area (rule of palms)
147.	(b)	V4: 192; EPC: 904, 1655	pediatric rule of nines
148.	(e)	V4: 192; EPC: 904, 1655	pediatric rule of nines
149.	(b)	V4: 192; EPC: 904, 1655	pediatric rule of nines
150.	(c)	V4: 192; EPC: 904, 1655	pediatric rule of nines
151.	(e)	V4: 193; EPC: 905	eschar
152.	(a)	V4: 181–183; EPC: 904–905	thermal burn systemic complications
153.	(e)	V4: 193, 200; EPC: 905, 910	circumferential burns
154.	(e)	V4: 43–46; EPC: 794	blast injuries
155.	(e)	V4: 43–46, 204–205; EPC: 795	inhalation injury
156.	(e)	V4: 206; EPC: 894–896	electrical burns

		PAGE KEY	SUBJECT
157.	(a)	V4: 199, 200; EPC: 911	burn severity categorization
158.	(c)	V4: 203–204; EPC: 913–914	fluid resuscitation for burns
159.	(a)	V4: 207–209; EPC: 896	acid burn pathology
160.	(e)	V4: 207–209; EPC: 896	alkali burn pathology
161.	(a)	V4: 208; EPC: 917	dry lime exposure management
162.	(b)	V4: 208; EPC: 917	phenol exposure management
163.	(c)	V4: 208; EPC: 918	sodium metal exposure management
164.	(c)	V4: 185–187; EPC: 897	types of radiation
165.	(c)	V4: 210–211; EPC: 899–900	radiation injury management

TEST SECTION 7 ANSWER KEY

Page Key texts:
Brady's *Paramedic Care: Principles & Practice*, AHA updated 2nd edition, 5-volume series of texts:

Brady's *Paramedic Care: Principles & Practice*, AHA updated 2nd edition, 5-volume series of texts:

  V1 indicates Volume 1: Introduction to Advanced Prehospital Care
  V3 indicates Volume 3: Medical Emergencies

EPC indicates Brady's *Essentials of Paramedic Care*, 2nd edition.
PEP indicates Brady's *Prehospital Emergency Pharmacology*, 6th edition.
TAC indicates Brady's *Taigman's Advanced Cardiology (In Plain English)*.
AHA indicates the 2005 American Heart Association Guidelines (*Circulation*, December 2005
  Supplement), and is followed by the PART number of the section to reference.

		PAGE KEY	SUBJECT
1.	(b)	V3: 74; EPC: 188	epicardium
2.	(d)	V3: 74; EPC: 188	myocardium
3.	(a)	V3: 74; EPC: 188	endocardium
4.	(e)	V3: 74; EPC: 188	myocardial muscle
5.	(d)	V3: 76–77; EPC: 191	normal flow of blood through the heart
6.	(d)	V3: 77; EPC: 191–192	coronary arteries
7.	(a)	V3: 16; EPC: 204–205	capillary gas, fluid, nutrient exchange
8.	(b)	V3: 79; EPC: 198	veins
9.	(b)	V3: 79; EPC: 193, 609	diastole
10.	(c)	V3: 79; EPC: 193, 609	systole
11.	(d)	V3: 79; EPC: 193, 609	systole
12.	(b)	V3: 79; EPC: 193, 609	diastole
13.	(a)	V3: 79; EPC: 193, 609	systole
14.	(b)	V3: 79; EPC: 193, 609	diastole
15.	(e)	V1: 207; V3: 79–80; EPC: 201, 1128	stroke volume
16.	(d)	V1: 207; V3: 80; EPC: 201, 1128	preload
17.	(c)	V1: 207; V3: 80; EPC: 202, 1128	afterload
18.	(c)	V3: 80–82; EPC: 193–194	cardiac nervous system control
19.	(e)	V1: 320–323; V3: 81–82; EPC: 340–342	alpha cardiac effects
20.	(d)	V1: 320–323; V3: 81–82; EPC: 340–342	beta cardiac effects
21.	(b)	V3: 82; EPC: 194, 1129	chronotropy and chronotropic

		PAGE KEY	SUBJECT
22.	(d)	V3: 82; EPC: 194, 1129	dromotropy and dromotropic
23.	(c)	V3: 82; EPC: 194, 1129	inotropy and inotropic
24.	(a)	V3: 82; EPC: 195	sodium
25.	(c)	V3: 82; EPC: 195	potassium
26.	(d)	V3: 82; EPC: 195	calcium
27.	(c)	V3: 84–85; EPC: 197, 1130	automaticity
28.	(a)	V3: 84–85; EPC: 197, 1130	excitability
29.	(e)	V3: 84–85; EPC: 197, 1130	conductivity
30.	(c)	V3: 84–85; EPC: 197	normal cardiac electrical conduction route
31.	(b)	V3: 84–85; EPC: 197	normal Purkinje fiber discharge rate
32.	(c)	V3: 84–85; EPC: 197	normal AV node discharge rate
33.	(d)	V3: 84–85; EPC: 197	normal SA node discharge rate
34.	(b)	V3: 88–89; EPC: 1133–1140	ECG time measurement
35.	(e)	V3: 88–89; EPC: 1133–1140	ECG time measurement
36.	(d)	V3: 90–95; EPC: 1330–1140	ECG voltage measurement
37.	(a)	V3: 90–95; EPC: 1133–1140	P wave, atrial depolarization
38.	(e)	V3: 90–95; EPC: 1133–1140	atrial repolarization
39.	(c)	V3: 90–95; EPC: 1133–1140	QRS, ventricular depolarization
40.	(b)	V3: 90–95; EPC: 1133–1144	T wave, ventricular repolarization
41.	(d)	V3: 90–95; EPC: 1133–1144	P-R interval
42.	(e)	V3: 90–95; EPC: 1133–1144	absolute refractory period
43.	(b)	V3: 90–95; EPC: 1133–1144	absolute refractory period of the T wave
44.	(e)	V3: 90–95; EPC: 1133–144	relative refractory period
45.	(c)	V3: 90–95; EPC: 1133–1144	relative refractory period of the T wave
46.	(d)	V3: 90–95; EPC: 1133–1144	normal P-R interval duration
47.	(b)	V3: 90–95; EPC: 1133–1144	normal QRS complex duration
48.	(b)	V3: 116–120; EPC: 1160–1164; TAC: 13–22	second-degree AV block/ I
49.	(c)	V3: 116–120; EPC: 1160–1164; TAC: 13–22	second-degree AV block/ II
50.	(d)	V3: 113–115; EPC: 1158–1159	uncontrolled A-fib with a BBB
51.	(c)	V3: 136–138; EPC: 1179, 1181–1182; TAC: 171–172	AV sequential (dual-chambered) pacemaker
52.	(a)	V3: 122–124; EPC: 1167	junctional rhythm

		PAGE KEY	SUBJECT
53.	(e)	V3: 98, 108, 129; EPC: 1144–1159	sinus rhythm with 1 PAC and 1 PVC
54.	(e)	V3: 98, 108, 129; EPC: 1144–1159	sinus rhythm with 1 PAC and 1 PVC
55.	(a)	V3: 112–113; EPC: 1144–1159	A-flutter with 4:1 & 3:1 ventricular response
56.	(c)	V3: 98, 122; EPC: 1144–1159	sinus rhythm with 1 PJC
57.	(a)	V3: 100–101, 215; EPC: 1144–1159	sinus tachycardia, QRS within normal limits, elevated S-T segment
58.	(a)	V3: 100–101, 227–228; EPC: 1144–1159	sinus tachycardia with a BBB, inverted T wave, no ectopy
59.	(a)	V3: 98, 215; EPC: 1144–1159	sinus rhythm, QRS WNL, S-T elevation indicative of AMI, no ectopy.
60.	(e)	AHA: 8; V3: 180–181; EPC: 1216, 1218	chest pain treatment
61.	(e)	AHA: 7.3	There is no mentation change, no "serious" signs or symptoms. Apart from oxygen, pharmacological treatment is not indicated.
62.	(a)	AHA: 7.3; V3: 164; EPC: 1204	Synchronized cardioversion is indicated. Because she's awake, she can be sedated.
63.	(a)	AHA: 7.3	Inverted T-waves and trigeminal PVCs, with c/o nausea and weakness = "silent" AMI. ASA is indicated; atropine is not, unless hypotensive.
64.	(d)	V3: 141, 142; EPC: 1185	pre-excitation syndromes
65.	(e)	V3: 141; EPC: 1185	Wolf-Parkinson-White syndrome
66.	(c)	EPC: 1185; TAC: 151–166:	Lown-Ganong-Levine syndrome
67.	(e)	V3: 142, 507; EPC: 1185; TAC: 200	J wave or Osborn wave
68.	(d)	V3: 142, 399; EPC: 1185; TAC: 27, 193	hyperkalemia and T waves
69.	(a)	V3: 142; EPC: 1185	hypokalemia and U waves
70.	(c)	V3: 139; EPC: 1184; TAC: 121	aberrant conduction
71.	(d)	V3: 141; EPC: 1184; TAC: 173	fusion beats
72.	(e)	V3: 141; EPC: 1184; TAC: 142	intermittent right bundle branch block
73.	(a)	V3: 111; EPC: 1156; TAC: 121–150	SVT vs. VT differentiation
74.	(c)	AHA: 7.3; V3: 111; EPC: 1156	SVT vs. VT treatment
75.	(b)	V3: 177–179; EPC: 1211–1212	angina
76.	(a)	V3: 177–179; EPC: 1211–1212	angina

		PAGE KEY	SUBJECT
77.	(c)	V3: 177–179; EPC: 1211–1212	angina
78.	(c)	V3: 177–179; EPC: 1214–1216	myocardial infarction
79.	(e)	V3: 177–179; EPC: 1214–1216	most common cause of AMI
80.	(a)	V3: 177–179; EPC: 1214–1216	most common complication of AMI
81.	(a)	V3: 177–179; EPC: 1214–1216	most common cause of death from AMI
82.	(d)	V3: 177–179; EPC: 1214–1216	most common site of AMI
83.	(a)	V3: 177–179; EPC: 1214–1216	subendocardial infarction
84.	(e)	V3: 177–179; EPC: 1214–1216	transmural infarction
85.	(b)	V3: 177–179; EPC: 1220–1222	left ventricular failure
86.	(c)	V3: 177–179; EPC: 1220–1222	bronchoconstriction and pulmonary edema
87.	(d)	V3: 183–188; EPC: 1220–1222	jugular vein distention
88.	(a)	V3: 183–188; EPC: 1220–1222	right ventricular failure
89.	(d)	V3: 183–188; EPC: 1220–1222	right ventricular failure
90.	(c)	V3: 183–188; EPC: 1220–1222	congestive heart failure management
91.	(e)	V3: 191–194; EPC: 1226–1229	cardiogenic shock
92.	(e)	V3: 191–194; EPC: 1226–1229	cardiogenic shock
93.	(a)	V3: 191–194; EPC: 1226–1229	cardiogenic shock management
94.	(c)	V3: 191–194; EPC: 1229	sudden death
95.	(d)	AHA: 5, 7.2; V3: 194–196; EPC: 1229–1230	cardiopulmonary arrest management
96.	(a)	AHA: 7.2; V3: 139–140; EPC: 1182–1183	pulseless electrical activity (PEA)
97.	(b)	V3: 139–140; EPC: 1182–1183; TAC: 27	PEA causes, hyperkalemia
98.	(e)	AHA: 7.4; V3: 139–140; EPC: 1182–1183	PEA management, sodium bicarb
99.	(d)	V3: 199–200; EPC: 1234, 1691	abdominal aneurysm
100.	(a)	V3: 199–200; EPC: 1234, 1691	thoracic aneurysm
101.	(b)	V3: 199–200; EPC: 1234, 1691	dissecting abdominal aneurysm
102.	(b)	V3:199–200; EPC: 1234, 1691	causes of aneurysm
103.	(d)	V3: 199–200; EPC: 1234, 1691	dissecting aortic aneurysm
104.	(b)	V3: 190–191; EPC: 1225–1226	hypertensive emergency
105.	(e)	V3: 190–191; EPC: 1225–1226	hypertensive emergency
106.	(b)	AHA: 5; V3: 162–166; EPC: 1199–1201	defibrillation
107.	(e)	AHA: 5; V3: 162–166; EPC: 1199–1201	defibrillation
108.	(c)	AHA: 5	monophasic & biphasic defibrillators

		PAGE KEY	SUBJECT
109.	(a)	AHA: 5, 7.2	monophasic defibrllation energy levels
110.	(c)	AHA: 5, 7.2	biphasic defibrillation energy levels
111.	(b)	AHA: 1, 5, 7.2	monophasic defibrillation energy levels
112.	(d)	AHA: 1, 5, 7.2	biphasic defibrillation energy levels
113.	(a)	AHA: 5, 7.2; V3: 162–163; EPC: 1199–1204	defibrillation
114.	(c)	AHA: 5, 7.3; V3: 164, 167, 168; EPC: 1204–1205	synchronized cardioversion
115.	(d)	AHA: 5, 7.3; EPC: 1204–1205	synchronized cardioversion contraindications
116.	(b)	AHA: 5, 7.3; V3: 140; EPC: 1156	unstable tachycardia patients
117.	(a)	AHA: 5, 7.3; V3: 140; EPC: 1156	stable tachycardia Tx levels
118.	(d)	AHA: 5, 7.3; V3: 164, 167; EPC: 1204–1205	synchronized cardioversion indications
119.	(e)	AHA: 5, 7.3; V3: 161, 164; EPC: 1205; PEP: 471, 288–292	synchronized cardioversion sedation
120.	(c)	AHA: 5, 7.3; V3: 164, 168; EPC: 1205	synchronized cardioversion PSVT, atrial flutter
121.	(a)	AHA: 5, 7.3; V3: 164, 168; EPC: 1205	synchronized cardioversion atrial fibrillation
122.	(a)	AHA: 5, 7.3; V3: 164, 168; EPC: 1205	synchronized cardioversion ventricular tachycardia
123.	(c)	V3: 110, 127; EPC: 1198	Valsalva maneuvers
124.	(a)	V3: 171–173; EPC: 1206, 1211	carotid sinus massage
125.	(b)	V3: 171–173; EPC: 1206, 1211	carotid sinus massage
126.	(b)	AHA: 5, 7.3; V3: 167; EPC: 1206	transcutaneous cardiac pacing
127.	(b)	V3: 167; EPC: 1206	transcutaneous cardiac pacing
128.	(a)	V3: 167; EPC: 1206	transcutaneous cardiac pacing energy levels
129.	(d)	V3: 167; EPC: 1206	transcutaneous cardiac pacing energy levels
130.	(e)	AHA: 5, 7.3; V3: 167; EPC: 1206	transcutaneous cardiac pacing

NAME-THAT-STRIP SECTION

[Test Section 7's Figure 7–14 = a Sample ECG]

FIGURE	DESCRIPTION
7–15	Runaway pacemaker with a capture rate of 100/min and accelerating.
7–16	Accelerated junctional rhythm, rate of 80/min, QRS WNL, with elevated S-T segments, no ectopy.
7–17	Complete (third-degree) AV block: junctional escape rate of 60/min, QRS WNL, elevated S-T segments, and an atrial rate of 110/min, no ectopy.
7–18	Sinus tachycardia (probable underlying rate of 120), QRS WNL, inverted T waves and two multifocal R-on-T PVCs; that degenerates into ventricular tachycardia. (Although similar in appearance at the start of the V-tach, this is not Ashman's phenomenon, because the underlying rhythm is not A-fib. See TAC, page 142.)
7–19	Mobitz I second-degree AV block (Wenckebach), at a rate of 60/min, QRS WNL, inverted T waves, in a 4:3 A:V ratio, no ectopy.
7–20	Sinus rhythm at a rate of 80/min, BBB, elevated S-T segments, and muscle-tremor ECG interference, no ectopy.
7–21	Pacemaker rhythm with effective capture at a rate of 72–80/min (underlying rhythm is atrial fibrillation), no ectopy.
7–22	Sinus dysrhythmia at a rate of 70/min, QRS WNL, inverted T waves, no ectopy.
7–23	Sinus bradycardia at a rate of 50/min, QRS WNL, elevated S-T segments, no ectopy.
7–24	Sinus rhythm with a BBB, elevated S-T segments, 2 unifocal PACs, and an underlying rate of 80/min.
7–25	Sinus rhythm with an underlying rate of 80/min, BBB, elevated S-T segments, 2 unifocal PVCs.
7–26	Sinus rhythm with an underlying rate of 80/min, inverted T waves, trigeminal unifocal PACs.
7–27	Controlled A-fib at a rate of 90/min, QRS WNL, elevated S-T segments, no ectopy.
7–28	SVT (likely an atrial tachycardia with 2:1 A:V ratio) at a rate of 180/min, QRS WNL, inverted T waves, no ectopy.
7–29	Sinus tachycardia with an underlying rate of 110/min, inverted T waves, 4 unifocal PACs.
7–30	Sinus rhythm with an underlying rate of 80/min, BBB, elevated S-T segments, 2 multifocal PVCs.
7–31	Sinus rhythm with a prolonged PRI (0.24), or first-degree AV block, with an underlying rate of 80/min, BBB, elevated S-T segments, and 1 PAC.

FIGURE	DESCRIPTION
7–32	Sinus bradycardia with a prolonged PRI (0.30), or first-degree AV block, at a rate of 60/min, QRS WNL, no ectopy.
7–33	Sinus bradycardia at a rate of 40/min, QRS WNL, inverted T waves, with 1 interpolated ventricular complex. (Because of the bradycardia, this can be considered an escape ventricular complex, rather than a PVC.)
7–34	Atrial flutter with a ventricular rate of 80/min, 4:1 A:V ratio, QRS WNL, elevated S-T segments, no ectopy.
7–35	Sinus tachycardia with an underlying rate of 110/min, QRS WNL, elevated S-T segments, bigeminal PACs that appear to have a BBB (likely due to premature depolarization of incompletely repolarized ventricles).
7–36	Sinus rhythm with an underlying rate of 80/min, QRS WNL, S-T WNL, 1 PVC and 1 PAC.
7–37	Sinus rhythm with an underlying rate of 80/min, BBB, elevated S-T segments, 1 PAC (QRS same BBB morphology as regular rhythm) and 2 unifocal PVCs.
7–38	Sinus rhythm with an underlying rate of 90/min, QRS WNL, bigeminal unifocal PJCs (QRS WNL).
7–39	Sinus rhythm with an underlying rate of 80/min, QRS WNL, elevated S-T segments, 1 couplet of unifocal PVCs.
7–40	Sinus rhythm with an underlying rate of 80/min, QRS WNL, 1 PJC (QRS WNL).
7–41	Sinus rhythm at a rate of 80/min and QRS WNL, interrupted by a premature atrial beat that initiates a regular atrial rhythm at a rate of 90/min, with a BBB, no other ectopy.
7–42	Atrial flutter with a variable ventricular response (4:1, 2:1, 3:1), QRS WNL, at an irregular rate ranging from 70/min to 120/min, no ectopy.
7–43	Sinus bradycardia with an underlying rate of 40/min, QRS WNL, inverted T waves, with 2 unifocal interpolated ventricular escape complexes.
7–44	Type II 2:1 AV Block (or Type II Second Degree Block), at a rate of 60/min, prolonged PRI (0.28) or first-degree block, BBB, elevated S-T segments, no ectopy. (Reference TAC pages 20 and 27, for Taigman's T wave Rule 1: "Teenagers have pimples. T waves do not." Those are nonconducted P waves on top of each T wave! Also reference TAC page 19 for Type II 2:1 AV Block discussion.)
7–45	Sinus rhythm with an underlying rate of 90/min, a prolonged PRI (0.24) or first-degree AV block, QRS WNL, inverted T waves, 4 unifocal PACs.
7–46	Atrial flutter with a 3:1 A:V response, at a rate of 100/min, QRS WNL, no ectopy.
7–47	Sinus rhythm with an underlying rate of 90/min, a prolonged PRI (0.26), or first-degree AV block, QRS WNL, quadrigeminal unifocal PVCs.
7–48	Uncontrolled atrial fibrillation at a rate of 110–120/min, QRS grossly WNL (borderline), elevated S-T segments, no ectopy.
7–49	Atrial fibrillation at a bradycardic ventricular response rate of 40–50/min, BBB, elevated S-T segments, no ectopy.

FIGURE	DESCRIPTION
7–50	Sinus rhythm with an underlying rate of 80/min, BBB, elevated S-T segments, bigeminal unifocal PVCs.
7–51	Sinus rhythm with an underlying rate of 70–80/min, QRS WNL, and a 3-beat "salvo" of unifocal V-tach (a triplet of unifocal PVCs).
7–52	Sinus rhythm with an underlying rate of 80/min, QRS WNL, inverted T waves, trigeminal unifocal PVCs.
7–53	Sinus rhythm with inverted T waves, 3 multifocal PVCs, and an underlying rate of 80/min.
7–54	Sinus tachycardia with an underlying rate of 100/min, QRS WNL, 1 couplet and 2 single multifocal PVCs.
7–55	Demand pacemaker rhythm with effective capture (pacer rate likely set at 72/min), that is overtaken by the patient's sinus dysrhythmia, with a slightly irregular underlying rate of 80/min, prolonged PRI (0.22), or first-degree AV block, underlying QRS WNL, inverted T waves, and muscle-tremor ECG interference.
7–56	Sinus rhythm with an underlying rate of 80/min, QRS WNL, inverted T waves, 2 unifocal bigeminal PACs.
7–57	Mobitz I second-degree AV block (Wenckebach) in a 4:3 A:V ratio, with a ventricular rate of 60/min, QRS WNL, elevated S-T segments, no ectopy.
7–58	Complete (third-degree) AV block with a junctional escape at a rate of 60/min, without ectopy.
7–59	Idioventricular escape rhythm at a rate of 40/min, sinus arrest.
7–60	Pacemaker rhythm (pacer rate likely set at 80/min), with periodically ineffective capture; ventricular capture rate of 30–40/min.
7–61	Junctional bradycardia at an underlying rate of 30/min, QRS WNL, with interpolated bigeminal ventricular escape beats; PVCs would come earlier in the cycle; if all complexes generate a pulse, the patient will have a heart rate of 60/min.
7–62	Unifocal ventricular tachycardia at a rate of 180–190/min.
7–63	Junctional rhythm at a rate of 50/min, inverted T waves, no ectopy.
7–64	Mobitz II second-degree (classical) AV block in a 2:1 ratio with a ventricular rate of 50/min, without ectopy.

TEST SECTION 8 ANSWER KEY

Page Key texts:

Brady's *Paramedic Care: Principles & Practice*, AHA updated 2nd edition, 5-volume series of texts:

> V1 indicates Volume 1: Introduction to Advanced Prehospital Care
> V3 indicates Volume 3: Medical Emergencies
> V4 indicates Volume 4: Trauma Emergencies

EPC indicates Brady's *Essentials of Paramedic Care*, 2nd edition.
PEP indicates Brady's *Prehospital Emergency Pharmacology*, 6th edition.
AHA indicates the 2005 American Heart Association Guidelines (*Circulation*, December 2005 Supplement), and is followed by the PART number of the section to reference.
MedD indicates that you should reference your Medical Dictionary.

		PAGE KEY	SUBJECT
1.	(d)	V1: 517–523; V3: 20; EPC: 213–217	hypoxia
2.	(b)	V1: 517–523; V3: 15; EPC: 213–217	CO_2 production
3.	(b)	V1: 517–523; V3: 15; EPC: 213–217	CO_2 elimination
4.	(d)	V1: 517–523; V3: 15; EPC: 213–217	CO_2 elimination
5.	(a)	V1: 517–523; V3: 15; EPC: 213–217	respiratory chemoreceptors
6.	(e)	V1: 517–523; V3: 15; EPC: 213–217	respiratory chemoreceptors
7.	(c)	V1: 517–523; V3: 15; EPC: 213–217	respiratory chemoreceptors
8.	(d)	V1: 517–523; V3: 20; EPC: 213–217	hypoxic drive
9.	(c)	V1: 528; EPC: 460	sighing
10.	(d)	V1: 528; EPC: 460	hiccuping
11.	(a)	V1: 528; EPC: 460	coughing
12.	(e)	V1: 528; EPC: 460	grunting
13.	(d)	V1: 528; EPC: 460	hiccupping and inferior AMI
14.	(a)	V3: 24; EPC: 1089	orthopnea
15.	(d)	V3: 24, 186; EPC: 1222	paroxysmal nocturnal dyspnea
16.	(c)	V3: 26; EPC: 1090	chest palpation
17.	(b)	V3: 10; EPC: 158	visceral pleura
18.	(c)	V3: 10; EPC: 158	parietal pleura

		PAGE KEY	SUBJECT
19.	(a)	V3: 28, 29; EPC: 1092	friction rub
20.	(d)	V3: 28, 29; EPC: 1092	rhonchi
21.	(b)	V3: 28, 29; EPC: 1092	stridor
22.	(e)	V3: 28, 29; EPC: 1092	rales
23.	(c)	AHA: 4; EPC: 1098–1099	obstructed airway management
24.	(d)	AHA: 4; EPC: 1098–1099	obstructed airway management
25.	(a)	AHA: 4; EPC: 1098–1099	obstructed airway management
26.	(e)	AHA: 4; EPC: 1098–1099	obstructed airway management
27.	(c)	AHA: 4; EPC: 1098–1099	obstructed airway management
28.	(c)	AHA: 4; EPC: 1098–1099	obstructed airway management
29.	(d)	AHA: 4; EPC: 1098–1099	obstructed airway management
30.	(b)	V3: 39–46; EPC: 1101–1108	chronic bronchitis
31.	(d)	V3: 39–46; EPC: 1101–1108	chronic bronchitis and asthma
32.	(a)	V3: 39–46; EPC: 1101–1108	pursed-lip breathing
33.	(e)	V3: 39–46; EPC: 1101–1108	COPD and wheezing
34.	(e)	V3: 50–51; EPC: 1109–1110	pneumonia
35.	(d)	V3: 50–51; EPC: 1109–1110	pneumonia
36.	(d)	V3: 50–51; EPC: 1109–1110	treatment of pneumonia
37.	(a)	V3: 55; EPC: 1114	carbon monoxide inhalation
38.	(d)	V3: 55; EPC: 1114	carbon monoxide inhalation
39.	(d)	V3: 55; EPC: 1114	carbon monoxide inhalation
40.	(e)	V3: 56–57; EPC: 1114–1116	pulmonary embolism
41.	(b)	V3: 56–57; EPC: 1114–1116	pulmonary embolism
42.	(e)	V3: 56–57; EPC: 1114–1116	pulmonary embolism
43.	(a)	V3: 56–57; EPC: 1114–1116	pulmonary embolism management
44.	(a)	V3: 58–59; EPC: 1116–1117	hyperventilation syndrome
45.	(d)	V3: 58–59; EPC: 1116–1117	hyperventilation syndrome
46.	(a)	V3: 58–59; EPC: 1116–1117	hyperventilation syndrome management
47.	(e)	V3: 249–251; EPC: 160–162	neuron
48.	(b)	V3: 249–251; EPC: 160–162	dendrite
49.	(c)	V3: 249–251; EPC: 160–162	axon
50.	(c)	V3: 249–251; EPC: 160–162	nerve impulse transportation
51.	(c)	V3: 253–256; EPC: 162–165	brain stem
52.	(c)	V3: 253–256; EPC: 162–165	cerebrum

		PAGE KEY	SUBJECT
53.	(a)	V3: 253–256; V4: 277; EPC: 162–165	cerebellum
54.	(c)	V3: 253–256; EPC: 162–165	cerebrum
55.	(a)	V3: 253–256; EPC: 162–165	cerebellum
56.	(b)	V3: 253–256; EPC: 162–165	medulla oblongata
57.	(a)	V3: 253–256; EPC: 162–165	speech center, temporal cerebrum
58.	(c)	V3: 253–256; EPC: 162–165	personality center, frontal cerebrum
59.	(a)	V3: 257, 258; EPC: 165, 167	spinal cord
60.	(b)	V3: 259–260; EPC: 172	12 cranial nerves
61.	(b)	V3: 256; EPC: 165	Circle of Willis
62.	(c)	V3: 261–262; EPC: 1245	altered LOC, structural causes
63.	(b)	V3: 261–262; EPC: 1246	altered LOC, toxic-metabolic causes
64.	(a)	V3: 264; EPC: 1248	mental status assessment AVPU
65.	(e)	V3: 264; EPC: 1248	cerebral function assessment
66.	(b)	V3: 273; EPC: 1256	AEIOU-TIPS
67.	(b)	V3: 273; EPC: 1256	AEIOU-TIPS
68.	(a)	V3: 274; EPC: 1256–1257	altered LOC management
69.	(b)	V3: 275–277; EPC: 1257–1262	cerebrovascular accident (CVA)
70.	(c)	V3: 275–277; EPC: 1257–1262	CVA predisposing factors
71.	(c)	AHA: 9; V3: 278; EPC: 1259	hemiplegia
72.	(b)	AHA: 9; V3: 278; EPC: 1259	paresthesia
73.	(e)	AHA: 9; V3: 278; EPC: 1259	hemiparesis
74.	(e)	AHA: 9; V3: 278; EPC: 1259	dysphasia
75.	(b)	AHA: 9; V3: 278; EPC: 1259	aphasia
76.	(a)	AHA: 9; V3: 278; EPC: 1259	dysarthria
77.	(c)	V3: 278; EPC: 1260	transient ischemic attack (TIA)
78.	(a)	V3: 278; EPC: 1260–1261	differentiating CVA and TIA
79.	(e)	V3: 279–280; EPC: 1261–1262	CVA/TIA management
80.	(d)	AHA: 9; V3: 279–280; EPC: 1261–1262	CVA/TIA management
81.	(e)	V3: 281–294; EPC: 1264–1266	seizures
82.	(c)	V3: 281–294; EPC: 1264–1266	grand mal seizures
83.	(e)	V3: 281–294; EPC: 1264–1266	petit mal seizures
84.	(b)	V3: 281–294; EPC: 1264–1266	psychomotor seizures

		PAGE KEY	SUBJECT
85.	(a)	V3: 281–294; EPC: 1264–1266	Jacksonian or focal motor seizures
86.	(a)	V3: 281–294; EPC: 1264–1266	Jacksonian or focal motor seizures
87.	(b)	V3: 281–294; EPC: 1264–1266	psychomotor seizures
88.	(a)	V3: 281–294; EPC: 1264–1266	Jacksonian or focal motor seizures
89.	(d)	MedD	gustatory aura
90.	(e)	MedD	tactile aura
91.	(c)	V3: 281; EPC: 1264; MedD	olfactory aura
92.	(b)	V3: 281; EPC: 1264	clonic phase of seizure
93.	(c)	V3: 281; EPC: 1264	tonic phase of seizure
94.	(e)	V3: 281; EPC: 1264	seizure event progression
95.	(c)	V3: 284; EPC: 1266	anticonvulsant medication
96.	(e)	V3: 283; EPC: 1266; PEP: 452–453	Dilantin®
97.	(b)	V3: 283; EPC: 1266	Tegretol®
98.	(e)	V3: 284–285; EPC: 1266	seizure management
99.	(c)	V3: 284–285; EPC: 1267	status epilepticus
100.	(e)	V3: 284–285; EPC: 1267	status epilepticus
101.	(b)	V3: 284–285; EPC: 1267	status epilepticus
102.	(d)	V3: 284–285; EPC: 1267	status epilepticus management
103.	(d)	V3: 285–286; EPC: 1267–1268	syncope
104.	(d)	V3: 335–338; EPC: 1296–1300	allergic reaction pathophysiology
105.	(e)	AHA: 10.6; V3: 338–344; EPC: 1296–1300	allergic reaction fatalities
106.	(d)	AHA: 10.6; V3: 338–344; EPC: 1296–1300	allergic reaction fatalities
107.	(e)	AHA: 10.6; V3: 334, 337; EPC: 1296–1300	*Hymenoptera* insect order
108.	(c)	V3: 337–339; EPC: 1296–1300	histamine effects
109.	(a)	V3: 337; EPC: 1296–1300	angioneurotic edema
110.	(b)	V3: 340; EPC: 1296–1300	urticaria
111.	(b)	V3: 340; EPC: 1296–1300	urticaria, hives
112.	(c)	AHA: 10.6; V3: 338–3403; EPC: 1296–1300	anaphylaxis signs and symptoms
113.	(e)	AHA: 10.6; V3: 338–349; EPC: 1296–1300	anaphylaxis signs and symptoms

		PAGE KEY	SUBJECT
114.	(d)	AHA: 10.6; V3: 338–340; EPC: 1296–1300	airway management of anaphylaxis
115.	(d)	AHA: 10.6; V3: 341–343; EPC: 1301–1303	IV fluids and anaphylaxis
116.	(b)	AHA: 10.6; V3: 341–343; EPC: 1301–1303	anaphylaxis management
117.	(a)	AHA: 10.6: V3: 334: EPC: 1301–1303	epinephrine 1:1000
118.	(c)	AHA: 10.6; V3: 334; EPC: 1301–1303	epinephrine 1:10000
119.	(d)	AHA: 7.2, 10.6, 28	epinephrine 1:10,000
120.	(d)	AHA: 7.4	epinephrine drip mix

TEST SECTION 9 ANSWER KEY

Page Key texts:

Brady's *Paramedic Care: Principles & Practice*, AHA updated 2nd edition, 5-volume series of texts:

    V2 indicates Volume 2: Patient Assessment
    V3 indicates Volume 3: Medical Emergencies
    V4 indicates Volume 4: Trauma Emergencies
    V5 indicates Volume 5: Special Considerations/Operations

EPC indicates Brady's *Essentials of Paramedic Care*, 2nd edition.

PEP indicates Brady's *Prehospital Emergency Pharmacology*, 6th edition.

AHA indicates the 2005 American Heart Association Guidelines (*Circulation*, December 2005 Supplement), and is followed by the PART number of the section to reference.

		PAGE KEY	SUBJECT
1.	(b)	V3: 303; EPC: 1281	exocrine glands
2.	(e)	V3: 303; EPC: 1281	endocrine glands
3.	(b)	V3: 303; EPC: 1281	exocrine glands
4.	(d)	V3: 310–311; EPC: 177–188	pancreas
5.	(c)	V3: 307–308; EPC: 177–188	pituitary gland
6.	(b)	V3: 312; EPC: 177–188, 1292–1293	adrenal glands
7.	(e)	V3: 309–310; EPC: 177–188	parathyroid glands
8.	(d)	V3: 309; EPC: 177–188, 1289	thyroid gland
9.	(b)	V3: 312–313; EPC: 177–188	estrogen/ovaries
10.	(b)	V3: 312–313; EPC: 177–188	progesterone/ovaries
11.	(c)	V3: 312–313; EPC: 177–188	testosterone/testes
12.	(b)	V3: 327; EPC: 1291–1292	Cushing's syndrome
13.	(c)	V3: 308; EPC: 177–188	antidiuretic hormone/pituitary gland
14.	(a)	V3: 312–313; EPC: 177–188	follicle-stimulating hormone/ovaries, testes
15.	(b)	V3: 308–309; EPC: 177–188	follicle-stimulating hormone/pituitary gland
16.	(c)	V3: 308; EPC: 177–188	oxytocin/pituitary gland
17.	(a)	V3: 312–313; EPC: 177–188	luteinizing hormone/ovaries, testes
18.	(d)	V3: 312; EPC: 177–188	adrenal glands
19.	(a)	V3: 307–308; EPC: 177–188	pituitary gland
20.	(c)	V3: 309; EPC: 177–188, 1289	thyroid gland
21.	(a)	V3: 310–311; EPC: 177–188	alpha cells, glucagon
22.	(d)	V3: 310–311; EPC: 177–188	beta cells, insulin

		PAGE KEY	SUBJECT
23.	(b)	V3: 310–311, 314–316; EPC: 185–186, 1282–1284	glycogen
24.	(c)	V3: 310–311, 314–316; EPC: 185–186, 1282–1284	glucose
25.	(a)	V3: 310–311, 314–316; EPC: 185–186, 1282–1284	glucagon
26.	(d)	V3: 310–311, 314–316; EPC: 185–186, 1282–1284	insulin
27.	(a)	V3: 310–311, 314–316; EPC: 185–186, 1282–1284	glycogenolysis
28.	(d)	V3: 310–311, 314–316; EPC: 185–186, 1282–1284	gluconeogenesis
29.	(d)	V3: 317–320, 323–324; EPC: 1282–1288	diabetes mellitus
30.	(c)	V3: 317–320, 323–324; EPC: 1282–1288	insulin absence
31.	(e)	V3: 317–320, 323–324; EPC: 1282–1288	ketone bodies
32.	(e)	V3: 317–320, 323–324; EPC: 1282–1288	Type I Diabetes Mellitus
33.	(e)	V3: 317–320, 323–324; 1282–1288	Type II Diabetes Mellitus
34.	(e)	V3: 317–320, 323–324; EPC: 1282–1288	hyperglycemic hyperosmolar nonketotic acidosis
35.	(e)	V3: 317–320, 323–324; EPC: 1282–1288	diabetic S/Sx: highest blood sugar levels
36.	(a)	V3: 317–320, 323–324; EPC: 1282–1288	altered LOC or unconscious patients
37.	(c)	V3: 317–320, 323–324; EPC: 1282–1288	diabetic S/Sx development time periods
38.	(e)	V3: 317–320, 323–324; EPC: 1282–1288	diabetic S/Sx development time periods
39.	(d)	V3: 317–320, 323–324; EPC: 1282–1288	diabetic S/Sx development time periods
40.	(b)	V3: 317–320, 323–324; EPC: 1282–1288	polydipsia, polyuria, polyphagia
41.	(e)	V3: 317–320, 323–324; EPC: 1282–1288	polydipsia, polyuria, polyphagia
42.	(c)	V3: 317–320, 323–324; EPC: 1282–1288	diabetic S/Sx: abdominal pain

		PAGE KEY	SUBJECT
43.	(d)	V3: 317–320, 323–324; EPC: 1282–1288	diabetic S/Sx: headache
44.	(a)	V3: 317–320, 323–324; EPC: 1282–1288	diabetic S/Sx: tachycardia
45.	(d)	V3: 317–320, 323–324; EPC: 1282–1288	diabetic S/Sx: cool, diaphoretic skin
46.	(b)	V3: 317–320, 323–324; EPC: 1282–1288	diabetic S/Sx: warm, dry skin
47.	(c)	V3: 317–320, 323–324; EPC: 1282–1288	diabetic S/Sx: Kussmaul's respirations
48.	(c)	V3: 317–320, 323–324; EPC: 1282–1288	diabetic S/Sx: fruity, acetone breath odor
49.	(d)	V3: 317–320, 323–324; EPC:1282–1288	diabetic S/Sx: seizure
50.	(d)	V3: 317–320, 323–324; EPC: 1282–1288	diabetic S/Sx: vomiting
51.	(a)	V3: 320; EPC: 1288	diabetic emergency management
52.	(d)	V3: 357; EPC: 1311	Mallory-Weiss syndrome
53.	(b)	V3: 358–359; EPC: 1313	esophageal varices
54.	(a)	V3: 358–359; EPC: 1313–1314	esophageal varices S/Sx
55.	(e)	V3: 358–359; EPC: 1313–1314	esophageal varices management
56.	(e)	V3: 356–358; EPC: 1311	upper GI bleed S/Sx
57.	(a)	V3: 357; EPC: 1312	hematemesis
58.	(e)	V3: 359–360; EPC: 1314	acute gastroenteritis
59.	(b)	V3: 359–360; EPC: 1314	acute gastroenteritis S/Sx
60.	(e)	V3: 364, 366; EPC: 1317	idiopathic inflammatory bowel disorders
61.	(a)	V3: 367–369; EPC: 1319	diverticulitis
62.	(c)	V3: 370; EPC: 1320–1321	bowel obstruction, infarction
63.	(a)	V3: 370; EPC: 1320–1321	bowel obstruction, hernia
64.	(e)	V3: 370; EPC: 1320–1321	bowel obstruction S/Sx
65.	(d)	V3: 409; EPC: 1346	pyelonephritis
66.	(b)	V3: 373–375; EPC: 1323	cholecystitis
67.	(c)	V3: 371–373; EPC: 1322	appendicitis S/Sx
68.	(e)	V3: 497; EPC: 1345	renal calculi S/Sx
69.	(d)	V3: 410; EPC: 1347	pyelonephritis S/Sx
70.	(b)	V3: 373–375; EPC: 1323	cholecystitis S/Sx

		PAGE KEY	SUBJECT
71.	(c)	V3: 353; EPC: 1333	posture suggestive of peritonitis
72.	(e)	V3: 407; EPC: 1345	posture suggestive of renal calculi
73.	(d)	V3: 409–410; EPC: 1347	posture suggestive of pyelonephritis
74.	(a)	V2: 99; V4: 443; EPC: 1067	rebound tenderness
75.	(e)	V3: 385–387; EPC: 224–225	kidney functions
76.	(e)	V3: 396–397, 400–401; EPC: 1335–1343	renal failure etiologies
77.	(a)	V3: 401; EPC: 1335–1343	uremia
78.	(c)	V3: 396; EPC: 1335–1343	oliguria
79.	(b)	V2: 18; EPC: 1065	hematuria
80.	(a)	V3: 396; EPC: 1335–1343	anuria
81.	(d)	V2: 18; EPC: 1284	polyuria
82.	(c)	V3: 396–404; EPC: 1335–1343	ascites
83.	(a)	V3: 396–404; EPC: 1335–1343	renal failure S/Sx
84.	(c)	V3: 396–404; EPC: 1335–1343	chronic renal failure S/Sx
85.	(c)	V3: 396–404; EPC: 1335–1343	renal failure management
86.	(e)	V3: 396–404; EPC: 1335–1343	renal dialysis methods
87.	(e)	V5: 277; EPC: 1335–1343	dialysis patient assessment and management
88.	(a)	V3: 396–404; EPC: 1335–1343	complications of renal dialysis
89.	(a)	V3: 408–409; EPC: 1345–1347	urinary tract infection
90.	(b)	V3: 432; EPC: 1365	gasoline ingestion management
91.	(d)	PEP: 413–414	adult dose of activated charcoal
92.	(c)	PEP: 413–414	pediatric dose of activated charcoal
93.	(a)	V3: 421; EPC: 1354; PEP: 313	activated charcoal with sorbitol
94.	(e)	V3: 424–425; EPC: 1357–1358	toxic inhalation management
95.	(b)	V3: 424–425; EPC: 1357–1358	toxic inhalation management
96.	(d)	V3: 426–427; EPC: 1360	S/Sx of organophosphate poisoning
97.	(d)	AHA 10.2; V3: 422; EPC: 1355; PEP: 342–344	organophosphate poisoning management
98.	(b)	V3: 436; EPC: 1368	N-Acetylcysteine
99.	(a)	V3: 436; EPC: 1368	acetylsalicylic acid OD S/Sx
100.	(b)	AHA 10.2; V3: 432–433; EPC: 1365–1366	tricyclic antidepressant OD

		PAGE KEY	SUBJECT
101.	(c)	V3: 441–444; EPC: 1373–1376	treatment of bites and stings
102.	(b)	V3: 441–444; EPC: 1373–1376	brown recluse spider
103.	(a)	V3: 441–444; EPC: 1373–1376	black widow spider
104.	(b)	V3: 441–444; EPC: 1373–1376	brown recluse spider
105.	(c)	V3: 441–444; EPC: 1373–1376	scorpion sting
106.	(a)	V3: 441–444; EPC: 1373–1376	black widow spider bite
107.	(d)	V3: 446–447; EPC: 1377–1379	coral snake
108.	(d)	V3: 446–447; EPC: 1377–1379	coral snake
109.	(c)	V3: 445–556; EPC: 1377–1379	snake bite management
110.	(c)	V3: 471–473; EPC: 1390–1392	blood type B
111.	(a)	V3: 471–473; EPC: 1390–1392	blood type O
112.	(b)	V3: 471–473; EPC: 1390–1392	blood type AB
113.	(e)	V3: 471–473; EPC: 1390–1392	universal blood donor
114.	(d)	V3: 471–473; EPC: 1390–1392	universal blood recipient
115.	(d)	V3: 472; EPC: 1391	Rh negative blood
116.	(b)	V3: 472–473; EPC: 1390–1392	transfusion reaction S/Sx
117.	(a)	V3: 472–473; EPC: 1390–1392	transfusion reaction management
118.	(e)	V3: 473; EPC: 1392	circulatory overload S/Sx
119.	(c)	V3: 481; EPC: 1399	polycythemia
120.	(e)	V3: 480–481; EPC: 1398	sickle cell disease

TEST SECTION 10 ANSWER KEY

Page Key texts:

Brady's *Paramedic Care: Principles & Practice*, AHA updated 2nd edition, 5-volume series of texts:

    V3 indicates Volume 3: Medical Emergencies
    V5 indicates Volume 5: Special Considerations/Operations

EPC indicates Brady's *Essentials of Paramedic Care*, 2nd edition.
PEP indicates Brady's *Prehospital Emergency Pharmacology*, 6th edition.
AHA indicates the 2005 American Heart Association Guidelines (*Circulation*, December 2005 Supplement) and is followed by the PART number of the section to reference.

		PAGE KEY	SUBJECT
1.	(e)	V3: 450; EPC: 1382; PEP: 336–341	opiates/narcotics
2.	(a)	V3: 450; EPC: 1382; PEP: 338–340	synthetic narcotic OD management
3.	(c)	V3: 450; EPC: 1380–1381	cocaine OD management
4.	(e)	V3: 450–451; EPC: 1380–1381	seizures and OD management
5.	(a)	V3: 450–455; EPC: 1381–1385	alcohol abuse related conditions
6.	(a)	V3: 454; EPC: 1385	conditions that mimic alcohol intoxication
7.	(b)	V3: 454–455; EPC: 1385–1386	alcohol withdrawal syndrome
8.	(b)	V3: 454–455; EPC: 1385–1386	alcohol withdrawal S/Sx & management
9.	(a)	V3: 454–455; EPC: 1385–1386	alcohol withdrawal S/Sx
10.	(b)	V3: 274–275; EPC: 1257	Korsakoff's psychosis
11.	(d)	V3: 274–275; EPC: 1257	Wernicke's syndrome
12.	(d)	V3: 274–275; EPC: 1257, 1386	thiamine administration
13.	(e)	V3: 495; EPC: 1409	thermoregulatory control center
14.	(d)	V3: 495; EPC: 1408	heat loss from evaporation
15.	(b)	V3: 495; EPC: 1408	heat loss from conduction
16.	(c)	V3: 495; EPC: 1408	heat loss from convection
17.	(b)	V3: 497–503; EPC: 1411–1414	heat-related-illness S/Sx
18.	(d)	V3: 497–503; EPC: 1411–1414	heat-related-illness S/Sx
19.	(c)	V3: 497–503; EPC: 1411–1414	heat-related-illness S/Sx
20.	(a)	V3: 497–503; EPC: 1411–1414	heat stroke and hyperkalemia
21.	(e)	V3: 497–503; EPC: 1411–1414	heat-related-illness management
22.	(d)	V3: 497–503; EPC: 1411–1414	heat-related-illness management

		PAGE KEY	SUBJECT
23.	(d)	V3: 497–503; EPC: 1411–1414	heat-related-illness management
24.	(d)	V3: 503; EPC: 1408–1409	normal oral temperature
25.	(a)	V3: 495–496; EPC: 1409	core temperature measurement
26.	(c)	V3: 503–504; EPC: 1415–1416	pyrexia
27.	(a)	V3: 503–504; EPC: 1415–1416	fever management
28.	(c)	AHA: 10.4; V3: 505–506; EPC: 1416–1420	mild hypothermia temperature
29.	(c)	AHA: 10.4; V3: 505–506; EPC: 1416–1420	moderate hypothermia temperature
30.	(c)	AHA: 10.4; V3: 505–506; EPC: 1416–1420	severe hypothermia temperature
31.	(d)	V3: 506; EPC: 1418	shivering cessation, pupillary dilation
32.	(b)	V3: 142, 507; EPC: 1185, 1419	J wave, Osborn wave
33.	(c)	V3: 142, 507; EPC: 1185, 1419	J wave, Osborn wave
34.	(b)	V3: 506; EPC: 1419	A-fib, moderate hypothermia
35.	(d)	V3: 507; EPC: 1419–1421	V-fib, hypothermia
36.	(e)	V3: 508–510; EPC: 1419–1421	mild hypothermia management
37.	(e)	V3: 508–510; EPC: 1419–1421	moderate/severe hypothermia
38.	(d)	AHA: 10.4; V3: 509; EPC: 1419–1421	hypothermic cardiac arrest management
39.	(a)	AHA: 10.4; V3: 509; EPC: 1419–1421	hypothermic cardiac arrest management
40.	(d)	AHA: 10.4; V3: 509; EPC: 1419–1421	hypothermic cardiac arrest management
41.	(a)	V3: 511–512; EPC: 1422–1421	pathophysiology of frostbite
42.	(c)	V3: 511–512; EPC: 1422–1421	treatment of frostbite
43.	(d)	AHA: 10.3; V3: 512–515; EPC: 1423	drowning
44.	(a)	AHA: 10.3; V3: 512–515; EPC: 1423	near drowning
45.	(c)	V3: 513–514; EPC: 1424	fresh water is hypotonic
46.	(a)	V3: 513–514; EPC: 1424	salt water is hypertonic
47.	(d)	V3: 513–514; EPC: 1424	fresh water aspiration
48.	(e)	V3: 514; EPC: 1425	successful resuscitation factors
49.	(d)	AHA: 10.3; V3: 515; EPC: 1425	treatment of drowning victims
50.	(a)	V3: 518; EPC: 1428	barotrauma

		PAGE KEY	SUBJECT
51.	(e)	V3: 516–518; EPC: 1427–1429	diving injuries
52.	(e)	V3: 516–518; EPC: 1427–1429	diving injuries
53.	(b)	V3: 519–520; EPC: 1428–1431	decompression illness
54.	(d)	V3: 519–520; EPC: 1428–1431	decompression illness management
55.	(b)	V3: 521–522; EPC: 1428–1431	pulmonary over-pressure accidents
56.	(e)	V3: 521–522; EPC: 1428–1431	arterial gas embolism
57.	(d)	V3: 521–522; EPC: 1428–1431	arterial gas embolism management
58.	(e)	V3: 522; EPC: 1428–1431	nitrogen narcosis
59.	(b)	V3: 523–526; EPC: 1433–1436	high altitude ascent
60.	(d)	V3: 523–526; EPC: 1433–1436	mild acute mountain sickness
61.	(c)	V3: 523–526; EPC: 1433–1436	severe acute mountain sickness
62.	(d)	V3: 523–526; EPC: 1433–1436	high altitude pulmonary edema
63.	(a)	V3: 523–526; EPC: 1433–1436	high altitude cerebral edema
64.	(e)	V3: 527; EPC: 897	forms of ionizing radiation
65.	(e)	V3: 528–529; EPC: 899–900	radiation exposure factors
66.	(a)	V3: 530; EPC: 899–900	radiation injury management risks
67.	(c)	V3: 546–548; EPC: 270–275	components of the immune system
68.	(c)	V3: 546–548; EPC: 270–275	lymphatic system
69.	(a)	V3: 546–548; EPC: 270–275	lymphatic system/spleen
70.	(d)	V3: 555; EPC: 69, 1453	The Ryan White Act
71.	(a)	V3: 540–541; EPC: 1443–1445	infectious diseases/antibiotics
72.	(c)	V3: 558–576; EPC: 1454–1471	infectious diseases & care provider risks
73.	(e)	V3: 568; EPC: 1463–1464	pneumonia
74.	(e)	V3: 571–572; EPC: 1466–1468	meningitis
75.	(d)	V3: 571–572; EPC: 1466–1468	meningitis transmission risks
76.	(e)	V3: 571–572; EPC: 1466–1468	meningitis S/Sx
77.	(a)	V3: 565–567; EPC: 1461–1462	tuberculosis
78.	(a)	V3: 565–567; EPC: 1461–1462	tuberculosis transmission protection
79.	(e)	V3: 565–567; EPC: 1461–1462	tuberculosis S/Sx
80.	(b)	V3: 562–564; EPC: 1458–1460	hepatitis
81.	(a)	V3: 562–564; EPC: 1458–1460	hepatitis transmission
82.	(b)	V3: 562–564; EPC: 1458–1460	hepatitis

		PAGE KEY	SUBJECT
83.	(c)	V3: 571–572; EPC: 1466–1468	meningitis S/Sx
84.	(d)	V3: 558–562; EPC: 1456–1462	tuberculosis/AIDS S/Sx
85.	(d)	V3: 558–562; EPC: 1456–1458	AIDS transmission
86.	(e)	V3: 558–562; EPC: 1456–1458	AIDS transmission
87.	(b)	V3: 560; EPC: 1457	Kaposi's sarcoma
88.	(d)	V3: 560; EPC: 1456–1458	pneumocystis carinii pneumonia
89.	(c)	V3: 570–571; EPC: 1465–1466	chickenpox
90.	(b)	V3: 584–588; EPC: 1478–1481	sexually transmitted diseases
91.	(b)	V3: 614; EPC: 1501	suicidal risk factors
92.	(c)	V5: 207–208; EPC: 1716–1718	partner abuse (domestic violence)
93.	(b)	V5: 209–211; EPC: 1718–1720	elder abuse
94.	(a)	V5: 211–216; EPC: 1657–1661	child abuse
95.	(e)	V5: 211–216; EPC: 1657–1661	child abusers
96.	(e)	V5: 211–216; EPC: 1657–1661	child abuse mistaken S/Sx
97.	(c)	V5: 211–216; EPC: 1657–1661	child abuse S/Sx
98.	(b)	V5: 211–216; EPC: 1657–1661	historical signs of child abuse
99.	(d)	V5: 211–216; EPC: 1657–1661	child abuse neglect S/Sx
100.	(b)	V5: 216–220; EPC: 1724–1727	sexual assault management
101.	(c)	V5: 229–230; EPC: 561	blind patient management
102.	(b)	V5: 279; EPC: 1769–1770	urinary tract medical devices
103.	(c)	V5: 241; EPC: 1742	poliomyelitis/pathological challenges

TEST SECTION 11 ANSWER KEY

Page Key texts:

Brady's *Paramedic Care: Principles & Practice*, AHA updated 2nd edition, 5-volume series of texts:

  V1 indicates Volume 1: Introduction to Advanced Prehospital Care
  V3 indicates Volume 3: Medical Emergencies
  V5 indicates Volume 5: Special Considerations/Operations

EPC indicates Brady's *Essentials of Paramedic Care*, 2nd edition.
PEP indicates Brady's *Prehospital Emergency Pharmacology*, 6th edition.
AHA indicates the 2005 American Heart Association (*Circulation*, December 2005 Supplement), and is followed by the PART number of the section to reference.
MedD indicates that you should reference your Medical Dictionary.

		PAGE KEY	SUBJECT
1.	(e)	V3: 636–637; EPC: 1512–1513	pelvic inflammatory disease (PID)
2.	(b)	V3: 636–637; EPC: 1512–1514	PID S/Sx
3.	(c)	V3: 627; EPC: 229–233	perineum
4.	(c)	V3: 629; EPC: 229–233	endometrium
5.	(a)	V3: 629–630; EPC: 229–233	myometrium
6.	(a)	V3: 653; EPC: 1524	gravida
7.	(b)	V3: 653; EPC: 1524	para/parity
8.	(d)	V3: 658–659; EPC: 1528–1529	abortion
9.	(d)	V3: 653–659; EPC: 1524–1529	gravida, parity, abortion/ miscarriage
10.	(c)	V3: 651, 653; EPC: 1522–1523	normal duration of pregnancy
11.	(e)	V3: 655; EPC: 1525–1528	pregnancy and medical disorders
12.	(a)	V3: 638, 660; EPC: 1530–1532	ectopic pregnancy
13.	(b)	V3: 660; EPC: 1530–1532	most common site of ectopic pregnancy
14.	(a)	VV3: 646–647; EPC: 1519	functions of the placenta
15.	(d)	V3: 661–662; EPC: 1530–1534	abruptio placentae
16.	(c)	V3: 660–661; EPC: 1530–1534	placenta previa
17.	(c)	V3: 660–661; EPC: 1530–1534	placenta previa
18.	(d)	V3: 661–662; EPC: 1530–1534	abruptio placentae
19.	(d)	V3: 662–663; EPC: 1530–1534	preeclampsia S/Sx
20.	(e)	V3: 663–664; EPC: 1530–1534	eclampsia S/Sx
21.	(d)	V3: 662; EPC: 1530–1534	preeclampsia management

		PAGE KEY	SUBJECT
22.	(e)	V3: 663–664; EPC: 1530–1534	eclampsia management
23.	(e)	V3: 662; EPC: 1530–1534	preeclampsia occurrence
24.	(b)	V3: 664–665; EPC: 1534	supine hypotensive syndrome
25.	(a)	V3: 664–665; EPC: 1534	supine hypotensive syndrome management
26.	(d)	V3: 664–665; EPC: 1534	supine hypotensive syndrome management
27.	(b)	V3: 666; EPC: 1536	Braxton-Hicks contractions
28.	(c)	V3: 668–670; EPC: 1538	stages of labor
29.	(b)	V3: 668–670; EPC: 1538	stages of labor
30.	(e)	V3: 668–670; EPC: 1538	stages of labor
31.	(a)	V3: 671–674; EPC: 1539, 1542	S/Sx of imminent field delivery
32.	(b)	V3: 671–674; EPC: 1539	field-delivery contraindications
33.	(c)	V3: 675–676; EPC: 1544	fetal distress
34.	(a)	V3: 672–673; EPC: 1544–1545	vertex position
35.	(b)	V3: 677–678; EPC: 1544–1545	breech presentation
36.	(a)	V3: 671; EPC: 1539	unruptured-amniotic-sac management
37.	(c)	V3: 671; EPC: 1539	wrapped umbilical cord management
38.	(e)	V3: 671; EPC: 1539, 1542	infant positioning, post-delivery
39.	(b)	V3: 674; EPC: 1542, 1549	placenta delivery, postpartum hemorrhage
40.	(c)	V3: 674; EPC: 1542	placenta transportation
41.	(e)	V3: 680; EPC: 1548	cephalopelvic disproportion
42.	(d)	V3: 681–683; EPC: 1544–1548	delivery complications
43.	(b)	V3: 681; EPC: 1549	meconium staining
44.	(e)	V3: 681; EPC: 1549	meconium staining
45.	(b)	V3: 680; EPC: 1548	multiple births
46.	(b)	V3: 682; EPC: 1550	uterine rupture
47.	(d)	V3: 682; EPC: 1550	uterine rupture S/Sx
48.	(e)	V3: 682–683; EPC: 1550	uterine inversion
49.	(b)	V3: 682–683; EPC: 1550	uterine-inversion management
50.	(e)	V3: 683; EPC: 1550	pulmonary embolus

		PAGE KEY	SUBJECT
51.	(d)	V5: 10–13; EPC: 1559–1563	newborn management
52.	(a)	V3: 674; EPC: 1642	umbilical clamp placement
53.	(a)	V3: 674; EPC: 1642	umbilical clamp placement
54.	(c)	V5: 6; EPC: 1557	neonatal breathing stimulus
55.	(d)	V3: 674–675; V5: 10; EPC: 1542	infant suctioning
56.	(b)	V5: 9–10; EPC: 1543	APGAR assessment times
57.	(c)	V5: 9–10; EPC: 1544	APGAR mnemonic
58.	(a)	V5: 9–10; EPC: 1544	APGAR mnemonic
59.	(d)	V5: 9–10; EPC: 1544	APGAR mnemonic
60.	(d)	V5: 9–10; EPC: 1544	APGAR mnemonic
61.	(b)	V5: 9–10; EPC: 1544	APGAR score maximum
62.	(a)	V5: 9–10; EPC: 1544	APGAR score minimum
63.	(a)	V3: 681; V5: 26; EPC: 1549, 1574	meconium staining
64.	(c)	V5: 13–21; EPC: 1563	neonatal distress vital sign
65.	(c)	V5: 13–21; EPC: 1564–1572	newborn assessment and management
66.	(e)	V5: 13–21; EPC: 1564–1572	newborn assessment and management
67.	(b)	V5: 13–21; EPC: 1564–1572	newborn assessment and management
68.	(e)	V5: 13–21; EPC: 1564–1572	newborn assessment and management
69.	(e)	V5: 18; EPC: 1566–1570	newborn ventilation rate
70.	(e)	V5: 21–22; EPC: 1570	newborn chest-compression rate
71.	(d)	V5: 21–22; EPC: 1570	newborn CPR compression/ ventilation ratio
72.	(d)	V5: 14–24; EPC: 1564	Inverted Pyramid for newborn resuscitation
73.	(d)	V5: 28–29; EPC: 1576	premature infant
74.	(d)	V5: 125–126; EPC: 1656–1657	SIDS
75.	(d)	V5: 125–126; EPC: 1656–1657	SIDS
76.	(b)	V5: 126; EPC: 1656–1657	SIDS management
77.	(e)	V5: 24, 85; EPC: 1572, 1622; PEP: 500	pediatric epinephrine IV/IO
78.	(e)	V5: 24, 85; EPC: 1572, 1622; PEP: 500	pediatric atropine IV/IO

458

			PAGE KEY	SUBJECT
79.	(a)		V5: 24, 85; EPC: 1572, 1622; PEP: 500	pediatric epinephrine ET
80.	(c)		AHA 12; V5: 85; EPC: 1622; PEP: 500	pediatric lidocaine
81.	(b)		AHA 12; V5: 85; EPC: 1622; PEP: 416	pediatric amiodarone
82.	(b)		V5: 50; EPC: 644	normal fontanelle
83.	(a)		V5: 50; EPC: 644	increased ICP fontanelle
84.	(c)		V5: 111; EPC: 644	dehydrated fontanelle
85.	(a)		V5: 108–109; EPC: 1640, 1642	pediatric seizures
86.	(a)		V5: 109; EPC: 1640, 1642	infant/child diazepam dose
87.	(b)		V5: 109; EPC: 1642	child diazepam dose
88.	(d)		V5: 110–111; EPC: 1643–1644	pediatric dehydration
89.	(b)		V5: 102–103; EPC: 1644	sepsis
90.	(c)		MedD	Reye's syndrome
91.	(e)		V5: 102, 109–110; EPC: 1643; MedD	meningitis, septicemia, Reye's
92.	(a)		V5: 109–110; EPC: 1643	meningitis
93.	(a)		V5: 109–110; EPC: 1643	meningitis
94.	(c)		MedD	Reye's syndrome
95.	(e)		V5: 102, 109–110; EPC: 1643; MedD	meningitis, septicemia, Reye's
96.	(c)		V5: 92–94; EPC: 1623–1632	croup (laryngotracheobronchitis)
97.	(a)		V5: 98; EPC: 1623–1632	bronchiolitis
98.	(d)		V5: 92, 94, 98; EPC: 1623–1632	bronchiolitis, epiglottitis, croup
99.	(c)		V5: 92–94; EPC: 1623–1632	croup
100.	(b)		V5: 93–95; EPC: 1623–1632	epiglottitis
101.	(e)		V5: 92–95; EPC: 1623–1632	croup, epiglottitis
102.	(e)		V5: 92–95; EPC: 1623–1632	croup, epiglottitis
103.	(e)		V5: 92, 94, 98; EPC: 1623–1632	bronchiolitis, epiglottitis, croup
104.	(c)		V1: 572–576; V5: 75–80; EPC: 1612–1619	pediatric airway and intubation
105.	(d)		V5: 32–33; EPC: 1644–1645	hypoglycemia in infants
106.	(a)		V5: 111–112; EPC: 1644–1645	hypoglycemia in children
107.	(d)		V5: 33; EPC: 1645; PEP: 422, 499	$D_{25}W$
108.	(b)		V5: 153–154; EPC: 1674	geriatric incontinence
109.	(e)		V5: 153–154; EPC: 1675	geriatric elimination difficulties
110.	(b)		V5: 196–197; EPC: 1708–1709	geriatric trauma

		PAGE KEY	SUBJECT
111.	(c)	V5: 196–197; EPC: 1684, 1700–1709	geriatric trauma
112.	(a)	V5: 185; EPC: 1701	osteoporosis
113.	(e)	V5: 200; EPC: 1713	spondylosis
114.	(b)	V5: 163; EPC: 625	kyphosis
115.	(a)	V5: 155–160; EPC: 1678–1692	geriatric CNS assessment
116.	(d)	V5: 172–173; EPC: 1690	geriatric myocardial infarction S/Sx
117.	(d)	V5: 184, 237; EPC: 1700	osteoarthritis
118.	(a)	V5: 157; EPC: 1676, 1677	cataracts
119.	(c)	V5: 157, 229; EPC: 594, 1676, 1733	glaucoma
120.	(b)	V5: 157; EPC: 1676	Meniere's disease

TEST SECTION 12 ANSWER KEY

Page Key texts:
Brady's *Paramedic Care: Principles & Practice*, AHA updated 2nd edition, 5-volume series of texts:

    V1 indicates Volume 1: Introduction to Advanced Prehospital Care
    V5 indicates Volume 5: Special Considerations/Operations

EPC indicates Brady's *Essentials of Paramedic Care*, 2nd edition.
PEP indicates Brady's *Prehospital Emergency Pharmacology*, 6th edition.
    AHA indicates the 2005 American Heart Association Guidelelines (*Circulation*, December 2005 Supplement), and is followed by the PART number of the section to reference.

		PAGE KEY	SUBJECT
1.	(b)	V5: 325; EPC: 1800–1801	calibrated equipment
2.	(e)	V5: 328–329; EPC: 1803	ambulance collisions
3.	(d)	V5: 330–331; EPC: 1804	emergency vehicle operation
4.	(e)	V5: 331–332; EPC: 1804–1805	lights and sirens
5.	(c)	V5: 331–332; EPC: 1804–1805	use of escorts
6.	(e)	V5: 331–332; EPC: 1804–1805	use of escorts
7.	(a)	V5: 334; EPC: 1806	intersections
8.	(e)	V5: 332; EPC: 1805	on-scene parking
9.	(d)	V5: 337; EPC: 1807–1808	air transportation
10.	(d)	V5: 337, 339; EPC: 1808–1809	preparation for helicopter transport
11.	(b)	V5: 339–340; EPC: 1809	helicopter approach
12.	(a)	V5: 339–340; EPC: 1809	helicopter approach
13.	(e)	V5: 352; EPC: 1809	MCI definition
14.	(d)	V5: 363–365; EPC: 1820–1821	START triage system
15.	(e)	V5: 363–365; EPC: 1820–1821	START triage system
16.	(e)	V5: 363–365; EPC: 1820–1821	START triage system
17.	(e)	V5: 363–365; EPC: 1820–1821	START triage system
18.	(e)	V5: 363–365; EPC: 1820–1821	START triage system
19.	(e)	V5: 363–365; EPC: 1820–1821	START triage system
20.	(d)	V5: 363–365; EPC: 1820–1821	START triage system
21.	(e)	V1: 38; EPC: 1829	critical incident stress management
22.	(a)	V1: 38; EPC: 1829	critical incident stress management
23.	(c)	V1: 38; EPC: 1829	critical incident stress management
24.	(b)	V5: 381; EPC: 1834	rescue awareness

		PAGE KEY	SUBJECT
25.	(d)	V5: 388–395; EPC: 1835–1838	rescue operations
26.	(d)	V5: 388–395; EPC: 1835–1838	rescue operations
27.	(c)	V5: 388–395; EPC: 1835–1838	rescue operations
28.	(d)	V5: 388–395; EPC: 1835–1838	rescue operations
29.	(a)	V5: 385–386; EPC: 1833	patient protection
30.	(b)	V5: 388–395; EPC: 1835–1838	rescue operations
31.	(c)	V5: 388–395; EPC: 1835–1838	rescue operations
32.	(c)	V5: 388–395; EPC: 1835–1838	rescue operations
33.	(b)	V5: 397; EPC: 1838–1844	water rescue
34.	(a)	V5: 400; EPC: 1838–1844	water rescue
35.	(c)	V5: 397; EPC: 1838–1844	water rescue
36.	(b)	V5: 402; EPC: 1844	mammalian diving reflex
37.	(a)	V5: 402; EPC: 1844	mammalian diving reflex
38.	(a)	V5: 402, 404; EPC: 1844	shallow-water rescue
39.	(b)	V5: 405–406; EPC: 1846	confined-space hazards
40.	(b)	V5: 408–410; EPC: 1847–1851	highway operations
41.	(b)	V5: 408–410; EPC: 1847–1851	highway operations
42.	(e)	V5: 408–410; EPC: 1847–1851	highway operations
43.	(a)	V5: 411–413; EPC: 1849–1850	vehicle-access safety
44.	(b)	V5: 411–413; EPC: 1849–1850	vehicle-access safety
45.	(b)	V5: 411–413; EPC: 1849–1850	vehicle-access safety
46.	(d)	V5: 416–417; EPC: 1852–1854	rough/hazardous terrain removal
47.	(c)	V5: 416–417; EPC: 1852–1854	rough/hazardous terrain removal
48.	(a)	V5: 429–430; EPC: 1857–1864	hazardous materials incidents
49.	(c)	V5: 429–430; EPC: 1857–1864	hazardous materials incidents
50.	(a)	V5: 353; EPC: 1812–1813, 1858	open versus closed incidents
51.	(a)	V5: 353; EPC: 1812–1813, 1858	open versus closed incidents
52.	(d)	V5: 430; EPC: 1857–1864	hazardous materials incidents
53.	(c)	V5: 430; EPC: 1857–1864	hazardous materials incidents
54.	(d)	V5: 430–436; EPC: 1857–1864	hazardous materials incidents
55.	(b)	V5: 438; EPC: 1857–1864	hazardous materials incidents
56.	(c)	V5: 435; EPC: 1857–1864	hazardous materials incidents
57.	(d)	V5: 438; EPC: 1857–1864	hazardous materials incidents

		PAGE KEY	SUBJECT
58.	(b)	V5: 444; EPC: 1867	pesticide exposure
59.	(d)	V5: 444; EPC: 1867	pesticide exposure
60.	(c)	AHA: 10.2; V5: 444; EPC: 1867; PEP: 341–344	pesticide exposure
61.	(d)	V5: 444; EPC: 1867	pesticide exposure
62.	(a)	V5: 444–445; EPC: 1867–1868	carbon monoxide inhalation
63.	(c)	V5: 444–445; EPC: 1867–1868	cyanide inhalation
64.	(c)	V5: 444–445; EPC: 1867–1868	carbon monoxide inhalation S/Sx
65.	(a)	V5: 444–445; EPC: 1867–1868	cyanide inhalation S/Sx
66.	(e)	V5: 444–445; EPC: 1867–1868	cyanide kit, amyl nitrite
67.	(e)	V5: 444–445; EPC: 1867–1868	cyanide kit, amyl nitrite
68.	(b)	V5: 459–461; EPC: 1873–1875	crime-scene safety
69.	(c)	V5: 459–461; EPC: 1873–1875	crime-scene safety
70.	(c)	V5: 459–461; EPC: 1873–1875	crime-scene safety
71.	(b)	V5: 459–461; EPC: 1873–1875	crime-scene safety
72.	(a)	V5: 469–470; EPC: 1882	body armor
73.	(c)	V5: 467–468; EPC: 1880	cover and concealment
74.	(d)	V5: 467–468; EPC: 1880	cover and concealment
75.	(c)	V5: 459–461; EPC: 1873–1881	crime-scene safety
76.	(e)	V5: 475; EPC: 1885	handling weapons
77.	(c)	V5: 465; EPC: 1879	clandestine drug labs
78.	(a)	V5: 465; EPC: 1879	clandestine drug labs
79.	(e)	V5: 462–463; EPC: 1877	violent incidents
80.	(e)	V5: 462–463; EPC: 1877	violent incidents
81.	(a)	V5: 468; EPC: 1881	distraction and evasion
82.	(d)	V5: 468; EPC: 1881	contact and cover
83.	(b)	V5: 472–475; EPC: 1883–1885	evidence preservation
84.	(a)	V5: 472–475; EPC: 1883–1885	evidence preservation
85.	(b)	V5: 472–475; EPC: 1883–1885	evidence preservation
86.	(a)	V5: 472–475; EPC: 1883–1885	evidence preservation
87.	(d)	V5: 472–475; EPC: 1883–1885	evidence preservation
88.	(e)	V5: 472–475; EPC: 1883–1885	evidence preservation
89.	(b)	V5: 475; EPC: 1885	crime-scene documentation
90.	(d)	V5: 475; EPC: 1885	crime-scene documentation

		PAGE KEY	SUBJECT
91.	(e)	AHA: 8, 9; PEP: 185–194	fibrinolytic administration
92.	(d)	AHA: 8; PEP: 185–194	fibrinolytic administration
93.	(c)	AHA: 8; PEP: 185–194	fibrinolytic administration
94.	(b)	AHA: 9; PEP: 185–194	fibrinolytic administration
95.	(a)	AHA: 9; PEP: 185–194	fibrinolytic administration
96.	(d)	AHA: 4	BLS ventilation initiation
97.	(b)	AHA: 4	BLS ventilation initiation
98.	(e)	AHA: 4	BLS & BVM ventilation
99.	(b)	AHA: 4	BVM ventilation
100.	(b)	AHA: 4	BVM ventilation
101.	(c)	AHA: 4, 11	CPR compression/ventilation ratios
102.	(c)	AHA: 4, 11	CPR compression/ventilation ratios
103.	(c)	AHA: 4, 11	CPR compression/ventilation ratios
104.	(c)	AHA: 4, 11	CPR compression/ventilation ratios
105.	(b)	AHA: 4, 11	CPR compression/ventilation ratios
106.	(a)	AHA: 13	CPR compression/ventilation ratios
107.	(a)	AHA: 13	CPR compression/ventilation ratios
108.	(e)	AHA: 11, 13	Infant CPR
109.	(e)	AHA: 11, 13	Infant CPR
110.	(a)	AHA: 4	CPR role-switching
111.	(b)	AHA: 4, 11, 13	CPR role-switching
112.	(b)	AHA: 4, 11, 13	CPR role-switching
113.	(c)	AHA: 4	CPR role-switching
114.	(a)	AHA: 4	chest compression rates
115.	(a)	AHA: 4	chest compression rates
116.	(d)	AHA: 4	effective chest compression
117.	(b)	AHA: 4	circulation checks during CPR
118.	(e)	AHA: 4, 7.2; EPC: 1231	defibrillation
119.	(c)	AHA: 4, 7.2; EPC: 1231	defibrillation
120.	(c)	AHA: 1, 4, 7.2	minimize CPR interruption

TEST SECTION 13 ANSWER KEY

Page Key texts:

Brady's *Paramedic Care: Principles & Practice*, AHA updated 2nd edition, 5-volume series of texts:

> V1 indicates Volume 1: Introduction to Advanced Prehospital Care
> V2 indicates Volume 2: Patient Assessment
> V3 indicates Volume 3: Medical Emergencies
> V4 indicates Volume 4: Trauma Emergencies
> V5 indicates Volume 5: Special Considerations/Operations

EPC indicates Brady's *Essentials of Paramedic Care*, 2nd edition.
MedD indicates that you should reference your Medical Dictionary.

		PAGE KEY	SUBJECT
1.	(d)	V1: 171; EPC: 92	anatomy
2.	(b)	V1: 171; EPC: 92, 241	physiology
3.	(a)	V1: 173–174; EPC: 241	pathophysiology
4.	(b)	V4: 339; EPC: 151–156	the sacrum
5.	(a)	V4: 339; EPC: 151–156	the coccyx
6.	(a)	V4: 339; EPC: 151–156	the vertebra
7.	(d)	V2: 67–72; EPC: 145–146	the ear
8.	(b)	V2: 67–72; EPC: 145–146	the cochlea
9.	(c)	V2: 67–72; EPC: 145–146	the semicircular canals
10.	(a)	V2: 67–72; EPC: 145–146	the tympanic membrane
11.	(e)	V5: 177; EPC: 1694	vertigo
12.	(e)	V2: 67–72; EPC: 596	otitis
13.	(e)	V2: 67–72; EPC: 139, 146	the mastoid process
14.	(b)	EPC: 205; MedD	rhinorrhea
15.	(e)	V2: 73; EPC: 828	epistaxis
16.	(a)	V2: 60, 64–67; EPC: 145–146	the iris
17.	(c)	V2: 60, 64–67; EPC: 145–146	the sclera
18.	(e)	V2: 60, 64–67; EPC: 145–146	the conjunctiva
19.	(b)	V2: 60, 64–67; EPC: 145–146	the cornea
20.	(a)	V2: 60, 64–67; EPC: 145–146	the retina
21.	(e)	V2: 60, 64–67; EPC: 145–146	the conjunctiva
22.	(d)	V2: 60, 64–67; EPC: 145–146	the pupil
23.	(a)	V2: 60, 64–67; EPC: 145–146	the iris

		PAGE KEY	SUBJECT
24.	(e)	V2: 60, 64–67; EPC: 145–146	the lens
25.	(c)	V2: 102; EPC: 120	206 bones in the body
26.	(e)	V1: 510; EPC: 206	hyoid bone
27.	(b)	V4: 220–221; EPC: 117–122	metaphysis
28.	(a)	V4: 220–221; EPC: 117–122	epiphysis
29.	(c)	V4: 220–221; EPC: 117–122	diaphysis
30.	(c)	V4: 225; EPC: 120–121	axial skeleton
31.	(d)	V4: 225; EPC: 120–121	appendicular skeleton
32.	(d)	V2: 126; EPC: 625	lordosis
33.	(c)	V4: 221; EPC: 119	red bone marrow
34.	(a)	V4: 221; EPC: 118	yellow bone marrow
35.	(b)	V4: 223; EPC: 120	ligaments
36.	(d)	V4: 133, 224; EPC: 121	tendons
37.	(c)	V4: 228; EPC: 129	medial malleolus
38.	(d)	V4: 228; EPC: 129	lateral malleolus
39.	(c)	V4: 227; EPC: 132	acetabulum
40.	(b)	V4: 226; EPC: 125, 127	clavicle
41.	(e)	V2: 110; EPC: 127, 157	scapula
42.	(d)	V4: 226; EPC: 126	radius
43.	(b)	V4: 226; EPC: 125, 127	clavicle
44.	(a)	V4: 226; EPC: 126	ulna
45.	(c)	V4: 227; EPC: 122	carpals
46.	(d)	V4: 227; EPC: 122–123	metacarpals
47.	(e)	V4: 227; EPC: 122–123	phalanges
48.	(a)	V2: 116; EPC: 122–129	tarsals
49.	(b)	V2: 116; EPC: 122–129	metatarsals
50.	(e)	V2: 116; EPC: 122–129	phalanges
51.	(b)	V4: 227–228; EPC: 221	ilium
52.	(a)	V4: 227–228; EPC: 131	patella
53.	(b)	V4: 227–228; EPC: 130	tibia
54.	(d)	V4: 227–228; EPC: 129–130	fibula
55.	(a)	V5: 234; EPC: 1736	paraplegia
56.	(c)	V3: 278, 521; EPC: 1259, 1431	hemiplegia

		PAGE KEY	SUBJECT
57.	(b)	V5: 234; EPC: 1736	quadriplegia
58.	(e)	V1: 514; EPC: 209	right lung anatomy
59.	(d)	V1: 514; EPC: 209	left lung anatomy
60.	(d)	MedD	medial
61.	(c)	MedD	lateral
62.	(a)	MedD	ventral
63.	(b)	MedD	dorsal
64.	(b)	MedD	posterior
65.	(a)	MedD	anterior
66.	(d)	MedD	superior
67.	(e)	MedD	inferior
68.	(d)	MedD	proximal
69.	(c)	MedD	distal
70.	(e)	MedD	abduction
71.	(b)	MedD	flexion
72.	(a)	MedD	extension
73.	(d)	MedD	adduction
74.	(b)	MedD	prone position
75.	(a)	MedD	Trendelenburg position
76.	(c)	MedD	laterally recumbent position
77.	(d)	V2: 205; MedD	semi-Fowler's position
78.	(e)	MedD	supine position
79.	(c)	V1: 510–512; EPC: 205–210	hypopharynx/laryngopharynx
80.	(a)	V1: 510–512; EPC: 205–210	nasopharynx
81.	(b)	V1: 510–512; EPC: 205–210	oropharynx
82.	(c)	V1: 510–512; EPC: 205–210	laryngopharynx
83.	(e)	V1: 510; EPC: 205–210	septum
84.	(d)	V1: 510; EPC: 205–210	cilia
85.	(d)	V1: 510; EPC: 205–210	tongue
86.	(d)	V1: 510–512; EPC: 205–210	epiglottis
87.	(c)	V1: 512–513; EPC: 205–210	carina
88.	(b)	V1: 512–513; EPC: 205–210	right mainstem bronchus
89.	(c)	V1: 512–513; EPC: 205–210	left mainstem bronchus

		PAGE KEY	SUBJECT
90.	(a)	V1: 514–516; EPC: 205–210	alveoli
91.	(a)	V1: 514–517; EPC: 205–210	alveoli
92.	(c)	V3: 653; EPC: 1510–1511	primagravida
93.	(d)	V3: 653; EPC: 1510–1511	primapara
94.	(d)	V3: 653; EPC: 1510–1511	multipara
95.	(e)	V3: 653; EPC: 1510–1511	multigravida
96.	(b)	V3: 317; EPC: 1284	polyphagia
97.	(e)	V3: 317; EPC: 1284	polyuria
98.	(d)	V3: 317; EPC: 1284	polydipsia
99.	(c)	V4: 419–420; EPC: 1057	needle thoracentesis
100.	(e)	V5: 151; EPC: 1693	dysphagia
101.	(d)	V5: 175; EPC: 1692	vasovagal syncope/valsalva maneuver
102.	(c)	V5: 175; EPC: 1692	orthostatic syncope
103.	(a)	V5: 177–179; EPC: 1694	Alzheimer's disease
104.	(e)	V5: 179; EPC: 1696	Parkinson's disease
105.	(e)	V2: 284; EPC: 742	decreased
106.	(d)	V2: 284; EPC: 742	increased
107.	(c)	V2: 284; EPC: 742	more than
108.	(a)	V2: 284; EPC: 742	less than
109.	(a)	V2: 284: ; EPC: 742	WNL
110.	(e)	V2: 285; EPC: 743	every
111.	(a)	V2: 284; EPC: 742	prn
112.	(c)	V2: 285; EPC: 742	NPO
113.	(a)	V2: 285; EPC: 742	prescription or therapy
114.	(b)	V2: 285; EPC: 742	treatment
115.	(c)	EPC: 115–118; MedD	fracture
116.	(e)	MedD	history
117.	(d)	MedD	diagnosis
118.	(b)	MedD	symptoms
119.	(c)	V2: 284; EPC: 742	secondary to
120.	(b)	V2: 285; EPC: 742	against medical advice

		PAGE KEY	SUBJECT
121.	(c)	V3: 292; EPC: 1359	hemiplegia
122.	(b)	V3: 292; EPC: 1359	paresthesia
123.	(e)	V3: 292; EPC: 1359	hemiparesis
124.	(e)	V3: 292; EPC: 1359	dysphasia
125.	(b)	V3: 292; EPC: 1359	aphasia
126.	(a)	V3: 292; EPC: 1359	dysarthria

TEST SECTION 14 ANSWER KEY

Page Key texts:

Brady's *Paramedic Care: Principles & Practice*, AHA updated 2nd edition, 5-volume series of texts:

   V3: indicates Volume 3: Medical Emergencies

EPC indicates Brady's Essentials of Paramedic Care, 2nd edition.
TAC indicates Brady's *Taigman's Advanced Cardiology (In Plain English)*.

		PAGE KEY	SUBJECT
1.	(b)	V3: 87–88; TAC: 62	bipolar limb leads
2.	(b)	V3: 205–206	bipolar limb leads
3.	(a)	V3: 205–206	bipolar limb leads
4.	(d)	V3: 205–206	bipolar limb leads
5.	(a)	V3: 205–206	bipolar limb leads
6.	(d)	V3: 205–206	bipolar limb leads
7.	(b)	V3: 205–206	bipolar limb leads
8.	(c)	V3: 206	bipolar limb leads
9.	(a)	V3: 209–212; TAC: 61–70	QRS axis determination
10.	(e)	V3: 209–212; TAC: 61–70	QRS axis determination
11.	(b)	V3: 209–212; TAC: 61–70	QRS axis determination
12.	(d)	V3: 209–212; TAC: 61–70	QRS axis determination
13.	(c)	V3: 209–212; TAC: 61–70	QRS axis determination
14.	(a)	V3: 227–230; TAC: 39, 58	bundle branch blocks
15.	(b)	V3: 227–230; TAC: 39, 58	bundle branch blocks
16.	(c)	V3: 227–230; TAC: 39, 58	bundle branch blocks
17.	(c)	V3: 230–231; TAC: 71–83	hemiblocks
18.	(b)	V3: 230–231; TAC: 71–83	hemiblocks
19.	(c)	V3: 230–231; TAC: 71–83	hemiblocks
20.	(e)	V3: 228–229; TAC: 37–60	left bundle branch block
21.	(b)	V3: 227–228; TAC: 37–60	right bundle branch block
22.	(b)	V3: 238; TAC: 42	angle of Louis
23.	(b)	V3: 209, 237–238; EPC: 1237	chest lead placement
24.	(c)	V3: 209, 237–238; EPC: 1237	chest lead placement
25.	(a)	V3: 209, 237–238; EPC: 1237	chest lead placement

		PAGE KEY	SUBJECT
26.	(b)	V3: 209, 237–238; EPC: 1237	chest lead placement
27.	(e)	V3: 209, 237–238; EPC: 1237	chest lead placement
28.	(a)	V3: 209, 237–238; EPC: 1237	chest lead placement
29.	(b)	V3: 214–219; TAC: 175–184	infarct or ischemia patterns
30.	(c)	V3: 214–215; TAC: 175–184	signs of ischemia
31.	(d)	V3: 214–215; TAC: 175–184	signs of ischemia
32.	(c)	V3: 214–215; TAC: 175–184	signs of injury
33.	(e)	V3: 214; TAC: 175–184	Q waves
34.	(c)	V3: 218–219; TAC: 175–184	anterior infarct
35.	(b)	V3: 218–219; TAC: 175–184	inferior infarct
36.	(d)	V3: 218–219; TAC: 175–184	lateral infarct
37.	(e)	V3: 221, 225; TAC: 175–184	right ventricular infarct
38.	(e)	V3:100–101, 210–212; TAC: 69–83	sinus tach; pathologic LAD
39.	(b)	V3: 230–231; TAC: 69–83	left anterior hemiblock
40.	(b)	V3: 94	PRI and QRS WNL
41.	(c)	V3: 210–212; TAC: 69	right axis deviation
42.	(e)	V3: 227–228; TAC: 49–50	right bundle branch block
43.	(d)	TAC: 1–60	bifascicular blocks
44.	(a)	AHA: 5, 7.3; TAC: 69, 121–149	V-tach
45.	(a)	TAC: 69, 121–149	morphology favoring V-tach

INTRODUCTION TO THE NATIONAL REGISTRY
and the NREMT-P ORAL and SKILLS Examination
EVALUATION PAGES

The National Registry of Emergency Medical Technicians (NREMT) has an informative Website that should be accessed by all paramedics preparing to take the NREMT-P test (or any similar test). Valuable information about the examination is available there.

The NREMT Paramedic Skills and Oral Station Evaluation Sheets represent the National Standard Performance order and content requirement for a paramedic to pass the NREMT-P Practical and Oral Examinations. These skills sheets are reproduced on the following 15 pages of this text.

To use these sheets in preparing for the exam, you have a couple of choices:

- Photocopy the pages provided here, OR
- Access, save, and print out the Adobe Acrobat files of the NREMT Paramedic Skills and Oral Station Evaluation Sheets, free of charge, from the NRMET Website. To do this:

  1. Go to the NREMT Website *(http://www.nremt.org)*.
  2. Click on the REGISTRANTS AND CANDIDATES link (at the top center of the NREMT home page).
  3. Click on the EXAM COORDINATOR DOCUMENTS link (at the left of that page).
  4. Click on the ADVANCED LEVELS SKILLS SHEETS link.
  5. Click on each of the Adobe Acrobat files containing paramedic Oral Stations and Skills Sheets, save them to your computer. Then, print copies as you need them.

The currently available NREMT Paramedic Skills and Oral Station Evaluation Sheets are reproduced on the following 15 pages of this text:

  1. Oral Station Scenario (an actual oral-exam scenario, as an example of this station)
  2. Oral Station Template (a guide for designing *your own* Oral Station Scenarios); there are two versions: a blank version you can fill in and a sample completed version, for your use as a reference.
  3. Oral Station Worksheet (used to score performers)
  4. Patient Assessment–Medical Skills Worksheet
  5. Patient Assessment–Trauma Skills Worksheet
  6. Spinal Immobilization (Seated Patient) Skills Worksheet
  7. Spinal Immobilization (Supine Patient) Skills Worksheet
  8. Pediatric (<2 yrs.) Ventilatory Management Skills Worksheet
  9. Dual Lumen Airway Device (Combitube® or PTL®) Skills Worksheet
  10. Ventilatory Management–Adult Skills Worksheet
  11. Bleeding Control/Shock Management Skills Worksheet
  12. Intravenous Therapy Skills Worksheet
  13. Pediatric Intraosseous Infusion Skills Worksheet
  14. Dynamic Cardiology Skills Worksheet
  15. Static Cardiology Skills Worksheet

BACKGROUND INFORMATION

EMS System description (including urban/rural setting)	Suburban EMS that responds to both emergency and non-emergency calls
Vehicle Type/response capabilities	2 person paramedic level transporting service
Proximity to and level/type of facilities	30 minutes to the attending physician s office 15 minutes to community hospital

DISPATCH INFORMATION

Nature of the call	Woman can t walk, requests transport to physician s office, non -emergent
Location	Well kept walk-up single family dwelling
Dispatch Time	1512 hours
Weather	68 F spring day
Personnel on scene	Daughter who is serving as primary care giver

SCENE SURVEY INFORMATION

Scene considerations	10 cement steps up to the front door No access for stretcher from any other doorway
Patient location	1st floor, back bedroom, narrow hallways & doorways
Visual appearance	Patient sitting in bed with multiple pillows holding her in an upright position, pale in color, does not respond to your presence in the room
Age, gender, weight	58 year old female, 200 pounds
Immediate surroundings (bystanders, family members present)	Clean, neat, well-kept surroundings Daughter is only family member present,

PATIENT ASSESSMENT

Chief Complaint	Altered level of consciousness
History of present illness	Daughter states My Mother just passed out a couple of minutes ago from the pain. Patient woke this morning with a painful left leg that has increased in pain, unable to walk without severe pain. Daughter states that her mother, Has a small sore on her left inner thigh that has gotten bigger over the past few hours and her doctor wants to see her in his office.
Patient responses, symptoms, and pertinent negatives	Patient opens her eyes to loud verbal stimulus but does not verbally respond

PAST MEDICAL HISTORY

Past Medical History	Adult onset diabetic controlled with diet and oral medication, hypertension, hernia repair years ago
Medications & Allergies	Glucophage bid, Lasix 20 mg qid, diltiazem qid, and Colace qid NKA
Social/family concerns	Patient lives alone after death of husband two years ago, daughter comes to her home each day to help her mother with daily chores

EXAMINATION FINDINGS	
Initial Vital Signs	BP 100/palpation P 130, rapid and weak R 8
Respiratory	Lung sounds are dimished bilaterally
Cardiovascular	Tachycardia, hypotensive
Gastrointestinal	---
Genitourinary	---
Musculoskeletal	---
Neurologic	Opens her eyes to loud verbal stimulus and withdraws to pain Utters incomprehensible sounds Pupils equal and responds sluggishly to light
Integumentary	Large ecchymotic area over the patient s entire left inner thigh extending into the groin, pelvis, and left lower abdomen Area is hot to touch with crepitation under the skin Skin is pale, hot, and moist to touch
Hematologic	---
Immunologic	---
Endocrine	Blood glucose 370 mg/dL
Psychiatric	---
PATIENT MANAGEMENT	
Initial stabilization	Assisted ventilation with high flow oxygen
Treatments	Assisted ventilation with high flow oxygen IV enroute
Monitoring	ECG sinus tachycardia, SpO$_2$ 85%
Additional Resources	Consider transportation to facility with immediate surgical capabilities and hyperbarics
Patient response to interventions	No change
TRANSPORT DECISION	
Lifting and moving patient	Place in Reeves stretcher to ambulance stretcher
Mode	Rapid
Facilities	Emergency department
CONCLUSION	
Field Impression	Septic shock
Rationale for Field Impression	Rapidly extending extremity infection, febrile, hypotension, and tachycardia, with altered LOC
Related pathophysiology	What is the basis for the septic shock in this case? Severe bacterial infection
Verbal Report	
MANDATORY ACTIONS	
Rapid identification of life-threat and immediate transportation to the emergency department High flow oxygen	
POTENTIALLY HARMFUL/DANGEROUS ACTIONS ORDERED/PERFORMED	
Delayed transportation for on scene interventions Taking the patietnt to the doctors office.	

BACKGROUND INFORMATION	
EMS System description (including urban/rural setting)	
Vehicle Type/response capabilities	
Proximity to and level/type of facilities	
DISPATCH INFORMATION	
Nature of the call	
Location	
Dispatch Time	
Weather	
Personnel on scene	
SCENE SURVEY INFORMATION	
Scene considerations	
Patient location	
Visual appearance	
Age, gender, weight	
Immediate surroundings (bystanders, family members present)	
PATIENT ASSESSMENT	
Chief Complaint	
History of present illness	
Patient responses, symptoms, and pertinent negatives	
PAST MEDICAL HISTORY	
Past Medical History	
Medications and allergies	
Social/family concerns	

EXAMINATION FINDINGS	
Initial Vital Signs	BP P R ; SpO$_2$ %
Respiratory	
Cardiovascular	
Gastrointestinal	
Genitourinary	
Musculoskeletal	
Neurologic	
Integumentary	
Hematologic	
Immunologic	
Endocrine	
Psychiatric	
PATIENT MANAGEMENT	
Initial stabilization	
Treatments	
Monitoring	
Additional Resources	
Patient response to interventions	
TRANSPORT DECISION	
Lifting and moving patient	
Mode	
Facilities	
CONCLUSION	
Field Impression	
Rationale for Field Impression	
Related pathophysiology	
Verbal Report	Please provide me with a verbal report on th is patient. Must include chief complaint, interventions, current patient condition, and ETA.

MANDATORY ACTIONS

POTENTIALLY HARMFUL/DANGEROUS ACTIONS ORDERED/PERFORMED

National Registry of Emergency Medical Technicians
Advanced Level Practical Examination

ORAL STATION

Candidate: _____ Examiner: _____

Date: _____ Signature: _____

Scenario: _____

Time Start: _____

	Possible Points	Points Awarded
Scene Management		
Thoroughly assessed and took deliberate actions to control the scene	3	
Assessed the scene, identified potential hazards, did not put anyone in danger	2	
Incompletely assessed or managed the scene	1	
Did not assess or manage the scene	0	
Patient Assessment		
Completed an organized assessment and integrated findings to expand further assessment	3	
Completed initial, focused, and ongoing assessments	2	
Performed an incomplete or disorganized assessment	1	
Did not complete an initial assessment	0	
Patient Management		
Managed all aspects of the patient's condition and anticipated further needs	3	
Appropriately managed the patient's presenting condition	2	
Performed an incomplete or disorganized management	1	
Did not manage life-threatening conditions	0	
Interpersonal relations		
Established rapport and interacted in an organized, therapeutic manner	3	
Interacted and responded appropriately with patient, crew, and bystanders	2	
Used inappropriate communication techniques	1	
Demonstrated intolerance for patient, bystanders, and crew	0	
Integration (verbal report, field impression, and transport decision)		
Stated correct field impression and pathophysiological basis, provided succinct and accurate verbal report including social/psychological concerns, and considered alternate transport destinations	3	
Stated correct field impression, provided succinct and accurate verbal report, and appropriately stated transport decision	2	
Stated correct field impression, provided inappropriate verbal report or transport decision	1	
Stated incorrect field impression or did not provide verbal report	0	
Time End: _____ **TOTAL**	15	

Critical Criteria

_____ Failure to appropriately address any of the scenario's "Mandatory Actions"

_____ Performs or orders any harmful or dangerous action or intervention

You must factually document your rationale for checking any of the above critical items on the reverse side of this form.

p308/8-003k

National Registry of Emergency Medical Technicians
Advanced Level Practical Examination

PATIENT ASSESSMENT - MEDICAL

Candidate: _____ Examiner: _____

Date: _____ Signature: _____

Scenario: _____

Time Start: _____

	Possible Points	Points Awarded
Takes or verbalizes body substance isolation precautions	1	
SCENE SIZE-UP		
Determines the scene/situation is safe	1	
Determines the mechanism of injury/nature of illness	1	
Determines the number of patients	1	
Requests additional help if necessary	1	
Considers stabilization of spine	1	
INITIAL ASSESSMENT		
Verbalizes general impression of the patient	1	
Determines responsiveness/level of consciousness	1	
Determines chief complaint/apparent life-threats	1	

Assesses airway and breathing
   -Assessment (1 point)
   -Assures adequate ventilation (1 point)
   -Initiates appropriate oxygen therapy (1 point)
	3	

Assesses circulation
   -Assesses/controls major bleeding (1 point)   -Assesses skin [either skin color, temperature, or condition] (1 point)
   -Assesses pulse (1 point)
	3	

Identifies priority patients/makes transport decision	1	
FOCUSED HISTORY AND PHYSICAL EXAMINATION/RAPID ASSESSMENT		

History of present illness
   -Onset (1 point)   -Severity (1 point)
   -Provocation (1 point)   -Time (1 point)
   -Quality (1 point)   -Clarifying questions of associated signs and symptoms as related to OPQRST (2 points)
   -Radiation (1 point)
	8	

Past medical history
   -Allergies (1 point)   -Past pertinent history (1 point)   -Events leading to present illness (1 point)
   -Medications (1 point)   -Last oral intake (1 point)
	5	

Performs focused physical examination [assess affected body part/system or, if indicated, completes rapid assessment]
   -Cardiovascular   -Neurological   -Integumentary   -Reproductive
   -Pulmonary   -Musculoskeletal   -GI/GU   -Psychological/Social
	5	

Vital signs
   -Pulse (1 point)   -Respiratory rate and quality (1 point each)
   -Blood pressure (1 point)   -AVPU (1 point)
	5	

Diagnostics [must include application of ECG monitor for dyspnea and chest pain]	2	
States field impression of patient	1	
Verbalizes treatment plan for patient and calls for appropriate intervention(s)	1	
Transport decision re-evaluated	1	
ON-GOING ASSESSMENT		
Repeats initial assessment	1	
Repeats vital signs	1	
Evaluates response to treatments	1	
Repeats focused assessment regarding patient complaint or injuries	1	

Time End: _____

CRITICAL CRITERIA   TOTAL   48

_____ Failure to initiate or call for transport of the patient within 15 minute time limit

_____ Failure to take or verbalize body substance isolation precautions

_____ Failure to determine scene safety before approaching patient

_____ Failure to voice and ultimately provide appropriate oxygen therapy

_____ Failure to assess/provide adequate ventilation

_____ Failure to find or appropriately manage problems associated with airway, breathing, hemorrhage or shock [hypoperfusion]

_____ Failure to differentiate patient's need for immediate transportation versus continued assessment and treatment at the scene

_____ Does other detailed or focused history or physical examination before assessing and treating threats to airway, breathing, and circulation

_____ Failure to determine the patient's primary problem

_____ Orders a dangerous or inappropriate intervention

_____ Failure to provide for spinal protection when indicated

You must factually document your rationale for checking any of the above critical items on the reverse side of this form.

p302/8-003k

National Registry of Emergency Medical Technicians
Advanced Level Practical Examination

PATIENT ASSESSMENT - TRAUMA

Candidate: _____   Examiner: _____

Date: _____   Signature: _____

Scenario # _____

Time Start: _____ NOTE: Areas denoted by "**" may be integrated within sequence of Initial Assessment	Possible Points	Points Awarded
Takes or verbalizes body substance isolation precautions	1	
SCENE SIZE-UP		
Determines the scene/situation is safe	1	
Determines the mechanism of injury/nature of illness	1	
Determines the number of patients	1	
Requests additional help if necessary	1	
Considers stabilization of spine	1	
INITIAL ASSESSMENT/RESUSCITATION		
Verbalizes general impression of the patient	1	
Determines responsiveness/level of consciousness	1	
Determines chief complaint/apparent life-threats	1	
Airway -Opens and assesses airway (1 point) -Inserts adjunct as indicated (1 point)	2	

Breathing
  -Assess breathing (1 point)
  -Assures adequate ventilation (1 point)
  -Initiates appropriate oxygen therapy (1 point)
  -Manages any injury which may compromise breathing/ventilation (1 point)
	4	

Circulation
  -Checks pulse (1point)
  -Assess skin [either skin color, temperature, or condition] (1 point)
  -Assesses for and controls major bleeding if present (1 point)
  -Initiates shock management (1 point)
	4	

Identifies priority patients/makes transport decision	1	
FOCUSED HISTORY AND PHYSICAL EXAMINATION/RAPID TRAUMA ASSESSMENT		
Selects appropriate assessment	1	
Obtains, or directs assistant to obtain, baseline vital signs	1	
Obtains SAMPLE history	1	
DETAILED PHYSICAL EXAMINATION		

Head
  -Inspects mouth**, nose**, and assesses facial area (1 point)
  -Inspects and palpates scalp and ears (1 point)
  -Assesses eyes for PERRL** (1 point)
	3	

Neck**
  -Checks position of trachea (1 point)
  -Checks jugular veins (1 point)
  -Palpates cervical spine (1 point)
	3	

Chest**
  -Inspects chest (1 point)
  -Palpates chest (1 point)
  -Auscultates chest (1 point)
	3	

Abdomen/pelvis**
  -Inspects and palpates abdomen (1 point)
  -Assesses pelvis (1 point)
  -Verbalizes assessment of genitalia/perineum as needed (1 point)
	3	

Lower extremities** -Inspects, palpates, and assesses motor, sensory, and distal circulatory functions (1 point/leg)	2	
Upper extremities -Inspects, palpates, and assesses motor, sensory, and distal circulatory functions (1 point/arm)	2	

Posterior thorax, lumbar, and buttocks**
  -Inspects and palpates posterior thorax (1 point)
  -Inspects and palpates lumbar and buttocks area (1 point)
	2	

Manages secondary injuries and wounds appropriately	1	
Performs ongoing assessment	1	

Time End: _____   **TOTAL**   43

CRITICAL CRITERIA

_____ Failure to initiate or call for transport of the patient within 10 minute time limit
_____ Failure to take or verbalize body substance isolation precautions
_____ Failure to determine scene safety
_____ Failure to assess for and provide spinal protection when indicated
_____ Failure to voice and ultimately provide high concentration of oxygen
_____ Failure to assess/provide adequate ventilation
_____ Failure to find or appropriately manage problems associated with airway, breathing, hemorrhage or shock [hypoperfusion]
_____ Failure to differentiate patient's need for immediate transportation versus continued assessment/treatment at the scene
_____ Does other detailed/focused history or physical exam before assessing/treating threats to airway, breathing, and circulation
_____ Orders a dangerous or inappropriate intervention

You must factually document your rationale for checking any of the above critical items on the reverse side of this form.

p301/8-003k

National Registry of Emergency Medical Technicians
Advanced Level Practical Examination

SPINAL IMMOBILIZATION (SEATED PATIENT)

Candidate:_____ Examiner:_____

Date: _____ Signature:_____

Time Start: _____	Possible Points	Points Awarded
Takes or verbalizes body substance isolation precautions	1	
Directs assistant to place/maintain head in the neutral, in-line position	1	
Directs assistant to maintain manual immobilization of the head	1	
Reassesses motor, sensory, and circulatory function in each extremity	1	
Applies appropriately sized extrication collar	1	
Positions the immobilization device behind the patient	1	
Secures the device to the patient's torso	1	
Evaluates torso fixation and adjusts as necessary	1	
Evaluates and pads behind the patient's head as necessary	1	
Secures the patient's head to the device	1	
Verbalizes moving the patient to a long backboard	1	
Reassesses motor, sensory, and circulatory function in each extremity	1	
TOTAL	12	

Time End: _____

CRITICAL CRITERIA

_____ Did not immediately direct or take manual immobilization of the head
_____ Did not properly apply appropriately sized cervical collar before ordering release of manual
immobilization
_____ Released or ordered release of manual immobilization before it was maintained mechanically
_____ Manipulated or moved patient excessively causing potential spinal compromise
_____ Head immobilized to the device **before** device sufficiently secured to torso
_____ Device moves excessively up, down, left, or right on the patient's torso
_____ Head immobilization allows for excessive movement
_____ Torso fixation inhibits chest rise, resulting in respiratory compromise
_____ Upon completion of immobilization, head is not in a neutral, in-line position
_____ Did not reassess motor, sensory, and circulatory functions in each extremity after voicing
immobilization to the long backboard

*You must factually document your rationale for checking any of the above critical items on the
reverse side of this form.*

p311/8-003k

National Registry of Emergency Medical Technicians
Advanced Level Practical Examination

SPINAL IMMOBILIZATION (SUPINE PATIENT)

Candidate:_____Examiner:_____

Date: _____Signature:_____

Time Start: _____	Possible Points	Points Awarded
Takes or verbalizes body substance isolation precautions	1	
Directs assistant to place/maintain head in the neutral, in-line position	1	
Directs assistant to maintain manual immobilization of the head	1	
Reassesses motor, sensory, and circulatory function in each extremity	1	
Applies appropriately sized extrication collar	1	
Positions the immobilization device appropriately	1	
Directs movement of the patient onto the device without compromising the integrity of the spine	1	
Applies padding to voids between the torso and the device as necessary	1	
Immobilizes the patient's torso to the device	1	
Evaluates and pads behind the patient's head as necessary	1	
Immobilizes the patient's head to the device	1	
Secures the patient's legs to the device	1	
Secures the patient's arms to the device	1	
Reassesses motor, sensory, and circulatory function in each extremity	1	
Time End: _____ **TOTAL**	14	

CRITICAL CRITERIA

_____ Did not immediately direct or take manual immobilization of the head
_____ Did not properly apply appropriately sized cervical collar before ordering release of manual immobilization
_____ Released or ordered release of manual immobilization before it was maintained mechanically
_____ Manipulated or moved patient excessively causing potential spinal compromise
_____ Head immobilized to the device **before** device sufficiently secured to torso
_____ Patient moves excessively up, down, left, or right on the device
_____ Head immobilization allows for excessive movement
_____ Upon completion of immobilization, head is not in a neutral, in-line position
_____ Did not reassess motor, sensory, and circulatory functions in each extremity after voicing immobilization to the device

You must factually document your rationale for checking any of the above critical items on the reverse side of this form.

National Registry of Emergency Medical Technicians
Advanced Level Practical Examination

PEDIATRIC (<2 yrs.) VENTILATORY MANAGEMENT

Candidate: _____ Examiner _____

Date: _____ Signature: _____

NOTE: If candidate elects to ventilate initially with BVM attached to reservoir and oxygen, full credit must be awarded for steps denoted by "**" so long as first ventilation is delivered within 30 seconds.

	Possible Points	Points Awarded
Takes or verbalizes body substance isolation precautions	1	
Opens the airway manually	1	
Elevates tongue, inserts simple adjunct [oropharyngeal or nasopharyngeal airway]	1	
NOTE: Examiner now informs candidate no gag reflex is present and patient accepts adjunct		
**Ventilates patient immediately with bag-valve-mask device unattached to oxygen	1	
**Ventilates patient with room air	1	
NOTE: Examiner now informs candidate that ventilation is being performed without difficulty and that pulse oximetry indicates the patient's blood oxygen saturation is 85%		
Attaches oxygen reservoir to bag-valve-mask device and connects to high flow oxygen regulator [12-15 L/minute]	1	
Ventilates patient at a rate of 12-20/minute and assures visible chest rise	1	
NOTE: After 30 seconds, examiner auscultates and reports breath sounds are present, equal bilaterally and medical direction has ordered intubation. The examiner must now take over ventilation.		
Directs assistant to pre-oxygenate patient	1	
Identifies/selects proper equipment for intubation	1	
Checks laryngoscope to assure operational with bulb tight	1	
NOTE: Examiner to remove OPA and move out of the way when candidate is prepared to intubate		
Places patient in neutral or sniffing position	1	
Inserts blade while displacing tongue	1	
Elevates mandible with laryngoscope	1	
Introduces ET tube and advances to proper depth	1	
Directs ventilation of patient	1	
Confirms proper placement by auscultation bilaterally over each lung and over epigastrium	1	
NOTE: Examiner to ask, "If you had proper placement, what should you expect to hear?"		
Secures ET tube [may be verbalized]	1	
TOTAL	**17**	

CRITICAL CRITERIA

_____ Failure to initiate ventilations within 30 seconds after applying gloves or interrupts ventilations for greater than 30 seconds at any time
_____ Failure to take or verbalize body substance isolation precautions
_____ Failure to pad under the torso to allow neutral head position or sniffing position
_____ Failure to voice and ultimately provide high oxygen concentrations [at least 85%]
_____ Failure to ventilate patient at a rate of 12-20/minute
_____ Failure to provide adequate volumes per breath [maximum 2 errors/minute permissible]
_____ Failure to pre-oxygenate patient prior to intubation
_____ Failure to successfully intubate within 3 attempts
_____ Uses gums as a fulcrum
_____ Failure to assure proper tube placement by auscultation bilaterally **and** over the epigastrium
_____ Inserts any adjunct in a manner dangerous to the patient
_____ Attempts to use any equipment not appropriate for the pediatric patient

You must factually document your rationale for checking any of the above critical items on the reverse side of this form.

p305/8-003k

National Registry of Emergency Medical Technicians
Advanced Level Practical Examination

DUAL LUMEN AIRWAY DEVICE (COMBITUBE® OR PTL®)

Candidate: _____ Examiner: _____

Date: _____ Signature: _____

NOTE: If candidate elects to initially ventilate with BVM attached to reservoir and oxygen, full credit must be awarded for steps denoted by "**" so long as first ventilation is delivered within 30 seconds.

	Possible Points	Points Awarded
Takes or verbalizes body substance isolation precautions	1	
Opens the airway manually	1	
Elevates tongue, inserts simple adjunct [oropharyngeal or nasopharyngeal airway]	1	
NOTE: Examiner now informs candidate no gag reflex is present and patient accepts adjunct		
**Ventilates patient immediately with bag-valve-mask device unattached to oxygen	1	
**Hyperventilates patient with room air	1	
NOTE: Examiner now informs candidate that ventilation is being performed without difficulty		
Attaches oxygen reservoir to bag-valve-mask device and connects to high flow oxygen regulator [12-15 L/minute]	1	
Ventilates patient at a rate of 10-12/minute with appropriate volumes	1	
NOTE: After 30 seconds, examiner auscultates and reports breath sounds are present and equal bilaterally and medical control has ordered insertion of a dual lumen airway. The examiner must now take over ventilation.		
Directs assistant to pre-oxygenate patient	1	
Checks/prepares airway device	1	
Lubricates distal tip of the device [may be verbalized]	1	
NOTE: Examiner to remove OPA and move out of the way when candidate is prepared to insert device		
Positions head properly	1	
Performs a tongue-jaw lift	1	

☐ USES COMBITUBE®	☐ USES PTL®		
Inserts device in mid-line and to depth so printed ring is at level of teeth	Inserts device in mid-line until bite block flange is at level of teeth	1	
Inflates pharyngeal cuff with proper volume and removes syringe	Secures strap	1	
Inflates distal cuff with proper volume and removes syringe	Blows into tube #1 to adequately inflate both cuffs	1	
Attaches/directs attachment of BVM to the first [esophageal placement] lumen and ventilates		1	
Confirms placement and ventilation through correct lumen by observing chest rise, auscultation over the epigastrium, and bilaterally over each lung		1	

NOTE: The examiner states, "You do not see rise and fall of the chest and you only hear sounds over the epigastrium."		
Attaches/directs attachment of BVM to the second [endotracheal placement] lumen and ventilates	1	
Confirms placement and ventilation through correct lumen by observing chest rise, auscultation over the epigastrium, and bilaterally over each lung	1	
NOTE: The examiner confirms adequate chest rise, absent sounds over the epigastrium, and equal bilateral breath sounds.		
Secures device or confirms that the device remains properly secured	1	

TOTAL **20**

CRITICAL CRITERIA

_____ Failure to initiate ventilations within 30 seconds after taking body substance isolation precautions or interrupts ventilations for greater than 30 seconds at any time

_____ Failure to take or verbalize body substance isolation precautions

_____ Failure to voice and ultimately provide high oxygen concentrations [at least 85%]

_____ Failure to ventilate patient at a rate of 10-12/minute

_____ Failure to provide adequate volumes per breath [maximum 2 errors/minute permissible]

_____ Failure to pre-oxygenate patient prior to insertion of the dual lumen airway device

_____ Failure to insert the dual lumen airway device at a proper depth or at either proper place within 3 attempts

_____ Failure to inflate both cuffs properly

_____ **Combitube** - failure to remove the syringe immediately after inflation of each cuff

_____ **PTL** - failure to secure the strap prior to cuff inflation

_____ Failure to confirm that the proper lumen of the device is being ventilated by observing chest rise, auscultation over the epigastrium, and bilaterally over each lung

_____ Inserts any adjunct in a manner dangerous to patient

You must factually document your rationale for checking any of the above critical items on the reverse side of this form.

p304/8-003k

National Registry of Emergency Medical Technicians
Advanced Level Practical Examination

VENTILATORY MANAGEMENT - ADULT

Candidate:_____ Examiner:_____

Date: _____ Signature: _____

NOTE: If candidate elects to ventilate initially with BVM attached to reservoir and oxygen, full credit must be awarded for steps denoted by "**" so long as first ventilation is delivered within 30 seconds.

	Possible Points	Points Awarded
Takes or verbalizes body substance isolation precautions	1	
Opens the airway manually	1	
Elevates tongue, inserts simple adjunct [oropharyngeal or nasopharyngeal airway]	1	
NOTE: Examiner now informs candidate no gag reflex is present and patient accepts adjunct		
**Ventilates patient immediately with bag-valve-mask device unattached to oxygen	1	
**Ventilates patient with room air	1	
NOTE: Examiner now informs candidate that ventilation is being performed without difficulty and that pulse oximetry indicates the patient's blood oxygen saturation is 85%		
Attaches oxygen reservoir to bag-valve-mask device and connects to high flow oxygen regulator [12-15 L/minute]	1	
Ventilates patient at a rate of 10-12/minute with appropriate volumes	1	
NOTE: After 30 seconds, examiner auscultates and reports breath sounds are present, equal bilaterally and medical direction has ordered intubation. The examiner must now take over ventilation.		
Directs assistant to pre-oxygenate patient	1	
Identifies/selects proper equipment for intubation	1	
Checks equipment for: -Cuff leaks (1 point) -Laryngoscope operational with bulb tight (1 point)	2	
NOTE: Examiner to remove OPA and move out of the way when candidate is prepared to intubate		
Positions head properly	1	
Inserts blade while displacing tongue	1	
Elevates mandible with laryngoscope	1	
Introduces ET tube and advances to proper depth	1	
Inflates cuff to proper pressure and disconnects syringe	1	
Directs ventilation of patient	1	
Confirms proper placement by auscultation bilaterally over each lung and over epigastrium	1	
NOTE: Examiner to ask, "If you had proper placement, what should you expect to hear?"		
Secures ET tube [may be verbalized]	1	
NOTE: Examiner now asks candidate, "Please demonstrate one additional method of verifying proper tube placement in this patient."		
Identifies/selects proper equipment	1	
Verbalizes findings and interpretations [compares indicator color to the colorimetric scale or EDD recoil and states findings]	1	
NOTE: Examiner now states, "You see secretions in the tube and hear gurgling sounds with the patient's exhalation."		
Identifies/selects a flexible suction catheter	1	
Pre-oxygenates patient	1	
Marks maximum insertion length with thumb and forefinger	1	
Inserts catheter into the ET tube leaving catheter port open	1	
At proper insertion depth, covers catheter port and applies suction while withdrawing catheter	1	
Ventilates/directs ventilation of patient as catheter is flushed with sterile water	1	
TOTAL	**27**	

CRITICAL CRITERIA

_____ Failure to initiate ventilations within 30 seconds after applying gloves or interrupts ventilations for greater than 30 seconds at any time
_____ Failure to take or verbalize body substance isolation precautions
_____ Failure to voice and ultimately provide high oxygen concentrations [at least 85%]
_____ Failure to ventilate patient at a rate of 10 - 12 / minute
_____ Failure to provide adequate volumes per breath [maximum 2 errors/minute permissible]
_____ Failure to pre-oxygenate patient prior to intubation and suctioning
_____ Failure to successfully intubate within 3 attempts
_____ Failure to disconnect syringe **immediately** after inflating cuff of ET tube
_____ Uses teeth as a fulcrum
_____ Failure to assure proper tube placement by auscultation bilaterally **and** over the epigastrium
_____ If used, stylette extends beyond end of ET tube
_____ Inserts any adjunct in a manner dangerous to the patient
_____ Suctions the patient for more than 10 seconds
_____ Does not suction the patient

**You must factually document your rationale for checking any of the above critical items on the reverse side of this form.**

p303/8-003k

BLEEDING CONTROL/SHOCK MANAGEMENT

Candidate: _____ Examiner: _____

Date: _____ Signature: _____

Time Start:_____

	Possible Points	Points Awarded
Takes or verbalizes body substance isolation precautions	1	
Applies direct pressure to the wound	1	
Elevates the extremity	1	
NOTE: The examiner must now inform the candidate that the wound continues to bleed.		
Applies an additional dressing to the wound	1	
NOTE: The examiner must now inform the candidate that the wound still continues to bleed. The second dressing does not control the bleeding.		
Locates and applies pressure to appropriate arterial pressure point	1	
NOTE: The examiner must now inform the candidate that the bleeding is controlled.		
Bandages the wound	1	
NOTE: The examiner must now inform the candidate that the patient is exhibiting signs and symptoms of hypoperfusion.		
Properly positions the patient	1	
Administers high concentration oxygen	1	
Initiates steps to prevent heat loss from the patient	1	
Indicates the need for immediate transportation	1	
Time End: _____ **TOTAL**	10	

CRITICAL CRITERIA

_____ Did not take or verbalize body substance isolation precautions
_____ Did not apply high concentration of oxygen
_____ Applied a tourniquet before attempting other methods of bleeding control
_____ Did not control hemorrhage in a timely manner
_____ Did not indicate the need for immediate transportation

You must factually document your rationale for checking any of the above critical items on the reverse side of this form.

National Registry of Emergency Medical Technicians
Advanced Level Practical Examination

INTRAVENOUS THERAPY

Candidate: _____ Examiner: _____

Date: _____ Signature: _____

Level of Testing: ☐ NREMT-Intermediate/85   ☐ NREMT-Intermediate/99   ☐ NREMT-Paramedic

Time Start: _____

	Possible Points	Points Awarded
Checks selected IV fluid for: -Proper fluid (1 point) -Clarity (1 point)	2	
Selects appropriate catheter	1	
Selects proper administration set	1	
Connects IV tubing to the IV bag	1	
Prepares administration set [fills drip chamber and flushes tubing]	1	
Cuts or tears tape [at any time before venipuncture]	1	
Takes/verbalizes body substance isolation precautions [prior to venipuncture]	1	
Applies tourniquet	1	
Palpates suitable vein	1	
Cleanses site appropriately	1	
Performs venipuncture -Inserts stylette (1 point) -Notes or verbalizes flashback (1 point) -Occludes vein proximal to catheter (1 point) -Removes stylette (1 point) -Connects IV tubing to catheter (1 point)	5	
Disposes/verbalizes disposal of needle in proper container	1	
Releases tourniquet	1	
Runs IV for a brief period to assure patent line	1	
Secures catheter [tapes securely or verbalizes]	1	
Adjusts flow rate as appropriate	1	

Time End: _____   **TOTAL**   21

CRITICAL CRITERIA
_____ Failure to establish a patent and properly adjusted IV within 6 minute time limit
_____ Failure to take or verbalize body substance isolation precautions prior to performing venipuncture
_____ Contaminates equipment or site without appropriately correcting situation
_____ Performs any improper technique resulting in the potential for uncontrolled hemorrhage, catheter shear, or air embolism
_____ Failure to successfully establish IV within 3 attempts during 6 minute time limit
_____ Failure to dispose/verbalize disposal of needle in proper container

NOTE: Check here (_____) if candidate did not establish a patent IV and do not evaluate IV Bolus Medications.

INTRAVENOUS BOLUS MEDICATIONS
Time Start: _____

Asks patient for known allergies	1	
Selects correct medication	1	
Assures correct concentration of drug	1	
Assembles prefilled syringe correctly and dispels air	1	
Continues body substance isolation precautions	1	
Cleanses injection site [Y-port or hub]	1	
Reaffirms medication	1	
Stops IV flow [pinches tubing or shuts off]	1	
Administers correct dose at proper push rate	1	
Disposes/verbalizes proper disposal of syringe and needle in proper container	1	
Flushes tubing [runs wide open for a brief period]	1	
Adjusts drip rate to TKO/KVO	1	
Verbalizes need to observe patient for desired effect/adverse side effects	1	

Time End: _____   **TOTAL**   13

CRITICAL CRITERIA
_____ Failure to begin administration of medication within 3 minute time limit
_____ Contaminates equipment or site without appropriately correcting situation
_____ Failure to adequately dispel air resulting in potential for air embolism
_____ Injects improper drug or dosage [wrong drug, incorrect amount, or pushes at inappropriate rate]
_____ Failure to flush IV tubing after injecting medication
_____ Recaps needle or failure to dispose/verbalize disposal of syringe and needle in proper container

You must factually document your rationale for checking any of the above critical items on the reverse side of this form.

p309/8-003k

National Registry of Emergency Medical Technicians
Advanced Level Practical Examination

PEDIATRIC INTRAOSSEOUS INFUSION

Candidate: _____ Examiner: _____

Date: _____ Signature: _____

Time Start:_____	Possible Points	Points Awarded

Checks selected IV fluid for:
    -Proper fluid (1 point)
    -Clarity (1 point)
	2	

Selects appropriate equipment to include:
    -IO needle (1 point)
    -Syringe (1 point)
    -Saline (1 point)
    -Extension set (1 point)
	4	

Selects proper administration set	1	
Connects administration set to bag	1	
Prepares administration set [fills drip chamber and flushes tubing]	1	
Prepares syringe and extension tubing	1	
Cuts or tears tape [at any time before IO puncture]	1	
Takes or verbalizes body substance isolation precautions [prior to IO puncture]	1	
Identifies proper anatomical site for IO puncture	1	
Cleanses site appropriately	1	

Performs IO puncture:
    -Stabilizes tibia (1 point)
    -Inserts needle at proper angle (1 point)
    -Advances needle with twisting motion until "pop" is felt (1 point)
    -Unscrews cap and removes stylette from needle (1 point)
	4	

Disposes of needle in proper container	1	
Attaches administration set to IO needle (with or without 3-way)	1	
Slowly injects saline to assure proper placement of needle	1	
Adjusts flow rate as appropriate	1	
Secures needle with tape and supports with bulky dressing	1	
Time End: _____ **TOTAL**	23	

CRITICAL CRITERIA
_____ Failure to establish a patent and properly adjusted IO line within the 6 minute time limit
_____ Failure to take or verbalize body substance isolation precautions prior to performing IO puncture
_____ Contaminates equipment or site without appropriately correcting situation
_____ Performs any improper technique resulting in the potential for air embolism
_____ Failure to assure correct needle placement
_____ Failure to successfully establish IO infusion within 2 attempts during 6 minute time limit
_____ Performing IO puncture in an unacceptable manner [improper site, incorrect needle angle, etc.]
_____ Failure to dispose of needle in proper container
_____ Orders or performs any dangerous or potentially harmful procedure

You must factually document your rationale for checking any of the above critical items on the reverse side of this form.

National Registry of Emergency Medical Technicians
Advanced Level Practical Examination

DYNAMIC CARDIOLOGY

Candidate: _____ Examiner: _____

Date: _____ Signature: _____

SET #_____

Level of Testing: ☐ NREMT-Intermediate/99   ☐ NREMT-Paramedic

Time Start:_____

	Possible Points	Points Awarded
Takes or verbalizes infection control precautions	1	
Checks level of responsiveness	1	
Checks ABCs	1	
Initiates CPR when appropriate [verbally]	1	
Attaches ECG monitor in a timely fashion [patches, pads or paddles]	1	
Correctly interprets initial rhythm	1	
Appropriately manages initial rhythm	2	
Notes change in rhythm	1	
Checks patient condition to include pulse and, if appropriate, BP	1	
Correctly interprets second rhythm	1	
Appropriately manages second rhythm	2	
Notes change in rhythm	1	
Checks patient condition to include pulse and, if appropriate, BP	1	
Correctly interprets third rhythm	1	
Appropriately manages third rhythm	2	
Notes change in rhythm	1	
Checks patient condition to include pulse and, if appropriate, BP	1	
Correctly interprets fourth rhythm	1	
Appropriately manages fourth rhythm	2	
Orders high percentages of supplemental oxygen at proper times	1	

Time End: _____   **TOTAL**   24

CRITICAL CRITERIA

_____ Failure to deliver any shock in a timely manner

_____ Failure to verify rhythm before delivering each shock

_____ Failure to ensure the safety of self and others [verbalizes "All clear" and observes]

_____ Inability to deliver DC shock [does not use machine properly]

_____ Failure to demonstrate acceptable shock sequence

_____ Failure to immediately order initiation or resumption of CPR when appropriate

_____ Failure to order correct management of airway [ET when appropriate]

_____ Failure to order administration of appropriate oxygen at proper time

_____ Failure to diagnose or treat 2 or more rhythms correctly

_____ Orders administration of an inappropriate drug or lethal dosage

_____ Failure to correctly diagnose or adequately treat v-fib, v-tach, or asystole

You must factually document your rationale for checking any of the above critical items on the reverse side of this form.